Natural Dyes

Natural Dyes

Dr. Rameshwar Dayal
University School of Biotechnology
Guru Gobind Singh Indraprastha University
New Delhi, India
[Funded By DST-USERS Scheme]

Formerly
Scientist-F and Head (Retd.), Chemistry Division
Forest Research Institute (ICFRE)
Dehra Dun, India

I.K. International Publishing House Pvt. Ltd.
NEW DELHI

Published by
I.K. International Publishing House Pvt. Ltd.
S-25, Green Park Extension
Uphaar Cinema Market
New Delhi–110 016 (India)
E-mail:info@ikinternational.com

ISBN: 978-93-84588-15-1

Published by Krishan Makhijani for I.K. International Publishing House Pvt. Ltd., S-25, Green Park Extension, Uphaar Cinema Market, New Delhi–110 016 and Printed by Rekha Printers Pvt. Ltd., Okhla Industrial Area, Phase II, New Delhi–110 020.

Preface

The concept of natural dyes is not new to the world as these have been used by people since time immemorial. The cave paintings made by prehistoric people clearly demonstrates their knowledge to make colours from natural materials. The use of natural dyes is deep rooted in the ancient history of India, and other equally primeval civilizations of the world. China, Egypt, Mexico, Peru, and other countries are not only known for their ancient civilization, but also for the use of natural dyes. Natural dyes were also used by people of Indus Valley Civilization, which existed during the 3300-1300 B.C. The relics from the excavations of Indus Valley Civilization (Harappa culture) have produced manifestation of fabrics and ropes dyed with natural colours, used by the people of Mohenjo-Daro. The pottery excavated also revealed presence of natural dyes. The earliest written record of the use of dyestuffs in 2600 B.C. can be traced to China. Indigo was used in India for about last 3500 years. The dyeing technology appears to have been originated in India during the ancient period, which later spread to other countries of the world. The credit goes to India and China for being the original homes not only for some dyes, but also for dyeing auxiliaries, used in the different countries of the world before the discovery of synthetic dyes. Plant breeding and isolation of dyes has been an important industry in the earlier times till the nineteenth century. The technique of dyeing and printing reached its zenith by the Medieval period and Indian textiles were known throughout the world.

Natural dyes were exclusively used for colouring till the middle of the nineteenth century. However, due to accidental discovery of the first synthetic dye mauveine by W.H. Perkin in 1856, and a number of other synthetic dyes in the later part of the nineteenth century, the use of natural dyes declined considerably by the beginning of the twentieth century. Research in the field of natural dyes never ceased completely, but continued at a very slow speed. It was realized during the last decade of the twentieth century that synthesis of synthetic dyes using aromatic chemicals and their subsequent use cause harmful effects on the environment. Moreover, the use of azo dyes, which released 22 aromatic amines, was prohibited. The safest and most eco-friendly option for dyeing is the natural dyes, which are produced in the nature's laboratory. Thus, there is a renewed interest in the study and revival of natural dyes during the last two decades. It will not be possible to replace synthetic dyes with natural dyes, but research and development in the field of natural dyes will help in reducing the environmental pollution to a certain limit.

Plants are the greatest organic chemists of the earth. In the nature's laboratory, they carry out complex organic syntheses with ease to manufacture complicated chemical compounds with an amazing degree of accuracy, structures, and characteristics. A number of compounds of the plants containing different functional groups and double bonds are coloured and can be used as dyes, either alone or in combination with other chemical constituents. The aim of this book is to provide a simple, concise, and collective description of the various aspects of natural dyes. The book consists of nine chapters, each explaining an aspect of natural dyes.

The first chapter provides a brief description of plants and forests, and the goods and services provided by them. Natural dye is one of the non-wood forest products (NWFPs) obtained from forests that has been used for dyeing since time immemorial. Some wood dyes are also known.

The second chapter gives a concise description about the history of natural dyes used by people since their appearance on the earth and their identification in ancient textiles. It also describes their use in India and different countries of the world. India played a key role in the development of famous natural dyes like indigo, lac dye, madder, and other dyes.

Chapter 3 deals with the classification of natural dyes, obtained mainly from plants, animals (insects), and minerals. Natural dyes can be classified on the basis of their botanical and common names, colours obtained from dye of plants, chemical classes, and application classes. A brief description about the discovery of synthetic dyes, azo dyes, and their harmful effects along with the merits and demerits of natural and synthetic dyes has been included in the chapter. The concise description about the future of natural dyes, research needs, and dyes export have also been given.

The plant sources of natural dyes have been described in Chapter 4 (Part A). These plants (392), which belong to different families, may be used for natural dyes. Different parts of the various plants contain the dye which is capable of imparting varying colours. The list of plant species on the basis of dye bearing parts and colours obtained using the dye have also been given separately. Chapter 4 (Part B) describes the two versatile dyes, namely lac and indigo. These dyes were used extensively during the ancient times. The use of synthetic colours for Holi celebration are harmful to people. Chapter 4 (Part C) describes the benefits of herbal colours and herbal gulal, as well as the toxic effects of synthetic colours.

Chapter 5 (Part A) deals with the chemistry of natural dyes. Natural dyes are mainly obtained from different parts of plants, namely, bark, flowers, fruits, leaves, seeds, roots, and wood. The dyes obtained from these parts consist of organic compounds belonging to different classes, namely, alkaloids, anthocyanins, carotenoids, flavonoids, indigoids, phenolics, quinones, and other classes. These compounds occur in different plants as secondary metabolites. Natural dyes are also obtained from animal and mineral sources. Chemical constituents isolated and characterized from dye-bearing parts of 69 prominent dye-yielding plant species and 4 insect species have also been described in the chapter. About 570 compounds were isolated in pure state by using the latest chromatographic techniques and their structures elucidated by using sophisticated spectroscopic techniques (IR, UV, NMR,MS, GC-MS,LC-MS, etc). Chapter 5 (Part B) describes the prominent constituents of some famous dyes. The major coloring constituents of some well-known natural dye sources used in the past were found to be curcumin from haldi or turmeric (*Curcuma longa*), crocin from saffron (*Crocus sativus*), lawsone from henna (*Lawsonia inermis*), indigotin from *Indigofera tinctoria*, alizarin from madder (*Rubia tinctoria*), and munjistin and alizarin from manjith (*Rubia cordifolia*), laccaic acids A, B, C, E, and lacaccic acid D from the lac dye (*Laccifer lacca),* carminic acid from cochineal insect (*Dactylopius coccus),* kermesic acid from kermes (*Kermes ilicis*), and 6,6 - dibromoindigotin from the shellfish *Murex brandis.*

Chapter 6 deals with the various types of mordants used for fixing the dye on the fabric, leading to an improvement in colourfastness of the dyed fabrics. A chemical compound that fixes the dye on the fabric by forming a coordination complex with the dye, which in turn attaches to the fabric, is known as mordant. The mordant has the affinity for both fibre and dyestuff and forms bond between them, there by fixing the dye on the material to be dyed. The metallic salts of some metals like aluminum, chromium, iron, and tin (alum, potassium dichromate, ferrous sulphate, and stannous chloride) are some of the examples. The mordanting methods of different types of mordants, namely, metal salts, natural mordants, and oil mordants, have been described. The limit of metallic mordants on the finished products after dyeing and a brief description of different fibres, namely, silk, wool, cotton, and polyester, has also been included in the chapter.

The use of natural dyes by humans started with their appearance on earth. The process of adhering colour to fibres, yarns, and fabrics by the application of dyes is known as dyeing. To start with, people might have started rubbing the dye-bearing parts of plants or their juices containing pigments onto the cloth. With the passage of time, improved dyeing methods were developed to obtain attractive colors on different fabrics. In Chapter 7, extraction of dyes from plants materials, methods of dyeing, mordanting, and dyeing trials by using the dye isolated from some abundantly occurring plant materials have been described. The colourfastness properties of the dyed fabrics is to be determined in order to find the suitability of the dye for its use. The brief description about ultrasound dyeing, microwave dyeing, supercritical carbon dioxide dyeing, and the process for the isolation of natural dyes in considerable amount have also been given.

The dyeing trials using natural dyes impart different colours to different fabrics. The acceptance of natural dye for its dyeing use depends on its colorfastness properties. Chapter 8 deals with the determination of the colourfastness properties of the dyed fabrics to washing, light, perspiration, and rubbing. The Blue Wool Scale, Grey Scale and determination of the colour coordinates of the dyed fabrics have also been described in brief.

The dyeing process has been in use since ancient times and has undergone numerous changes for its betterment. The low dye adsorption by the fibre from the dyeing bath leads to weak colourfastness properties, indicating the non-suitability of the dye for dyeing the fibre. The dye adsorption on the fibre (wool, silk, cotton and others) can be increased by using mordants. The dyeing process is an interaction between the fibre and the dye solution at particular set of conditions and occurs due to the selective dye adsorption on the fibre. In Chapter 9, a brief description about the theory of dyeing of fibres have been described. A concise narration of the valence bond and molecular orbital application to colour has also been included.

The book will be useful to academicians, scientists, industrialists, researchers, students, and people who are interested in the use and the scope of natural dyes, and contribute to environmental protection.

Dr. Rameshwar Dayal

Acknowledgements

I owe a deep sense of gratitude to the Almighty God for providing me the intuition, perseverance, guidance, inspiration, faith and strength to carry out this work. I am thankful to Prof. P.C. Sharma, Dean and Prof. Rajinder K. Gupta, ex-Dean University School of Biotechnology, Guru Gobind Singh Indraprastha University for providing the facility during the book writing work. Thanks are also due to Swati Singh for editing the manuscript besides her very busy schedule, Sunil Sharma for typing the manuscript, incorporation of the corrections a number of times to my satisfaction and other secretarial assistance, and Dr. Ravindra Singh, Research officer (Chemistry), Central Council for Research in Ayurvedic Sciences, Janakpuri, New Delhi for going through the manuscript and photographs of the plants. The financial assistance for the book writing project from the Science and Engineering Research Board, Department of Science and Technology under the Utilization of Scientific Expertise of Retired Scientists (USERS) Scheme New Delhi, is gratefully acknowledged. I wish to thank all the authors of the available books, research papers on Natural Dyes and Forest Research Institute, Dehra Dun, where I worked from November 23, 1977 till my superannuation on August 31, 2009. It is not possible to convey my indebtedness to the authors of the research papers, review articles, proceedings, monographs, books, etc; which were consulted by me for obtaining the useful information. An attempt had been made to convey some magnitude of my gratitude by citing the reference of their works. I am grateful to my parents for their blessings. I am also thankful to my wife for her angelic tolerance, inspiration and support, and to Rahul, Priyanka, Deepak and Parul for their encouragement. Sincere efforts have been made to give concise and the correct information, however I will be happy to receive useful suggestions if any to improve further editions of this book.

Dr. Rameshwar Dayal

Contents

Tables

Author

Dr. Rameshwar Dayal obtained his B.Sc. (1969) from Delhi University, M.Sc. (1971) from Kurukshetra University, and Ph.D (1976) from Delhi University, specializing in the field of Natural Products Chemistry. Dr. Dayal joined Forest Research Institute, Dehra Dun (1977) as Research Officer, and worked there in various positions and superannuated as Scientist-F and Head, Chemistry Division in August, 2009.

Dr. Dayal has about 38 years of research experience in the field of natural products chemistry. The main thrust of his research activities was directed towards the "Phytochemistry of Indian Forest Trees", particularly of medicinal, aromatic, dye bearing, and insecticidal plants. The phytochemical examination of a number of plant species was undertaken which led to the characterization of numerous chemical compounds using latest separation and spectroscopic techniques. A number of compounds exhibited significant pharmacological and biopesticidal activities.

During the long research career at Forest Research Institute (FRI), Dr. Dayal executed a number of projects with financial support from, Indian Council of Forest Research and Education (ICFRE) and various national funding agencies such as Ministry of Environment and Forests (MoEF), Department of Scientific and Industrial Research (DSIR), Department of Biotechnology (DBT), Department of Science and Technology (DST), New Delhi.

The work on the development of technologies for value addition of non- wood forest products, as a team leader needs special mention which includes the development of know-how for the isolation of natural dyes from abundantly available plants on a pilot plant scale, a process for the preparation of compost from readily available biomass, and development of eco-friendly biopolymer based hair dye. The work on natural dyes is of significant importance in providing additional sources of natural dyes, demand of which is on increase due to adverse environmental and health risk of synthetic dyes. The simple processes were developed for the isolation of bioactive bryonolic acid and ursolic acid from *Tetrameles nudiflora* (Bark and leaves) and *Eucalyptus* hybrid (leaves), respectively. The contributions made by Dr. Dayal in synthetic organic chemistry include the synthesis of a number of new prenylated flavonoids, and macrocylic lactones of perfumery value from pine needles wax.

Dr. Dayal disseminated his research work through 89 research papers published in several national and international journals, presentation of a large number of research papers in several national and international conferences/symposia/workshop/seminar, a number of technical reports, six patents and transfer of three technologies to the interested entrepreneurs. He supervised ten students for the award of Ph.D degree.

In recognition of his research work, he was conferred Vishist Vaigyanik Purskar award (1999-2000) of the Ministry of Environment and Forests, Government of India, for the outstanding contribution in the field of "Phytochemistry of Forest Trees"; ICFRE-Cash Award of Indian Council of Forestry Research and Education, Dehra Dun (1989-99) for outstanding contribution to research in the field of "Forest Utilization"; letter of appreciation from Director General, ICFRE for award of Vishist Vaigyanik Purskar, and Incentive Cash Award (1998) from National Research Development Corporation, New Delhi. Dr. Dayal joined the Department of Science and Technology project of book writing on Natural Dyes awarded to him under the Utilization of Scientific Expertise of Retired Scientists (USERS) Scheme at University School of Biotechnology, Guru Gobind Singh Indraprastha University, New Delhi in October, 2011.

1 Introduction

INTRODUCTION

Human civilization has its roots in the forests since the prehistoric days. During the early days of civilization, humans had only limited needs like food, clothing, and shelter. However, with the advancement of civilization, their requirements also grew. India has a rich variety of soils, climate, and other ecological factors, which have endowed it with vast and rich forests. According to the Global Forest Resource Assessment (GFRA, 2010), done by the Food and Agriculture Organisation of the United Nations (FAO), world's total forest area is about 4 billion hectares, which covers 31% of total land area [Forest Survey of India (FSI), 2011]. A list of top ten countries with largest forest area in the world is given in Table 1.1. India ranks tenth among the most forested countries of the world, with 21.05% (6,92,027km^2) of its geographical area, covered by trees and forests (FSI, 2011). The status of forest cover in India is given in Table 1.2. The average per capita availability of forest land in India is approximately ~0.08 hectare, supporting about 16% of the world's population and 18% of world's livestock, as compared to the world's average of 0.60 hectare.

Table 1.1 Top ten countries with the largest forest areas in 2010

S.No.	*Name of country*	*Area in million ha*
1.	Australia	149
2.	Brazil	519
3.	Canada	310
4.	China	206
5.	Democratic Republic of the Congo	154
6.	India	68
7.	Indoensia	94
8.	Russian Federation	809
9.	Sudan	69
10.	USA	304
	Total World	4,033

Table 1.2 Status of forest cover in India

S.No.	*Class of forest cover*	*Area (km)²*	*Geographical area (%)*
1.	Moderately thick forest (cover density between 40% to 70%)	3,20,736	9.76
2.	Open forests (cover density less than 40%)	2,87,820	8.75
3.	Very thick forest (more than 70% cover density)	83,471	2.54
4.	Total forest cover	6,92,027	21.05
5.	Scrub	42,176	1.28
6.	Non forest	25,53,060	77.67
7.	Total geographical area	32,87,263	100.00

*Includes 4662 km² area under mangroves.

India is the second most populous and seventh largest country of the world with an area of 32,87,263 million km². It has a land frontier of 15,200 km and a coastline of 7500 km. It is, in fact, the meeting zone of three major biogeographic realms, namely the Indo-Malayan, the Eurasian, and the Afro-tropical (Singh and Khanna, 2008). All these features provide India with a tremendous 2.50% of the global land mass and an amazing range of biodiversity. India represents one of the 12 mega-diversity regions in the world and houses four of the 34 globally identified biodiversity hotspots, namely the Himalayas, Indo-Burma, Western Ghats and Sri Lanka and Sundaland, which contain thousand of varieties of plants and animals found nowhere else on the planet. India comprises around 91,200 species of animals and 45,500 species of plants. It has 99 national parks and 515 wildlife sanctuaries. Wild fauna of India includes about 390 species of mammals, 458 species of birds, 521 species of reptiles, 231 species of amphibians, 5749 species of fishes, 4000 species of mollusks, and 5000 species of insects in addition to other invertebrates. Among the flora, India houses Angiosperms (17,527), Gymnosperms (67), Pteridophytes (1,200), Bryophytes (2,500), Algae (7,157), Fungi (14,500), Lichens (2,223) and Bacteria (850) which amounts to about 11% of the known floral diversity of the world. (Anon, 2009; Bawa et al., 2009). Table 1.3 represents forest cover belonging to different forest type groups, of which tropical dry deciduous forest (41.87%) and tropical moist deciduous forest (19.73%) and tropical semi-evergreen forest (13.79%) are the major ones. These plant species occur on India's land mass of 32,87,263 km² (FSI, 2011).

Biological diversity or biodiversity is the total variability within all living organizations and the ecological complexes they inhabit. It includes bacteria, viruses, fungi, and other microorganisms, as well as animals and plants. Presently, all organisms are grouped into five kingdoms, namely, Animalia, Fungi, Monera, Plantae (plants) and Prostista. Kingdom Plantae (plants) covers multicellular, photosynthetic plants, which occur on land, on sea shores, in lakes, and in streams. They are mainly land plants and aquatic, often marine, red, brown and green algae. Land plants have been divided into two major groups, namely, Bryophytes, consisting of the mosses and liverworts,

Table 1.3 Increasing forest cover in different forest type groups

S. No.	*Group name*	*% of Forest Cover*
1.	Subtropical Dry Evergreen Forest	0.03
2.	Tropical Dry Evergreen Forest	0.13
3.	Littoral and Swamp Forest	0.69
4.	Montane Wet Temperate Forest	0.69
5.	Himalayan Dry Temperate Forest	0.84
6.	Tropical Thorn Forest	2.25
7.	Sub-Alpine and Alpine Forest	2.55
8.	Subtropical Pine Forest	2.63
9.	Subtropical Broad leaved Hill Forest	2.69
10.	Tropical Wet Evergreen Forest	2.92
11.	Himalayan Moist Temperate Forest	4.12
12.	Plantation/TOF	5.07
13.	Tropical Semi-Evergreen Forest	13.79
14.	Tropical Moist Deciduous Forest	19.73
15.	Tropical Dry Deciduos Forest	41.87
	Total	100

which survive only in moist places, and a large group of vascular plants. Plants are the major producer of biomass on the earth. The term "plants" generally refers to vascular plants, viz., the green plants around us. They consist of distinctive organs, namely, leaf, stem, and root. The vascular system, composed of xylem and phloem, regulates water and, inorganic and organic substances in these organs. Vascular plants (mainly flowering plants) may be classified into three groups, namely Angiosperms (plants forming fruit and seed), Gymnosperms (plants with seeds but no fruit) and Ferns (plants without flower and seeds, but produce spores in sporangia on their leaves). There are about 2,50,000 species of Angiosperms and form the largest group of existing plants on the earth. Angiosperms may be annuals, biennials or perennials, some of which survive for hundreds of years. They also vary in their sizes from 1 mm *(Wolffia)* to 100 m *(Eucalyptus)*. The leaves, stem, and roots of Angiosperms may be modified for different functions. These plants yield timber, food, fibre, medicine, dyes, pesticides, and a number of other useful products for the existence of humans and animals.

Seed plants without flowers are known as Gymnosperms and survive for several years. There are about 650 species of living Gymnosperms, of which cone-bearing plants (conifers) form the largest group. Pine, fir, spruce, cedar, cypress, and larch species are some of the examples. Conifers produce adequate amounts of soft wood, which is used in making plywood, particle board, construction, packing, paper, and other useful products. Vascular plants without seeds are called ferns or pteridophytes and consist of about 10,000 plant species. These are mainly found in humid tropics, and a few ancient ferns grow to over 20 m. Majority of ferns, have pinnately compound leaves, i.e., the blade is subdivided into leaflets arranged in the style of a feather. Most of the ferns, except tree ferns have horizontally-growing rhizomes. Ferns are often the first plants to appear after forest fire (Mahadevan et al., 1988).

Small plants that grow densely together in moist and shady places are known as Bryophytes. They include mosses and liverworts and comprise about 14,500 species. In the rainy season, they also form green carpets or mats on damp soil, rock walls, and barks of trees. Because of the absence of vascular tissue in bryophytes, water and food materials are transported from cell to cell. Moreover, in the absence of true roots, their cells absorb moisture directly from the ground and atmosphere, and thus, these can thrive only in the moist places. A moss plant is composed of an erect stem-like structure, densely dressed with small green, leaf-like outgrowths arranged spirally. Mosses are attached like outgrowths arranged spirally. They rarely grow more than 15 cm in height and help in binding soil and thus preventing soil erosion.

Forests provide a wide range of goods and services. Goods include timber, fuel-wood, fodder, and other non-timber forest products (NTFPs) (i.e., products of vegetable, mineral, and animal origin). The important services include conservation of ecosystems, maintenance of quality and supply of water at a sustained rate, release of nutrients from soils, removal of harmful chemicals from the air, climate control and prevention or reduction of the severity of avalanches, droughts, erosion, and floods. Forests also provide a wide range of economic and social benefits, such as forest products, employment, and protection of places of cultural value (FAO, 2006). Presently, forests have very vital roles to play due to increased environmental pollution and services rendered by them in reducing the same. Forests conservation also protects the biodiversity, watershed and landscape beauty. The increasing awareness about the services provided by forests all over the world, will help the forests owners (Government and communities of poor people) in increased revenue generation leading to their better livelihoods.

Forests, like other ecosystems, are affected by climate changes. They absorb carbon dioxide (CO_2) from the atmosphere, and store it in wood, leaves litter, roots and soil by acting as "carbon sinks"; however, burning of forests releases carbon back into the atmosphere. Forests also moderate the global climate by acting as sinks. It is estimated that the ecosystems store more carbon than the entire atmosphere (FAO, 2006; Kishwan et al., 2009). The country reports of FAO indicate that global forest vegetation stores 283 Gt of carbon in its biomass, and an additional 38 Gt in dead wood, for a total of 321 Gt, and IPCC (2000) assumed 359 Gt of carbon in these pools.

The total forests cover (21.05%) of India acts as a prominent sink of carbon dioxide, and according to estimates, is capable of removing 11.25% of India's CHG emissions (CO_2 equivalent) at 1994 levels. The comparative recent values for developing countries are available as per their separate National Communications (NATCOMS) to the United Nations Framework Convention on Climate Change (UNFCCC). This equals to 100% emissions from houses and in transportation or agricultural emissions of 40%.

It is, thus, evident that India's forest and tree covers serve as a major mode of carbon mitigation for India and the entire world. According to an estimate, the total carbon storage in Indian forests is likely to increase to 6,998 mt by 2015 (MOEF, 2009). Therefore, plants are of immense importance for the survival of mankind. Normally, a tree lives for about 50 years and its value is priceless as it generates oxygen for human survival, recycles fertility, contributes to soil erosion control, air pollution control, shelter for birds and animals, regulates groundwater table and number of other ecosystem services. Dr. Tarak Mohan Das of calcutta University quantified the services rendered by a tree, under eight heads, for the first time in 1979 (Das, 1979). The total price of these service was estimated to be Rs. 15,70,000/- at that time. It now amounts to about Rs. 3,55,13,000/-. According to Delhi Greens, the economic value with respect to oxygen production of a healthy tree about Rs. 23,72,50,000/- per year (Delhi Green, 2013)

Forest products are divided into two groups, namely, wood and non- wood forest products (NWFPs). Wood includes timber and fuelwood, and all other products obtained from forests are covered under NWFPs, which are defined as the goods of biological origin other than wood, derived from forests, other wooded land and trees outside forests. These include mainly medicinal plants, essential oil-yielding species, spices, gums, resins, gum-oleo-resins, fatty oils, tanning and natural dye materials, edible plants; fodder and forage plants, beverage-yielding plants, saponin-yielding plants, fish, insect, and rodent poisons, green manures, rubber other latex yielding plants, beedi leaves, bamboos, canes, fibres, and flosses. Animal products such as lac, honey beeswax, tussar, ivory, wild animal hides, skins, bones and horns, bird feathers, and, musk are also classified as NWFPs (Murthy, 1974; Anon, 2009a). Due to ban on cutting of forest trees, NWFPs are now the main source of revenue generation from the forests. Besides meeting the indigenous demands, they also earn valuable foreign exchange.

Natural dye is an important NWFP. Woods of some forest plants also yield natural dyes. The use of natural dyes began with the dawn of civilization. In the past, natural dyes were obtained from different parts of plants, viz., fruits, leaves, berries, seeds, flowers, and roots. Later on, natural dyes were also obtained from minerals and animals. Primitive people used these dyes for colouring the walls of caves, improving their appearance by painting their bodies, and for other uses. The extensive use of these dyes began after the development of the art of weaving. These dyes were also used for making food more attractive by colouring it. But, after the discovery of first synthetic dye by the William Henry Perkin in 1856, the use of natural dyes started decreasing and by beginning of the 20th century, their use decreased to minimum after the synthesis of dyes such as alazirin, indigo and a number of other azo dyes by cost-effective synthetic methods. As a result of cost-effectiveness of synthetic dyes, natural dyes almost disappeared from the commercial trade. However, it is difficult to replace the food dyes like turmeric, and annatto, which are of medicinal value. Synthetic dyes cause environmental pollution during their production and use, which is a matter of concern. Due to the growing concern about the health of people and safety of environment, there is a need to find alternatives to synthetic dyes. The safest may be the use of the natural dyes, which are not only eco-friendly but also prevents environmental pollution. These dyes are mostly obtained from plants and no major laboratory reactions are required to prepare them, thereby keeping a check on air and water pollution. Cultivation of plants for natural dye on degraded land will also help in improving the environment. But, very little organized effort has been made to revive and promote the use of natural dyes.

References

1. Anon (2009). *India's Fourth National Report on the Convention of Biological Diversity,* Ministry of Environment and Forests, Government of India, New Delhi.
2. Anon (2009a). *Training Module on Capacity Building of Communities Involved in Sustainable Forest Management.* Published by Forest Research Institute, Dehra Dun, India, pp. 123-128 (Katha), pp. 129-143 Natural Dye, pp. 145-150 (Tannins)
3. Bawa, R, Gupta, R, Rai, K, Dutt, B, Brahmi, M.K. [Editors] (2009). *Advance Training on Non-timber Forest Produce.* Published by Dr. Rajan Bawa, Coordinator, Regional Centre, Nature Afforestation and Eco Development Board, Dr. Y.S. Parmar University of Horticulture and Forestry, Nauni, Solan, Himachal Pradesh, India. 180-184
4. Das, Tarak Mohan (1979). *Indian Biologist,* Vol. XI, No 1-2.
5. Delhi Greens (2013). Report on economic valuation of oxygen supplying ecosystem service of healthy trees, August 13, 2013. *The Times of India,* August 20, 2013.
6. FAO (2006). *Global Forest Resources Assessment 2005: Progress Towards Sustainable Forest Management.* FAO Forestry Paper 147. Food and Agriculture Organization of the United Nations, Rome.
7. FSI (2011). *India State of Forest Report,* 2011. Published by Forest Survey of India, Ministry of Environment and Forests, Government of India, Dehra Dun, India.
8. IPCC (2000). *Land-Use Change and Forestry.* Special report, Cambridge University. Intergovernmental Panel on Climate Change, Geneva.
9. Kishwan, J., Pandey, R., Dadhwal, V.K. (2009). *India's Forest and Tree Cover: Contribution as a Carbon Sink,* Technical Paper. Indian Council of Forestry Research and Education, Dehra Dun, India.

9a. Mahadevan; Futehally, Z., Lahiry, D., Biology (Part II). National Council of Educational Research and Training, New Delhi, pp. 159-168, 217-240.

10. Murthy, K.T. (1974). *Minor Forest Products of India: Their Availability and Demand, Utilization and Exploitation and Suggestions for Their Development.* Paper presented in Tenth Commonwealth Forestry Conference. pp. 1-18.
11. MOEF (2009). *India's Forest and Tree Cover. Contribution as a Carbon Sink.* Ministry of Environment and Forests, Government of India, New Delhi.
12. Singh, R.V., Khanna, D. (2008). *The Role of Civil Society in the Forestry Sector in India.* Published by The Director; Forest Research Institute, Dehra Dun, India. p. 56.

2 History of Natural Dyes

INTRODUCTION

The concept of natural dyes is not new to the world as these have been used by people since time immemorial. The stains left behind by juices on the hands of primitive people while collecting bark, flowers, fruits, leaves, roots, and other parts of various plants for foods and other daily needs led to the discovery of natural dyes. Some mineral sources and materials isolated from a few animals were also used as dyes. Cave paintings made by prehistoric people about 25,000 years ago clearly demonstrate their knowledge of making colours from natural materials (Tortora, 1978). It is difficult to predict the exact period of the start of dye art in the world. In the beginning, the plant materials might have been applied directly on the items to be dyed, viz., walls of caves, baskets, pottery, body, garments, and many other items of daily use, by primitive people. Later on, the juices of plants were extracted and used for dyeing. The use of natural dyes is deep rooted into ancient history of India and other equally primeval civilizations of the world. China, Egypt, Mexico, Peru, and other countries are not only known for their ancient civilizations, but also for the use of natural dyes. The earliest written record of the use of dyestuffs in 2600 B.C. can be traced to China (Siva, 2007). The dyeing technology appears to have been originated in India during the ancient period, which later spread to Egypt, Persia, and Phoenicia (Wali, 1989). The credit goes to India and China for being the original homes not only for some dyes, but also for dyeing auxiliaries, notably those used in Europe before the discovery of synthetic dyes (Gulati, 1989). Archaeological evidence indicates that dyeing was an industrial business venture in Egypt, India, and Mesopotamia around the third millennium B.C. (Roy, 1978). Numerous different natural sources of dyes have been documented in the ancient records. Natural dyes were acquired from plants, animals, and mineral sources by all countries of the world since time immemorial. Breeding of plants and isolation of dye from their different parts have been an important industry from earlier times till the nineteenth century.

In ancient days, a dye, either alone or with a mordant, or a pigment mixed with a binding material was used for colouring textiles. Most of the dyes are of organic nature, while a pigment consists of inorganic substances derived from minerals (ores). Dyes were not only used for imparting different colours to textiles, but also by manufacturers of cosmetics, drugs, foods, furs, leather, paints, paper, perfumes, printing inks, and photographic films. Nowadays, dyes are also used as stains in histological and bacteriological works, as well as indicators. Natural dyes were exclusively used for colouring till the middle of the nineteenth century. However, due to accidental discovery of the first synthetic dye by W.H. Perkin in 1856, azo dyes in 1864, and a number of other synthetic dyes in the later part of the nineteenth century, the use of natural dyes declined to minimum by the beginning of the twentieth century, and almost completely vanished from the scene of dyeing and printing. Ten dyestuffs, namely, indigo, woad blue, cochineal, lac, madder red, logwood black, fustic, weld, safflower, and saffron yellow, were, however, most prominent due to their trade in the ancient world. These dyes are now available in very limited amounts. It is surprising to know that by the twentieth century, natural dyes stood at the verge of extinction. However, God is omnipotent and nothing is permanent in this world, and the same happened with natural dyes.

HISTORY OF NATURAL DYES IN INDIA

Dyeing and paintings are primitive traditions in India. Indigo has been in use in India for the last 3500 years (Thrope, 1921; Gulati, 1989). Pitamber, a yellow garment, was used in the Indian Epic period. Natural dyes have been described in *Atharvaveda*, and *Bhrigu Samhita* was written using natural dye (Srivastava, 1989). The details of the possible use of natural dyes starting from vedic period (particularly in atharvavedic and succeeding period) in chronological order are given in Table 2.1 (Mohanty, 1989; Roy, 1978).

Natural dyes were also used by people during the period of Indus Valley Civilization (IVC). The IVC, one of the three primeval civilizations, existed during the period 3300-1300 B.C. and now has its ruins in Pakistan, northwest India, and eastern Afghanistan. The civilization covered around 1,260,000 km^2, constituting the largest ancient civilization. The other two ancient civilizations of Mesopotamia and Egypt also existed along with IVC. The IVC is also known as the Harappan civilization as the excavations in 1921-1922 led to the discovery of the first city of IVC at Harappa by Sir John Hubert Marshall, Rai Bahadur Daya Ram Sahni, and Madho Sarup Vats. It was declared as a UNESCO World Heritage Site in 1980 due to its utmost significance (UNESCO, 2010). A research carried out by an international research team (in 2003 and 2008) led by Dr. Liviu Giosan, a geologist with the Woods Hole Oceanographic Institution, suggested that the decline in monsoon rains led to weakened river dynamics, which might have played a key role in both the development and fall of the Harappan culture. The research findings published in the proceedings of National Academy of Sciences also indicated that the mythical Saraswati River was actually not fed by Himalayan glaciers, as believed earlier. Instead, it was found to be a perennial monsoon-supported watercourse, and aridification reduced it to short seasonal flows. The digital maps of the landforms constructed by the Indus and other neighbouring rivers were made by using satellite photos and topographic data. The field studies were carried out by drilling, coring, and digging trenches. The samples collected were used to determine the sediments' origins, whether brought in by rivers or wind, and their age, in order to develop the chronology of landscape changes. The latest research findings (2014)

Table 2.1 Natural dyes used during ancient period

S.No.	*Period*	*Year*	*Dye used*	*Nature*	*Part used*
1.	Vedic period	c. 1500 B.C.-500 B.C.	The tinctorial properties of the following recognised (a) Asikni (probably indigo from *Indigofera tinctoria*)	Vegetable	Whole plant
	(Atharvavedic and succeeding periods)		(b) Haridra (Turmeric, *Curcuma longa*)	Vegetable	Rhizomes
			(c) Lodhra (*Symplocos racemosa*)	Vegetable	Bark, Wood
			(d) Mahāranjana (Safflower, *Carthamus tinctorius*)	Vegetable	Flowers
			(e) Manjistha (Madder, *Rubia cordifolia*)	Vegetable	Whole plant
			(f) Lac (*Laccifer lacca*)	Animal	Exudate from female insect body
2.	Post-Vedic period	c. 500 B.C-c. 3rd century A.D.	The following dye-stuffs were introduced. (a) Nila (Indigo, *Indigofera tinctoria*)	Vegetable	Whole plants
			(b) Saffron (*Corcus sativus*)	Vegetable	Flowers
			(c) Krmi (Kermes)	Animal	Insects
			(d) Rocāna	Animal product, cow urine	Cow urine as a source of yellow substance
			(e) Gairika (Red – Ochre)	Mineral	Mineral
			(f) Khanjana (Carbon black)	Carbon	Carbon black
3.	Classical age upto Medieval period	320-550 A.D.	A number of natural dyes used. Some prominent ones are: (a) Kampillaka (*Mallotus philippensis*)	Vegetable	Seeds
			(b) Jatuka (*Oldenlandia species*)	Vegetable	Roots
			(c) Pattanga (*Caesalpinia sappan*)	Vegetable	Wood

Contd...

S.No.	Period	Year	Dye used	Nature	Part used
			(d) Indragopa (*Cochineal*) Black dyes were also developed using following sources	Animal	Insect
			(e) Abhaya (*Terminalia chebula*)	Vegetable	Rind of the fruits
			(f) Amalaka (*Emblica officinalis*)	Vegetable	Fruits
			(g) Bhringaraja (*Eclipta prostrata*)	Vegetable	Whole plant
			(h) Nila (*Indigofera tinctoria*)	Vegetable	Whole plant particularly leaves and flowers
			(i) Pippala (*Piper longum*)	Vegetable	Fruits
			(j) Ayas (Iron)	Mineral	Mineral metal
			(k) Kasisa (Iron sulphate)	Inorganic salt	Salt
			(l) Tuttha (Copper sulphate)	Inorganic salt	Salt
			(m) Anjana (Antimony sulphate)	Inorganic salt	Salt
4	Medieval period	5-15th century A.D.	(a) Process for the extraction of colouring principles from the dye yielding material and mordanting properties of tuvari (alum) were discovered.	Inorganic salt	Salt
			(b) Process for the use of iron mordant and alum for fixing different colours were discovered in the late medieval period(Eighteenth century A.D.).	Inorganic salt	Salt

by a team of scientist of the University of Cambridge and Banaras Hindu University suggested the destruction of IVC by climatic changes (Sinha, 2014).

The relics from the excavations of IVC (Harappan culture) have produced manifestation of fabrics and ropes dyed with natural colours by the people of Mohenjo-Daro around 3000 B.C. A purple-coloured cotton piece was found sticking to silver vase, indicating that people of Mohenjo-Daro civilization used natural dyes. Chemical examination of the piece found it to be dyed with Indian madder or manjith *(Rubia cordifolia)* (Marshal, 1931; Gulati, 1989). The pottery excavated also revealed the presence of natural dyes.

More than 300 dye-producing plants are listed in various historical manuals, and one can find a number of references to them in Sanskrit works, such as the *Arthashastra* and the *Atharvaveda*. Indigo was the most valuable dye that originated from Indian subcontinent. Safflower, which imparted bright pink colour, was cultivated as an agriculture crop in India during the Chandragupta Maurya's kingdom in 300 B.C. Around 350 dye-yielding plant species were used by talented Indian craftsmen, as early as three years before the Christian era (Gulati, 1989). Many earlier textiles items dyed using natural dyes do not exist today due to their biodegradable nature. However, exhibition of wall paintings at Ajanta, Bagh, Ellora, and other sites show that dyeing and paintings were well known in India in the past (Agarwal and Tiwari, 1989). India took a lead role in perfecting the art of dyeing in the ancient times. The Ajanta Caves, situated in Aurangabad district of Maharashtra in India, are 30 rock-cut cave monuments, which date from the second century B.C.E. to the 600 C.E. The caves comprise paintings and sculptures considered to be the masterpieces of Buddhist religious art (Ajanta Caves, 2006) and the reminiscent frescos of the Sigiriya paintings in Sri Lanka (UNESCO). The caves were built in two phases. The construction of the first phase was started in the second century B.C.E. and the second group of caves was constructed around 600 C.E. (Ajanta Caves, 1982). The palace decorations and the ceilings of the temples of Hallebid and Bellur in Karnataka strongly affirm to the comprehensive skills of the Indian craftsmen in using natural dyes (Srivastava, 1989).

Natural dyes were also used during the pre-Muslim and Muslim periods in India, and the indication of their use is preserved in the form of paintings, manuscripts, and dresses of that period. Some of the records of court historians and biographers like Firdausi were written and illustrated with natural dyes. The naturally dyed silk and muslin fabrics of India were famous in different countries of the world during the sixteenth and seventeenth centuries (Srivastava, 1989). The garments found in Egyptian tombs were found to be dyed with indigo and safflower from India (Chandramouli, 1989). The technique of dyeing and printing reached its zenith by the Medieval period, and Indian textiles were known throughout the world (Agarwal and Tiwari, 1989).

Natural dyes were not only used for dyeing, printing, and paintings, but also as cosmetics by womenfolk. Henna (mehndi powder) was introduced in India around 712 A.D. by Muslims at the time of conquering Sind for the first time (Gulati, 1989). Mehndi powder was used by women, especially brides, for colouring palms and soles.

Catechu with lime and betel leaf was used for tinting lips, "Missi" for gums, and Surma to sharpen the effect of eyelashes. Natural dyes were also used, and are still being used, for colouring food items. Turmeric (haldi), saffron (kesar), and Kashmiri chilli are some of the examples of food colours (Srivastava, 1989). According to the *Encyclopedia Britannica,* turmeric was, and is still being used as a colouring matter and condiment in India for a very remote period. There is no synthetic substitute for these dyes. Tattooing is also as old as the use of natural dyes. It was used for decoration, to have a distinctive mark or to ward off evils. A tattoo on the human body can be made by using a pigment, namely, charcoal powder, gunpowder, indigo, soot, ultramarine, or vermillion.

India took lead for the introduction of widely used dyes, namely Al, indigo and manjith, turmeric, annatto, and other dyes throughout the world in the past. The roots of Saranguy or "Al" (*Morinda citrifolia* or *Morinda tinctoria*) were used for dyeing during the seventeenth century A.D. The "Al" roots were used as a cheap substitute of madder in many parts of India. The blue dress of British navy was dyed with Indigo (Leggett, 1944). Alexander the Great used the red dye obtained either from Kermes or madder in the battle against Persians. A number of Indian dyes rose to international fame till the middle of the nineteenth century, and due to their immense importance, these were exhibited at industrial exhibition held at London in the year 1851 (Table 2.2, Royle, 1851; Bhardwaj and Jain, 1982).

ANCIENT EXTRACTION METHODS

(a) General methods of extraction of vegetable dyes

Water, the universal solvent, was used for the extraction of vegetable dyes due to their solubility in it. Hot water is usually preferred for this purpose. In the past, aqueous alkaline or acidic solution was also used for the extraction of dyes, depending upon the presence of functional groups. The aqueous dye solution was separated from plant material by filtration through a muslin cloth. Removal of water yielded the solid dye in powder form or as a hygroscopic solid, depending upon the nature of chemical compounds present in the dye.

(b) Extraction of indigo

The extraction of the blue dye indigo from *Indigofera tinctoria* around 200 years back was carried out in two types of vats, namely, *khari* or alkaline vat and *mitha* or sweet vats. Caustic lime, gur, and sajji were the constituents of the alkaline vat, whereas slaked lime and sugar (in the form of *sira*) were used in the sweet vat. The latter method was more predominant due to higher dye yield (Bhardwaj and Jain, 1982). The leaves of *Indigofera tinctoria* were soaked in water in a vat to obtain the soluble glucoside indican in the liquid extract. The enzymes present hydrolyzed indican to glucose and indoxyl. After the completion of fermentation process, the greenish-yellow, sweet-smelling liquid was separated and agitated with sticks for about three hours for its oxidation. This converted indoxyl into grains of insoluble blue coloured indigo (indigotin) coagulated and settled at the bottom, which was separated, washed, and

Table 2.2 Indian vegtable dyes in London exhibition (1851)

S. No.	*Botanical name*	*Family*	*Dye name*	*Place of origin at that time*
1.	*Morinda citrifolia* Linn.	Rubiaceae	Al	Budelkhand and Sagar
2.	*Bixa orellana* Linn.	Bixaceae	Annatto	Dacca
3.	*Oldenlandia umbellata Linn.* syn. *Hedyotis umbellata* (Linn.) Lam.	Rubiaceae	Chay root	Dyndygal, Nellore, Masulipatam
4.	*Indigofera tinctoria* Linn.	Fabaceae	Indigo	Bengal, Madras
5.	*Terminalia belerica* Roxb.	Combretacae	Lodh	Himalayas
6.	*Haematoxylon compechianum* Linn.	Fabaceae	Logwood	Calcutta Botanical Garden
7.	*Rubia cordifolia* Linn.	Rubiaceae	Manjith	Assam, Bombay, Nepal, Quetta and Scindiah's territory etc.
8.	*Terminalia chebula* Retz.	Combretaceae	Myrobalan	Bengal
9.	*Pterocarpus santalinus* Linn.f.	Fabaceae	Red Saunders	Madras
10.	*Carthamus tinctorius* Linn.	Asteraceae	Safflower	Dacca
11.	*Caesalpinia sappan* Linn.	Fabaceae	Sappan Wood	Mergui, Malabar
12.	*Curcuma longa* Linn. syn. *C. domestica* Val.	Zingiberaceae	Turmeric	Bengal, Malabar

pressed into cakes for use. At times, some people bypassed the fermentation step and instead added a decoction of tannin-rich *Syzygium cuminii* bark in the vat containing indigo liquor. The acidic nature of decoction probably assisted in the decomposition of indican into indoxyl and indigotin without undergoing fermentation. The gums or juices of some plants, such as butea gum,were found to be helpful in precipitating the dye (Watt, 1972). The high-quality indigo contained 50% to 70% of indigotin, which varied from 6% to 9% in the inferior quality (fig indigo).

(c) Extraction of the animal dye: Lac dye

The twigs of the host tree, on which the lac insect *Laccifera lacca* occurred as natural pests, were cut into pieces and crushed to separate the resinous material. This dye-containing material was put into water-containing tubs and beaten with a wooden pestle or trodden underfoot. The red-coloured water solution was then filtered to remove the foreign material. Complete removal of water by evaporation yielded the dye, which was compressed into cakes (Watt, 1972).

(d) Extraction of the animal dye: Cochineal dye

Insects *Dactylopius coccus,* which host on *Opuntia species,* yielded the cochineal dye (Leggett, 1944).The insects after collection from plants were either dried in sun or killed by putting them in hot water. The dye can be extracted by hot water. An insect yields only small quantity of dye; therefore, a large number of insects were needed to produce the dye in considerable amount. According to an estimate, one pound of cochineal dye could be obtained from about 70,000 insects.

(e) Extraction of minerals dyes

Simple procedures were involved in the use of mineral dyes for dyeing. For example, the two mineral dyes, namely, Red ochre (hirmāji) and Yellow ochre (ramraj), used for dyeing were obtained by grinding the ochreous clay into powder and mixing this powder with water.

MORDANT USED FOR DYEING

Most of the natural dyes are mordant dyes. A mordant fixes the dye on the fibre due to its affinity for both. Some inorganic salts and plant mordants were used for dyeing in the eighteenth and nineteenth centuries and even earlier. Alum, blue vitriol, iron, tin, and chromium salts were used as inorganic mordants, while myrobalan, lodh bark, and lemon juice constituted plant mordants. Sometimes, a few dyestuffs in conjunction with other dyes served as mordants. Alum, a non-toxic mordant, was used with chay, dhak, dhau, kamela, madder, patang, and turmeric dyes. Myrobalan was used as mordants for dyeing with al, bel, haldi, manjith, safflower, and tesu. Lodh bark was used as a mordant with al, dhak, manjith, and patang, while imli was found suitable as a mordant for dyeing silk with safflower. Post-mordanting or simultaneous mordanting methods were used for fixing dyes (Bhardwaj and Jain, 1982).

DYEING METHODS

Different colours, such as blue, black, green, red, and yellow were produced on cotton, silk, and wool in the past. A number of dyes, mordants, and chemical assistants were used for dyeing these fabrics. The general methods used earlier (about 150 years back) for dyeing of cotton, silk, and wool are described in Table 2.3 (Bhardwaj and Jain, 1982). The methods used in the past for dyeing cotton by using different natural dyes as given in *Nuskha Khulasatul Mujarrrebat*, and some of them later translated and published (Naqvi, 1980), of which a few are described in Table 2.4. The quantities of the plant material used for dyeing have been mentioned as nearest ratio which was calculated by taking 1 seer (933.10 g), 1 tola (11.664 g) and 1 dām (20 g). Some important natural dyes of vegetable (plant species) origin used for dyeing cotton, silk, and wool in the past (18–19th century) are given in Table 2.5.

Table 2.3 General dyeing methods used in the past

Type of fibre	*Dyes used*	*Dyeing*
Cotton	*Indigo, madder, Turmeric, Safflower, Harsinghar, Patang, Dhak, Al, Pomegranate rind, Myrobalans and other dyes*	(i) Impregnation of cotton with an aqueous solution of dung. (ii) Washing of cloth and its drying in the sun. (iii) Immersing of cloth in oil and alkali for making the fibre soft and removal of dirt if any. Castor oil and *Sajjikhār* or *papadkhār* were the frequently used oil and alkali. (iv) Dyeing and mordanting of the cloth was done.
Silk	*Catechu, Cochineal, Indigo, Kamala, Lac, Red Ochre, Safflower, Sappan Turmeric, Yellow ochre and other dyes*	(i) Dewaxing of raw silk by washing with alkali solution. (ii) Mordanting of the washed silk with alum. (iii) Dyeing of the mordanted silk in a particular dye solution. The acidic pH of the dye solution was maintained as per the need.
Wool	*Babul bark, Catechu, Cochineal, Dhak, Indigo, Lac, madder, Myrobalans, Pomegranate rind, Tamarind, Turmeric and other dyes*	(i) Washing of yarn with an alkali mainly reh and *sajji*. (ii) Mordanting of cleaned yarn with mordants like alum, gall nuts, lodh bark, tamarind and other mordants depending on the colour shades required to be produced. (iii) Dyeing of the mordanted yarn in a particular dye solution to obtain the desired colour/shade (Pim, 1898).
Chintz printing	*By using different dyes*	A speciality cloth item having a variegated or spotted design known as Chintz was developed in India. Its printing involved four steps, namely, oiling, making of faint lines on cloth to help the printers, producing fast the marks previously stamped, and making of the different colours and shades. The method was varied as per the requirement (Hadi, 1896).

NATURAL DYE TRADE

Some Indian dyes, viz., indigo, lac, safflower, and madder, had a very good commercial value due to their superior quality and were exported for revenue generation during

S. No.	*Constituents used for dyeing (Ratio)*	*Procedure*	*Colour imparted*
12.	(i) Turmeric (ii) Alum (iii) Pomegranate rind (iv) Henna (v) Bright madder (vi) Catechu *Paparya* (3:1.5:2:2:1:1)	Crush the turmeric in water and strain it through a piece of cloth. Immerse the material to be dyed in the liquid for about one and a half hours. Wash with clean water and dry it. Dip it into alum solution and dry it. Pound the pomegranate rind, henna, catechu and madder together with water, and separate the dye solution through a piece of a clean cloth. Dye the material in this solution and dry it. Repeat the process once more. There after add *ahar* in the same solution and again dip the material in it, dry it and finally, beat it smooth.	Green
13.	(i) Turmeric (ii) Alum (iii) *Shahab i khasa* (iv) Water (v) Lemon (vi) Starch- As per requirement (3:1:6:38:3)	Dye the material in the liquor obtained on pounding the turmeric with water, and wash twice. Submerge the material in alum solution and dry it. Add *shahab* in about three litres of water, steep and leave the material in the solution for a while. Rub the lemon with water, strain it, and add it in the *Shahab* solution, again dip the material, rub it with hand, take it out. Add starch in the same solution and keep the material for some time to absorb it, dry the material.	Like Amber
14.	(i) Vermillion (ii) Saffron (iii) Starch-according to requirment (41:1)	Crush the vermillion with water, dye the material and dry it. Boil saffron in water, filter it, dye the material in the solution and dry it. Add starch in the same liquid, dip the material and dry it.	Sandali (Sandal) colour

Table 2.5 Some Important Indian sources of vegetables dyes used in the past

S.No.	Name of plant species	Family	Common name	Part used	Pattern	Colour
1.	*Acacia catechu* Willd.	Mimosaceae	Khair	Wood	Tr	Brown
2.	*Acacia concinna* (Willd) D.C	Mimosaceae	Rutta	Leaf, fruit	Sh	Brown
3.	*Acacia leucophloea* Willd.	Mimosaceae	Safed Kikar	Wood	Tr	Brownish red
4.	*Adhatoda zeylancia* Medic.	Acanthaceae	Arusa	Leaves	Sh	Yellow, grey
5.	*Aegle marmelos* (Linn.) Correa ex. Roxb	Rutaceae	Bel	Rind of fruit	Tr	Yellow
6.	*Alpinia galanga* (Linn.) Willd.	Zingiberaceae	Kulanjan, Barakulanjar	Roots, stalk	H	Yellow, brown
7.	*Anogeissus latifolia* (Roxb.) Wall. ex Bedd.	Combretaceae	Dhaura	Leaves, bark	Tr	Black
8.	*Areca catechu* Linn.	Arecaceae	Supari	Nuts	Tr	Red, brown
9.	*Artocarpus heterophyllus* Lam.	Myrtaceae	Kanthal, Kathal	Wood	Tr	Yellow
10.	*Berberis lycium* Linn.	Berberidaceae	Kashmal	Roots	H	Yellow
11.	*Bixa orellana* Linn.	Bixaceae	Annatto	Seeds	Tr	Orange, red
12.	*Bombax ceiba* Linn.	Bombacaceae	Semal	Crimson flowers	Tr	Reddish brown
13.	*Butea monosperma* (Lam.) Taub.	Fabaceae	Dhak	Flowers	Tr	Yellow
14.	*Caesalpinia sappan* Linn.	Fabaceae	Patang,Sappan Wood	Wood	Tr	Red
15.	*Carthamus tinctorius* Linn.	Asteraceae	Kusum (Safflower)	Flowers	H	Red
16.	*Cassia auriculata* Linn.	Fabaceae	Tarwar, Tanner's Cassia	Flowers	Sh	Yellow
17.	*Casurina equisetifolia* Linn.	Casurarinaceae	Jangli saru	Bark	Tr	Brown
18.	*Cassia fistula* Linn.	Fabaceae	Amaltas	Bark, sapwood	Tr	Red
19.	*Cedrela toona* Roxb. ex Rottl. & Willd. syn. *Toona ciliata* M.Roemer	Meliaceae	Tun	Flower, sawdust	Tr	Red, pink

Contd...

S.No.	Name of plant species	Family	Common name	Part used	Pattern	Colour
20.	*Ceriops tagal (Perr.)* C.B. Robins	Rhizophoraceae	Goran	Bark	Mangrove	Brown
21.	*Cinnamomum tamala* (Buch.-Ham.) Nees & Eberm.	Lauraceae	Dalcini, Tejpat	Leaves	Tr	Brown
22.	*Crocus sativus* Linn.	Iridaceae	Kesar	Flowers	H	Yellow
23.	*Curcuma amada* Roxb.	Zingiberaceae	Am-haldi	Rhizomes	H	Yellow
24.	*Curcuma longa* Linn.	Zingiberaceae	Haldi	Rhizome	H	Yellow
25.	*Curcuma zedoria* (Christm.) Rosc.	Zingiberaceae	Kachura	Rhizome	H	Yellow
26.	*Datisca cannabina* Linn.	Datiscaceae	Akalbir	Root	H	Yellow
27.	*Diospyros embryopteris* Linn.	Ebenaceae	Makartendi	Fruit	Tr	Brown
28.	*Emblica officinalis* Gatern.	Euphorbiaceae	Avala, Amla	Fruits, bark	Tr	Blue, black
29.	*Syzygium cumini (L.)* Skeels.	Myrtaceae	Jamun	Bark, leaf	Tr	Red
30.	*Garcinia morella* Desr.	Clusiaceae	Tamal	Bark(exudate)	Tr	Yellow
31.	*Hedychium spicatum* Buch.-Ham.	Zingiberaceae	Kacri, Bagaldu	Rhizome	H	Yellow
32.	*Indigofera tinctoria* Linn.	Fabaceae	Nil, Indigo	Leaves	Sh	Blue
33.	*Lawsonia inermis* Linn.	Lythraceae	Mehendi	Leaves	Sh	Yellow-orange, red
34.	*Mallotus philippensis* Muell. Arg.	Euphorbiaceae	Kamala, Kamela	Fruits	Tr	Red
35.	*Mammea longifolia* Planch. & Triana syn.	Clusiaceae	Suringi, Nagkesar	Flower buds	Tr	Red
36.	*Mangifera indica* Linn.	Anacardiaceae	Am	Bark, leaves	Tr	Yellow
37.	*Memecylon umbellatum* Burm. f	Melastomataceae	Anjani, Iron Wood Tree	Leaves	Sh or small tree	Yellow
38.	*Morinda citrifolia* Linn.	Rubiaceae	Al	Root, bark	Sh	Red, yellow
39.	*Myrica esculenta* Buch.-Ham.	Myricaceae	Kaiphal	Bark	Tr	Yellow
40.	*Nyctanthes arbortristis* Linn.	Oleaceae	Harsinghar	Flowers (Corolla)	Sh	Bright orange

41.	*Oldenlandia umbellata* Linn.	Rubiaceae	Cherivelu Chay or Chay root	Root	Woody annual	Red
42.	*Pistacia integerrima* Stewart. ex. Brandis	Anacardiaceae	Kakar-singi	Galls of leaves and petioles	Tr	Yellow
43.	*Pterocarpus marsupium* Roxb.	Fabaceae	Bijasal	Bark	Tr	Brownish red
44.	*Punica granatum* Linn.	Lythraceae	Anar	Rind of fruit	Tr	Yellowish brown, light red
45.	*Quercus infectoria* Olivier	Fagaceae	Majuphal	Gall nuts	Tr	Yellow
46.	*Rubia cordifolia* Linn.	Rubiaceae	Manjit	Root, stem	Cl	Orange, scarlet-red
47.	*Shorea robusta* Gaertn.f.	Diptocarpeaceae	Sal	Bark	Tr	Red, black
48.	*Symplocos racemosa* Roxb.	Symplocaceae	Lodh	Leaf, bark, wood	Tr	Yellow
49.	*Tagetes erecta* Linn.	Asteraceae	Genda	Petal	H	Yellow
50.	*Tamarindus indica* Linn.	Fabaceae	Imli	Leaves, flowers, fruits	Tr	Yellow
51.	*Tamarix troupii* Hole	Tamaricaceae	Jhau	Galls	Small Tree	Yellow, brown
52.	*Tectona grandis* Linn.f.	Verbenaceae	Sagwan	Leaves	Tr	Yellow, reddish
53.	*Terminalia belerica* Roxb.	Combretacae	Bahera	Bark, fruit	Tr	Blue
54.	*Terminalia chebula* Retz.	Combretaceae	Harra	Bark, ripe and Unripe fruit	Tr	Yellow, brown
55.	*Ventilago madraspatana* Gaertn.	Rhamnaceae	Surabi, Rakatvali	Root, bark	Cl	Chocolate, red
56.	*Woodfordia fruticosa* Kurz.	Lythraceae	Dhau, Dawi	Leaves, flowers	Sh	Pink, red

(Cl = Climber; H = Herb; Sh = shrub; Tr = Tree)

the eighteenth and nineteenth centuries. Besides the foreign trade, natural dyes also played a significant role in internal trade within the country. The useful dyes used for internal trade in different divisions of North-Western Provinces and Oudh during 1885-1886 to 1894-1895 were al, catechu, harsinghar, indigo, lac, madder, pomegranate, safflower, sappan, turmeric, etc. Lucknow was an important center of dyes (Bhardwaj and Jain, 1982).

After the discovery of synthetic dyes, the trade in natural dyes started declining. The Magenta dye was the first synthetic dye imported in India by a Bombay company "Cama Ramji" in 1867. In later years, Germans were successful in introducing synthetic alizarin in Indian market, followed by other synthetic dyes, thus leading virtually to the closure of indigenous dye industry. This is evident from the fact that the import of synthetic dyestuffs rose from Rs. 38.14 lakhs in the year 1887 to around Rs. 45.00 lakh in the year 1900 (Bhardwaj and Jain, 1982).Moreover, the export of Indian dye indigo declined from about 41.25 lakh pounds to 38.50 lakh pounds during the years 1896-1897 to 1901-1902, whereas the import of synthetic dyes increased from 2.5 lakh pounds in 1897 to 12.50 lakh pounds in 1903 (Ghosh, 1907). The use of natural dye reduced to minimum in the beginning of the twentieth century.

India was the leading manufacturer of natural dyes up to mid-nineteenth century and earned very good profit from this trade. The English rulers helped to grow the dye industry, especially the indigo production, during that period due to their own interest in exporting it and transferring the money to their country. No sooner did the trade started becoming ineffective due to introduction of synthetic dye in the foreign market and so they immediately switched over to the import of synthetic dyes to India instead of exporting natural dyes. Their sole interest was to benefit their own country. The Indian natural dye industry collapsed mainly due to introduction of synthetic dyes in the Indian market at reduced price by foreign countries like United Kingdom and Germany. The export of Indian dyes was also affected as these were taxed heavily in England and other European countries. Moreover, English people took very little interest in establishing chemical industries in India and revenue generated by export of various items, including natural dyes, was invested in the industrial revolution in foreign countries.

NATURAL DYES IN OTHER COUNTRIES

Natural dyes were also used in other parts of the world since time immemorial. A brief description of the dyes that were used in different countries of the world during ancient period has been described by Druding. The first written record of the use of natural dyes was found in China, dated 2600 B.C. (Druding, 1982). Mollusk was used for making Tyrian purple dye, which was discovered by Phoenicians in 1500 B.C., and was used for colouring on entire Mediterranean coast (Srivastava, 1989). The textiles dyed by using this dye were so expensive that only royal family could afford them. Purple dye was extracted from three sources, namely a small gastropod mollusk found in all seas or from a crustacean called trumpet shell or purple fish, found near Tyre, on the Mediterranean coast. A deep-purple fluid secreted by their body was harvested by cracking the shell and digging out a vein located near the shellfish head with a small, pointed utensil. The mucus-like materials of the veins were mixed together to get the dye. According to an estimate, around 8500 shellfish produced 1 g of the dye, thereby making it a very costly dye. Alexander the Great, on conquering the Persian capital Susa, found purple robes dating to 541 B.C. in its royal treasury. The chemical examination of red fabrics collected from the tomb of King Tutankhamen in Egypt showed the presence of alizarin, a pigment isolated from madder (Driessen, 2014).

The insect dyes like kermes and cochineal became more common by the fifteenth century. The wood dye isolated from logwood was introduced in England for dyeing of clothes in the seventeenth century. The cultivation of indigo was started in England, and cudbear, a natural dye prepared from a variety of lichens, was patented. In addition, a vegetable dye quercitron isolated from the inner bark of North American Oak was patented in 1775. Until the middle ages, European used the leaves of abundantly growing shrub "woad" for making a blue dye for colouring textiles/fabrics. This dye was replaced by indigo, an ancient shrub well known to Indians and Egyptians, and isolated from the leaves/branches of *Indigofera tinctoria* by fermentation process. Indigo was a preferred dye as the dyed fabric possesses excellent washing and light-fastness properties. Manufacturing of natural indigo continued till the beginning of the twentieth century. In 1905, Adolf von Baeyer was awarded the Noble Prize in chemistry for discovering the molecular structure of indigo and developing a process to produce it synthetically on a commercial scale. Natural indigo was replaced by synthetic, thus ending an ancient and honoured botanical history (Driessen, 2014). There is again an interest in natural indigo as one tonne of synthetic indigo produces considerable amount of waste material (pollutants). Because of continuous advancement, dyes like Brazilwood, indigo, madder, weld, woad, and a dark-reddish-purple dye were known in dyeing. Some useful information about the history of natural dyes is depicted in Table 2.6.

Table 2.6 Dyes used during ancient period

S.No.	*Period*	*Samples*	*Natural dye*	*Ref.*
1.	4th-12th century	Coptic-textiles from National Museum in Warsaw (Early Christian Art)	Anthraquinone, indigoid and flavonoid dyes	Orska-Gawrys et al., 2003
2.	6th-7th century	Coptic museum textiles in Cairo	Madder and indigoid dyes	Abdel- Kareem et al., 2011
3.	15th-19th century	Post-Byzantine era	Cochineal dye	Karapanagiotis et al., 2005
4.	16th-early 20th century	Mount Athos	Cochineal, madder, indigo and logwood	Mantzouris et al., 2011.
5.	19th-20th century	Romanian textiles	Cochineal and yellow dye	Petroviciu and Wouters, 2002

The dyed textile samples collected from the sites of a number of archaeological excavations were subjected to chemical examination in order to know the dyes used at that time. The findings are given below (Taylor, 1986):

(a) Chemical examination of natural dyes used in Egyptian textiles collected from different sites, from the fourteenth century B.C. till the medieval period, indicated that madder and ocherous earth were used for dyeing of flax in 1400 B.C. (Taylor, 1986). In 100 A.D., madder was also used for dyeing wool, while red dyes isolated from kermes and lac were found on the wool samples of 1500 A.D. A bluish-purple dye made from shellfish (*Murex trunculus*) was used for colouring the four tassels in the corners of the Tallit and the Jewish ritual garment and prayer shawl (Elliott, 1976; Zideman, 1981)

(b) Wool samples (Vindolanda Textiles) of around 90 A.D., constituting mainly of the Roman textiles, were tested for the presence of dyes used in the dyeing of these samples at that time. Some of the samples were found to be dyed with madder (Wild and Ryder, 1977; Taylor, 1986). The Saxon wool samples (London) of 900 A.D. to 1100 A.D. were found to be dyed either with lichen purple, indigotin, or madder dyes (Taylor, 1986). The wool samples (Viking Textile) collected from Dublin, dating back to the tenth and eleventh centuries, were chemically examined, which also indicated the presence of madder, indigotin, and lichen purple on them (Taylor, 1986).

(c) Wool and a few silk dyed specimens (Anglo-Scandinavian Samples of York) of the ninth to eleventh centuries were examined and found to have been dyed with madder, indigotin, and lichen purple (Taylor, 1986; Kok, 1966; Walton, 1986).

(d) Around 500 woven fabric samples dyed with yellow dyes (1500 A.D. to 1850 A.D.) collected from the museums of Europe and America were examined by chromatography techniques. The result indicated that 80% of the samples were dyed with the dye isolated from weld (Taylor, 1986). The main constituent of the dye was luteolin.

(e) A number of red-coloured silk and wool samples (1450 A.D. to 1600 A.D.) collected from different museums were analyzed by thin layer chromatography (TLC) for the identification of the dyes (Holenk de Graaff, 1976). The samples were found to be dyed with kermes and polish cochineal, and their use was diminished and substituted by American cochineal by the end of the sixteenth century. The use of madder also varied and decreased due to the introduction of cochineal during the sixteenth century. The dyes used in the well-known Flemish tapestry-weaving centre of Tourani were found to be indigotin, madder, and weld (Taylor, 1986). Samples of Netherland's tapestry of the fifteenth and sixteenth centuries were also dyed with indigotin, madder, and weld (Taylor, 1986). Alum and iron salt were used as mordants.

(f) Textiles dating back to the fifteenth and sixteenth centuries (Newcastle samples) were found to be dyed mainly with madder (Harbottle and Ellison, 1981; Taylor,

1986), while the samples of the seventeenth century contained mainly indigotin on them (Ellison and Harbottle, 1983; Taylor, 1986).

(g) Textiles samples of the fourteenth to the middle of the sixteenth century unearthed from Baynard Castle site, London, on testing indicated the presence of madder as the main dye used at that time (Harbottle and Ellison, 1981).

The above results indicated that indigo, madder, cochineal, kermes, weld, and purple dyes were used in dyeing of the textiles in the ancient times in different countries of the world. These dyes imparted different colours, namely blue, red, yellow, and purple.

CHARACTERIZATION OF DYES IN ANCIENT TEXTILE SAMPLES

Beautiful colours were obtained in the past by using natural dyes obtained from different sources. Researchers all over the world were eager to characterize the dyes and the sources from which these were obtained. The books on the chemistry of natural colourants (Perkin and Everest, 1918; Mayer and Cook, 1943; Thomson, 1957; 1971) are of immense importance. Reviews on natural colourants are also useful (Kroschwitz and Howe-Grant, 1993; Taylor, 1986). Some useful basic information about the chemistry of blue and red colouring matters occurring in plants was provided by Robinson, a few of which were used as dyes (Robinson, 1942).

The general analytical approach for the identification of dye from ancient textiles is to first extract the dye into a solution, followed by its characterization using chromatographic techniques (TLC, CC, HPLC and HPTLC) and spectral data (IR, UV, ^{1}HNMR, 2DNMR, ^{13}CNMR, and MS). With the advent of modern spectroscopic and very sensitive microanalytical techniques in the past three decades (Ferreria, 2004; Wouters and Verkecken, 1989a, 1989b), it has now become possible to characterize the dye present in very small amount (0.1 mg) of the sample (thread). For example, high-performance liquid chromatograph (HPLC) fitted with diode array detector permits dyestuff detection from very small quantity of dyed sample by comparison with authentic samples. It is possible to identify the unknown compounds extracted from a dyed sample with the help of HPLC coupled with mass spectrometer (HPLC-MS). With the development of latest extraction techniques like supercritical fluid extraction and other mild extraction techniques, it is possible to isolate the dye components in more natural state; their chromatographic studies with authentic samples and other studies may lead to the identification of natural sources (Zhang and Laursen, 2005; Guinot and Andary, 2006). Most of the natural dyes are mordant type. It is now possible to quantify the metal ion present in the dyed fabric by inductively coupled plasma separation with atomic emission spectrometric (ICP-AES) or inductively coupled plasma separation with mass spectrometric detection (ICP-MS) of the dyed samples (~0.25 mg) digested with nitric acid (Taylor, 1986; Melo et al., 2005). The chemical constituents of some dyes responsible for imparting different colours to some ancient textiles were characterized and are given below.

(1) Plant dyes

(a) Blue dye: Indigotin isolated from *Indigofera tinctoria* (indigo) and *Isatis tinctoria* (Woad) was mainly used to produce blue colour on textiles. It can be extracted from textiles under hot conditions with pyridine/water (1:1) (Taylor, 1986), dimethyl formamide (Schweppe, 1980), or acetic acid (Hofenk de Graaff, 1974).

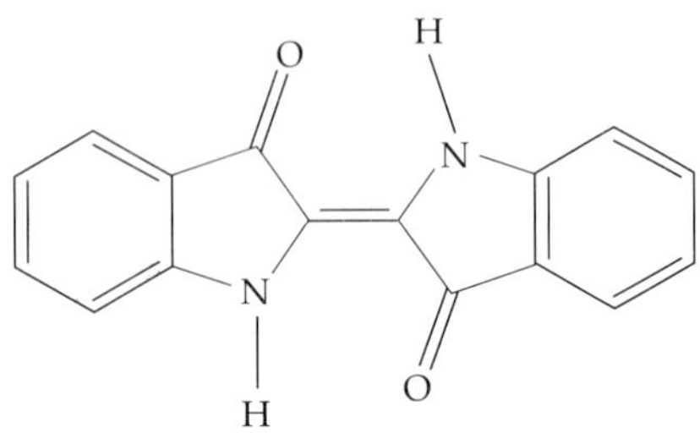

Indigotin

(b) Red dyes: Roots of *Rubia cordifolia* (the Indian madder, Manjith) and *Rubia tinctorium* (the European madder) constitute the plant red dye. The chemical analysis of the madder roots indicated that the colourants consists of hydroxy anthraquinones, of which munjistin, purpurin, purpuroxanthin, and pseudopurpurin occurred in the former, while ruberythric acid, rubiadin glucoside, purpurin, alizarin, pseudopurpurpin, purpuroxanthin, and munjistin occurred in the latter, alongwith other constituents (Anon, 2009; Bunnett and Thomson, 1968; Thomson, 1971). The dye consists of hydroxy anthraquinone glycosides, which readily break down to give hydroxy anthraquinones. These are mordant dyes and are extracted from textiles using acidic alcohol.

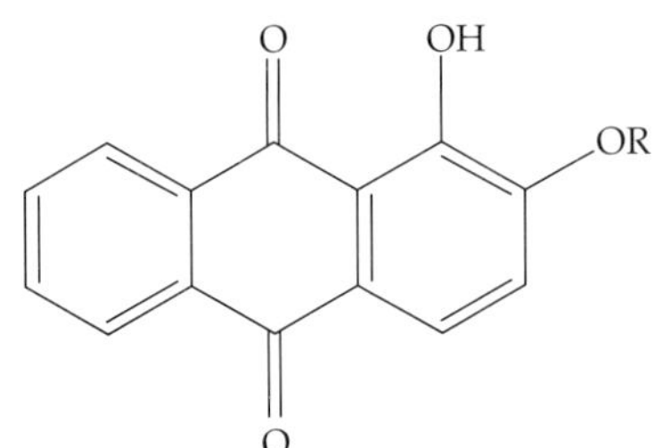

	R
Alizarin	H
Ruberythric acid	Primeverose

(c) Yellow dyes: Yellow dyes are mainly mordant type dyes and can be removed from textiles by treatment with acid/alcohol mixture. These dyes basically consist of flavonoid type compounds, which on TLC over silica gel, polyamide, and cellulose in different solvent systems indicate the number of constituents present in the dyed fabric. Luteolin was found to be the major colouring compound. A few yellow dyes, such as saffron and turmeric, are direct dyes and are sufficiently coloured to be detected by recording their UV spectra in the visible region. Crocin and curcumin were found to be the active ingredients, respectively. The chemical constituents of a number of yellow dyes have been complied (Taylor, 1986). As these dyes have tendency to fade, thus unable to survive on old textiles.

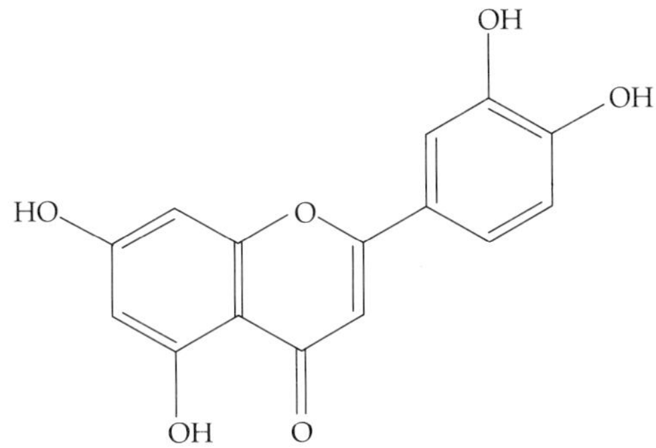

Luteolin

(2) Insect dyes

(a) Red dyes: The insect red dyes are mordant dyes and can be isolated from textiles by heating with methanolic sulphuric acid. Removal of the solvent yielded the dye, which can be characterized by chromatographic and spectroscopic techniques. The prominent red dye was isolated from the insect *Kermes vermilio*. Kermesic acid was found to be the main colourant of the dye. In the sixteenth century, it was superseded by cochineal (*Dactylopius coccus*), which contained carminic acid as its main colourant. Some more *Dactylopius* species were also used on the isolation of dye (Rodd, 1979; Taylor, 1986; Thomson, 1971).

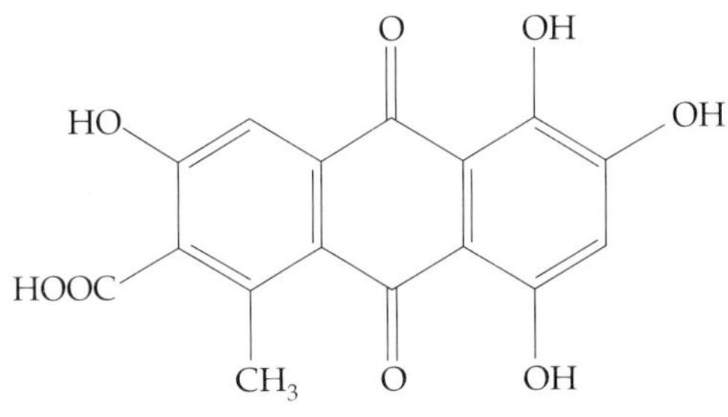

Kermesic acid

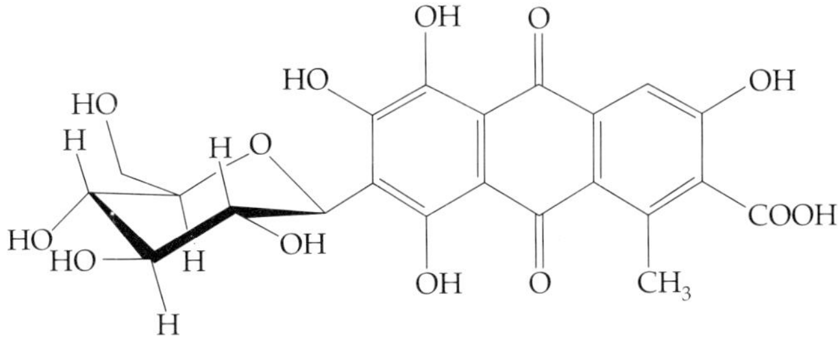

Carminic acid

The lac insect *Laccifer lacca* was and is still being used to produce the lac dye in India and other countries of the world. The main constituents of the dye were characterised as laccaic acids A, B, C, E, and laccaic acid D (Rodd, 1979; Thomson, 1971).

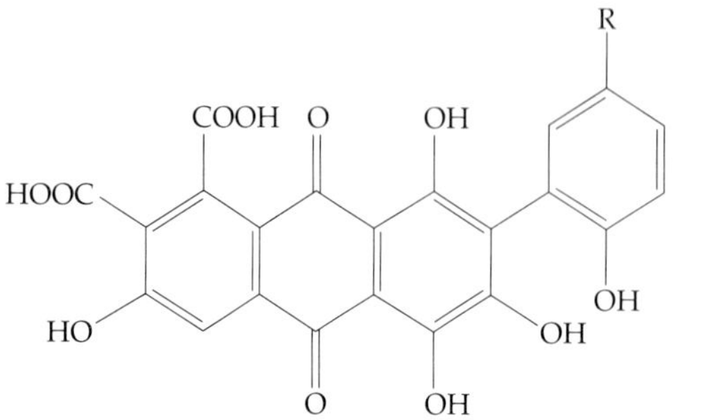

	R
Laccaic acid A:	$CH_2CH_2NHCOCH_3$
Laccaic acid B:	CH_2CH_2OH
Laccaic acid C:	$CH_2CH\ (NH_2)\ COOH$
Laccaic acid E:	$CH_2CH_2NH_2$

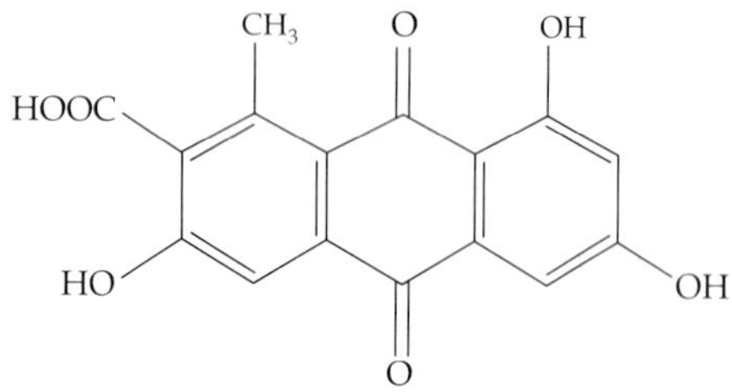

Laccaic acid D

(b) Purple dye: The chief colourant of shellfish purple dye was 6, 6′- dibromoindigotin (Baker, 1974), which can be extracted from textile by heating with pyridine (Taylor, 1986) or acetic acid (Hofenk de Graaff, 1974).

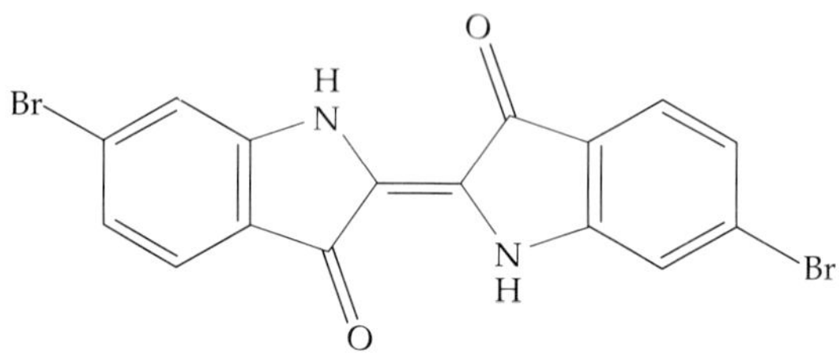

6,6′-dibromoindigotin

References

1. Abdel-Kareem, O, Alawi, M.A. Mubarak, M.S. (2011). Identification of natural dyes in selected Museum textiles using HPLC with photodiode detection. *Research Journal of Textile and Apparel*, 15(2), 84-94
2. Agarwal, O.P. and Tiwari, R. (1989). Mineral pigments in India. Paper Presented at the First National Seminar on Natural Dyes, Jaipur. 27-32.
3. Ajanta Caves. (1982). *Advisory Body Evaluation*. UNESCO International Council of Monuments and Sites. Retrieved 27 October, 2006. (http://whc.unesco.org/archive/advisory body evaluation/242.pdf), p. 2.
4. Ajanta Caves, India: Brief Description, UNESCO World Heritage Site. Retrieved 27 October, 2006.(http:// whc.Unesco.org/en/list/242).
5. Anon. (Reprinted) (2009). The *Wealth of India, Raw Materials.* Council of Scientific and Industrial Research, New Delhi, India, pp. 82-84.
6. Baker, J.T. (1974). *Endeavour*, 33, 11-17.

7. Bhardwaj, H.C. and Jain, Kamal, K. (1982). Indian dyes and dyeing industry during 18-19th century. *Indian Journal of History of Science,* 17(1), 70-81.
8. Bunnett, A.R. and Thomson, R.H. (1968). *JCS(C),* 2437-2441.
9. Chandramouli, K.V. (1989). The primary colours of natural dyes. *Presented at the National Seminar on Natural Dyes,* Jaipur.
10. Driessen, K. (2014). The earliest dyes (Quilt History). http://www.quilthistory.com/dye.htm.
11. Druding, S.C. (1982). *Dye History from 2600 B.C. to 20th Century.* (http.//www.Straw.com.sig/dyehist.html).
12. Ellison, M. and Harbottle, B. (1983). *Arch. Aeliana* 5th ser, 11, 217, 240, 262-263.
13. Elliott, R.L. (1976). *J.S.D.C,* 92, 303-304.
14. Encyclopedia Britannica. (A new survey of knowledge). *Natural Dyes.* Encylopedia Britannica Ltd William Berton-Publisher, London, Vol. 7, pp. 813-814.
15. Ferreria, E., Hulme, A., McNab, H., Quye, A. (2004). The natural constituents of historical textiles dyes. *Chem. Soc. Rev,* 33, 329-336.
16. Ghose, D.R. (1907). *The Indian Industrial Guide,* S.K. Lahiri & Co, Calcutta, pp. 95-96.
17. Gulati, A.N. (1989). *Natural Indian Dyes and the Art of Their Applications* National Seminar on the Natural Dyes, held at Jaipur on October 20-21, pp. 111-117.
18. Guinot, P. and Andary, C. (2006). Molecules involved in the dyeing process with flavoniods, in the *25th Meeting on Dyes in History and Archaeologly,* Suceava, Romania, 21-22 September.
19. Hadi, Saiyid Muhammad (1896). *A Monograph on Dyes and Dyeing in the Northern-Western Provinces and Oudh,* pp. 52-54.
20. Harbottle, B.and Ellison, M. (1981). *Arch. Aeliana,* 5th ser, 9 190, 228, 248-249.
21. Hofenk de Graaff, J.H. (1974). *Studies in Conservation,* 19, 54-55.
22. Hofenk de Graaff, J.H. and Roeloffs, W.T. (1976). *Amer Dyestuffs Rep,* 65(3), 32-34.
23. Karapanagiotis, L; Daniilia; Tsakalof, A; Chryssoulakis, Y. (2005). Identification of red natural dye in post-Byzatine icons by HPLC. *Journal of Liquid Chromatography and Related Technologies,* 28(5), 739-749.
24. Kok, A. (1966). *The Lichenologist,* 3, 248-272.
25. Kroschwitz, J.I and Howe-Grant, M. (1993) [Editors]. Encylopedia of Chemical Technology, Vol. 8, Fourth edition, A Wiley-Interscience Publication, pp. 784-809.
26. Leggett, W.F. (1944). *Ancient and Medieval Dyes.* Chemical Publishing Company, Inc. U.S.A., pp. 6-7, 30-31, 83-84.
27. Mantzouris, D; Karapanagiotis, L; Valianou, L; Panayiotou, C. (2011). HPLC-DAD-MS analysis of dyes identified in textiles from Mount Athos. *Analytical and Bioanlytical Chemistry,* 399 (9), 3065-3079.
28. Marshall, J. (Editor) (1931). *Mohenjo-Daro and Indus valley Civilisation 1,* London, p. 33.
29. Mayer, F. (1943). *Chemistry of Natural Colouring Matters,* ACS Monograph No. 89, translated and revised by A.H. Cook. Reinhold, Publishing Corp. New York.

30. Melo, M.J., Aguiar-Ricardo, A., Cruz, P. (2005). A green approach to antique textile cleaning. *Preprints of the 14th Triennial Meeting of the ICOM Committee for Conservation*, Vol. 2, pp. 994-954, James & James.
31. Mohanty, B.C. (1989). A brief historical note on the use of natural dyes in India and a case study on kum (khum, khuma) dyeing in Manipur. Presented at the National seminar on Natural dyes, Jaipur, p. 7.

31a. Naqvi, H.K. (1980). Colour making and dyeing of cotton textiles in medieval Hindustan. Indian Journal of History of Science, 15(1), 58-70.

32. Orska-Gawrys, J; Surowiec, I; Kehl, J; Rejniak, H; Urbaniak-Walczak, K; Trojanowicz, M. (2003). *Journal of chromatography*, 98(2), 239-248.
33. Perkin, A.G. and Everest, A.E. (1918). *The Natural Organic Colouring Matters*, London, Longmans Green.
34. Petroviciu, I. and Wouters, J. (2002). Analysis of natural dyes from Romanian 19th- and 20th century ethnographical textiles by DAD-HPLC. *Dyes in History and Archaeology*; 18, 57-62
35. Pim, A.W. (1898). *A Monograph on Wollen Fabrics in the Northern-Western Provinces and Oudh*, p. 13.
36. Robinson, R. (1942). *Endeavour*, I, 92-101.
37. Rodd (1979). *Chemistry of Carbon Compounds*, Second Edition, S. Coffey, Vol. 3H, p. 79.

37a. Roy, Mira. (1978). Dyes in ancient and medieval India. Indian Journal of History of Science, 13(2), 83-112.

38. Royle, J.F. (1851). *On the Culture and Commerce of Cotton in Indian and Elsewhere*, p. 600.
39. Schweppe, H. (1980). *JAIC*, 19(1), 14-23.
40. Sinha, K. (2014). Climate change killed Indus cities. *Times of India,* p. 23. (www.timesofindia.com)

40a. Siva, R. (2007). Status of natural dyes and dye-yielding plants of India. Current Science, 92(7), 916-925.

41. Srivastava, M.G. (1989). Natural dyes: An overview. Paper *Presented in the National Seminar on Natural Dyes,* held at Jaipur on October 20-21, pp. 1-6.
42. Taylor, G.W.(1986).Natural dyes in textile applications. *Rev. Prog. Colouration*, vol. 16, 53-61.
43. Thomson, R.H. (Butterworth, 1957). *Naturally Occurring Quinones.* (a) First Edition, (b) Second Edition (Academic Press, 1971), pp. 373, 443-459, 462-472.
44. Thrope. (1921). *Dictionary of Applied Chemistry*, pp. 590-591.
45. Tortora, P.G. (1978). *Understanding Textiles.* New York, Macmillan. Co. Inc. pp. 163-180.
46. "UNESCO page-Ancient City of Sigiriya." (http://whc.unesco.org/en/list/202). UNESCO.org. http://whc unesco. org/en/list/202. Retrieved 2010-09-17.

47. Wali, B. (1989). *Technology of Dyeing: An Introspection. National Seminar on the Natural Dyes*, held at Jaipur on October 20-21, pp. 85-90.

48. Walton, P. (1986). The Textiles from 16-22 Coppergate. *Archaeologly of York*, 17, Fascicule 5.

49. Watt, G. (1972). *A Dictionary of the Economic Products of India*. Cosmo Publications, Delhi, 1972, Vol. 4, pp. 597, 435-445, 455.

50. Wild, J. and Ryder, M.C. (1977). *Vindolanda, III. The Textiles* (Vindolanda Trust).

51. Wouters, J. (1985). *Studies in Conservations*, 30, 119-128.

52. Wouters, J. and Verhecken, A. (1989a). The coccid insect dyes: HPLC and computerized diode-array analysis of dyed yarns. *Studies in Conservation*, 34, 189-200; Wouters, J. and Verhecken A. (1989b). The scale insect dyes (*Homoptera: Coccoidea*). Species recognition by HPLC and diode-array analysis of the dyestuff, *Annales de la Soxiete Entomlogique de France*, 25(4), 339-410.

53. Zhang, X. and Laursen, R.A. (2005). Development of mild extraction methods for the analysis of natural dyes in textiles of historical interest using LC-diode array detector-MS. *Anal. Chem*; 77(7), 2022-2025.

54. Zideman, I.I. (1981). J.S.D.C; 97, 362-364. Historic textiles and paper materials. *Adv. Chem. Ser; 212* Ed. H.L Needles and S.H. Zeroman, 1986, 175-186.

3 Classification of Natural Dyes

INTRODUCTION

A substance used to change the colour of fabric, hair, wood, skin, food, and other items is known as dye. It should have a suitable colour and have the capacity to fix itself to the fabric naturally or with the help of a chemical (organic or inorganic) known as a mordant. The dyed fabric should have good to excellent colour-fastness properties to light, washing, rubbing, and perspiration. Numerous natural dyes have been known since ancient times. These were mainly extracted from plant, animal, and mineral sources. However, with the discovery of a number of synthetic dyes in the later part of the nineteenth century, their use started decreasing and reduced to minimum in the twentieth century. The dyes can be classified into two types:

(1) Natural dyes
(2) Synthetic dyes or artificial dyes or coal tar dyes

NATURAL DYES

Natural dyes can be defined as the colourants (dyes and pigments) obtained from plants, animals, or minerals without any chemical modification. The manuscript dating back to 2600 B.C. mentions the use of dyestuff in China. In India, the use of natural dyes dates back to the Indus Valley Civilization between the fourth and second millennium B.C. More than 300 dye-producing plants are listed in various historical manuals and a number of references are available in Sanskrit works, such as the *Arthashastra* and the *Atharvaveda* (Srivastava, 1989). Different parts of the plants, namely flowers, fruits and their rinds, leaves, bark, wood, roots, seeds, plant gall, all kinds of plant extracts, lichens, and fungi, along with lac insect, beetles, snails, and minerals, are some sources of natural dyes. Our forefathers through their cleverness and continuous and repeated experimentation were able to develop bright colours from natural dyes, which remained fast for centuries. These dyes were used for colouring the textiles (silk, wool, cotton, etc.), in food, and for paintings works, and for other purposes, without carrying out chemical modifications on them. Indigo was one of the most valuable dyes that originated from the Indian subcontinent. Plants are the greatest organic chemists on the earth. They

Punica granatum

Cassia fistula

Bixa oralina

Tamarindus indica

Cinnamomum tamala

Curcuma longa

Areca catechu

Butea monosperma

Mallotus philippensis

Curcuma amada

Curcuma longa

Lawsonia inermis

Toona ciliata

synthesize complex dye-yielding and others organic molecules essential for sustaining life on the earth. Both dyes and pigments are colouring matters with different properties and techniques of their use. Some differences between them are given below.

Dyes	*Pigments*
(i) Dyes are soluble in the vehicle with which it is mixed.	(i) Pigments are insoluble in the vehicle (or binder) with which it is mixed to make a paint, varnish, lacquer, printing ink, or other coating composition.
(ii) Dyes are applied generally in aqueous solution by impregnation of an object.	(ii) Pigments are applied by painting and staining a material body.
(iii) Some dyes form complexes with metal salts to form chelates known as lakes. Coloured lake formed is known as aluminum lake or calcium lake or any other lake, depending on the salt used.	

Natural dyes, when used as textile dyes, are mainly mordant dyes; however, some dyes belong to other groups (direct, vat, solvent, acid, and pigment). The majority of the prominent historically important dyes (colourants) are the members of anthraquinone, naphthoquinone, indigoid, carotenoid, and flavonoid groups. There appears to be no natural dyes belonging to sulphur, azoic, disperse, or ingrain types (Hancock and Boxworth, 1997).

SOURCES OF NATURAL DYES

Natural dyes are obtained from plants, animals (insects), and mineral sources. The dyes obtained from different parts of plants are known as vegetable dyes. The dyes obtained from other sources, viz., animals (insects) and minerals, are called animal and mineral dyes, respectively.

(i) Vegetable dyes

Natural dyes obtained primarily from different parts of plants, namely, leaves, bark, fruits, flowers, wood, roots, etc., are known as vegetable dyes. The majority of natural dyes are vegetable dyes. Some examples of vegetables dyes are given in Table 3.1, which indicates that these can be obtained from different parts of plants, viz. wood (*Artocarpus heterophyllus*), bark (*Mangifera indica*), roots (*Rubia cordifolia*), fruits (*Bixa orellana*), flowers (*Crocus sativus*), and leaves (*Tectona grandis*). Vegetable dyes can be divided into three classes, namely, direct dyes, mordant dyes, and vat dyes, depending on their acceptance by the material to be dyed.

Table 3.1 Sources of some vegetable dyes

S.No.	*Species*	*Family*	*Popular Name*	*Dye bearing parts*	*Pattern*	*Vital colour*
1.	*Acacia catechu* Willd.	Mimosaceae	Khair, Cutch Tree	Heartwood	Tr	Red & brown
2.	*Adina cordifolia* (Roxb.) Hook.f.ex Brandis	Rubiaceae	Haldu	Wood	Tr	Yellow
3.	*Allium cepa* Linn.	Liliaceae	Piyaz	Skin	Bulbous biennial	Orange
4.	*Arnebia densiflora* (Nordm.) Ledeb.	Boraginaceae	Ratanjot	Root	H	Red
5.	*Artocarpus heterophyllus* Lam.	Moraceae	Kathal, Jackfruit	Wood, roots	Tr	Yellow
6.	*Bischofia javanica* Blume	Euphorbiaceae	Bishop Wood	Bark, seeds	Tr	Black
7.	*Butea monosperma* (Lam.) Taub.	Fabaceae	Palas, Flame of the Forest	Flowers	Tr	Yellow-orange
8.	*Butea superba* Roxb.	Fabaceae	Palas lata	Roots	H	Red
9.	*Carthamus tinctorius* Linn.	Asteraceae	The Safflower, False Saffron	Flower	H	Red
10.	*Curcuma longa* Linn. syn *C. domestica* Val.	Zingiberaceae	Haldi, Turmeric	Rhizome	H	Yellow
11.	*Haematoxylon campechianum* Linn.	Fabaceae	Logwood, Patang	Heartwood	Tr	Black
12.	*Indigofera tinctoria* Linn.	Fabaceae	Nil, Indigo	Leaf, whole plant		Blue
13.	*Juglans regia* Linn.	Juglandaceae	Akhrot	Bark, fruits, leaves unripe fruits	Tr	Camel, brown, black, red
14.	*Lawsonia inermis* Linn.	Lythraceae	Mehndi, Henna	Leaves	Sh	Yellow-red, orange
15.	*Mallotus philippensis* Muell.Arg.	Euphorbiaceae	Kamala Tree, Sindur	Fruits	Tr	Orange
16.	*Mangifera indica* Linn.	Anacardiaceae	Aam, The Mango	Bark	Tr	Yellow
17.	*Memecylon umbellatum* Burm.f.	Melastomataceae	Iron Wood Tree	Leaves	Sh, Tree	Yellow

18.	*Morinda citrifolia* Linn.	Rubiaceae	Al, Ach	Root bark	Tr	Red, yellow
19.	*Nardostachys jatamansi* DC.	Valerianaceae	Jatamansi, Spikenard	Inflorescence	H	Red
20.	*Pterocarpus santalinus* Linn. f.	Fabaceae	Red Sandal wood, Raktachandan	Wood	Tr	Red
21.	*Punica granatum* Linn.	Punicaceae	Anar, Pomegranate	Rind, flowers	Tr	Yellowish brown, light red
22.	*Quercus infectoria* Olivier	Fagaceae	Majuphal, Gall Oak	Galls	Tr	Yellow, dark grey
23.	*Reseda luteola* Linn.	Resedaceae	Weld, Dyer's Rocket	Whole plant	H	Deep yellow
24.	*Rheum emodi* Wall.ex Meissn.	Polygonaceae	Dolu, Himalayan Rhubarb	Rhizome	H	Reddish yellow
25.	*Rhizophora mucronata* Lam.	Rhizophoraceae	True mangrove	Bark	Tr	Chocolate
26.	*Rubia cordifolia* Linn.	Rubiaceae	Indian madder, Manjit	Roots	Cl	Scarlet-red, orange
27.	*Semecarpus anacardium* Linn.f.	Anacardiaceae	Bhilawa, The Marking Nut Tree	Bark, fruits	Tr	Black
28.	*Tagets erecta* Linn.	Asteraceae	Genda, Azetac or African Marigold	Flowers	H	Yellow
29.	*Tectona grandis* Linn.f.	Verbenaceae	Sagwan, Teak	Leaves, bark, roots	Tr	Reddish, yellowish brown
30.	*Terminalia chebula Retz.*	Combretaceae	Harra, Myrobalan	Fruits, barks	Tr	Yellow, brown
31.	*Ventilago denticulata* Willd.	Rhamnaceae	Pitti	Bark, root	Sh	Red

(Cl = Climber; H = Herb; Sh = Shrub; Tr = Tree)

(ii) Animal dyes

(a) The dyes obtained from insects come under this group. Animal dyes were obtained mainly from the following insects and insects groups.

(b) Lac scales *(Laccifer lacca)* found on lac host plants, viz., *Butea monosperma* (Palas) and other hosts.

(c) Gall-like coccids (*Kermes* sp.) found on oaks.

(d) Cochineal scales (*Dactylopius* spp.) found on Opuntia.

(e) Sea snails *(Murex brandaris, Murex trunculus)* found in the Eastern Mediterranean.

Some of the examples are Tyrian purple dye from sea snails (*Murex brandaris)* and red dyes obtained from kermes (*Kermes* species*)*, lac insects (*Laccifera lacca*), and cochineal scales (*Dactylopius* species), respectively.

(iii) Mineral colouring matters

These consist of naturally occurring minerals and form the third source of natural dyes. The khaki colour obtained on cotton by using the mixture of iron and chromium salts, is an example of mineral colouring matter (Tewari, 1994). The other mineral colourants used for dyeing and painting are cinnabar (*sengraf*), red lead (*sindur*), redochre (*geru*), zinc-white (*safeda*), yellow ochre (*ramraj*), etc. Many minerals also act as good mordants. The important naturally occurring mineral colouring matters are given in Table 3.2 (Agarwal and Tiwari, 1989). Minerals dyes (pigments) have also been used in paintings and dyeing since ancient time, as is evident from the wall paintings of Ajanta, Ellora, Bagh, and other sites. A coloured substance used as paint is known as pigment. Although synthetic pigments appeared in the market in the twentieth century, yet naturally occurring minerals have their own advantages. Chemical examination of the ancient wall paintings, cloths, and papers indicated the use of a number of minerals in the earlier times.

CLASSIFICATION OF NATURAL DYES

Natural dyes have been classified in a number of ways. In the beginning, these dyes were classified according to the alphabetical order. With the passage of time, natural dyes were classified on the basis on their chemical structures belonging to different classes of organic compounds, and the grouping within each structure class was made according to the hue. Classification of dyes in the colour index is made according to its chemical constitution and major application classes. The dyes are arranged according to the hue within the application class. The natural dyes constitute a separate section in the colour index. The classification of dyes was also made on the basis of their botanical names, common names, and traditional colours (Gulrajani and Gupta, Paul et al., 1996).

(1) Classification Based on Botanical Names

Natural dyes can be classified according to the botanical names of the plants from which these were and are being isolated. The names of some plant species yielding prominent vegetable dyes from their different parts are given in Table 3.3.

Table 3.2 Important Mineral Dyes

S. No.	*Name*	*Constituents*	*Colour*	*Source*	*Occurrence*	*Uses*
1.	Chalk	Calcium carbonate ($CaCO_3$)	White	(a) Found with limestone deposit. (b) Also obtained by grinding shells, or calcining them. (c) Conch shell white favoured in India	Orissa, Bihar, Madhya Pradesh, Rajasthan, Karnataka and Gujarat	Paintings
2.	Charcoal black	Charo black	Black	(a) Characoal prepared in India by controlled burning of twigs and wood of tamarind tree, powdered to make black pigment. (b) Shells of almonds and coconuts after charring and powdering also used for black pigment.	–	Black pigment
3.	Cinnabar or vermillion (Sangraf)	Mercury sulphide (HgS)	Bright red	(i) Made from mercury main ore cinnabar by grinding (ii) Synthetically produced by combining mercury and sulphur.	Central India	Paintings
4.	Gypsum	Calcium sulphate ($CaSO_4.2H_2O$)	White	Plaster of Paris is prepared by heating gypsum	Brazil, Pakistan, Jamacia, Iran, Spain, Thailand, Germany, Italy and other countries	Important pigment
5.	Malachite	Copper carbonate hydroxide mineral $Cu_2(CO_3)(OH)_2$	Green	Found associated with copper ore. The stone carefully selected, ground and sieved to prepare the pigment	Rajasthan, Bihar, Andhra Pradesh, Sikkim, and Nepal	Pigment

Contd...

S.No.	Name	Constituents	Colour	Source	Occurrence	Uses
6.	Red lead (*sindur*)	Pb_3O_4 (on long exposure to light and air turns to brown or black.	Bright orange red	(i) Does not occur in mineral form. Prepared by heating Litharge (PbO) or white lead ($PbCO_3$) (ii) Commercially prepared by oxidation of lead	Lead ores occur in Bihar, Himalayas, Rajasthan and Peninsular India	Paintings
7.	Red ochre or Indian red (*geru*)	Ferric oxide Fe_2O_3 (anhydrous); $Fe_2O_3H_2O$ (hydrous) stable compound no effect of light or alkalies	Red Red to dull yellow	(i) Deposit found in almost all parts of India. (ii) Also synthetically made	All parts of India	Paintings
8.	Terre-Verte (*green earth*)	Glauconite and celadonite (hydrous iron, magnesium and aluminium potassium silicate)	Yellow green to greenish grey	(i) Few varieties are used as pigments	All over India	Widely used
9.	Verdigirs	Copper acetate, $(CH_3COO)_2$ Cu	Bright and deep green	Prepared by action of vinegar on copper foils	—	Textiles and miniature paintings
10.	Yellow ochre (yellow)	Iron oxide in various hydrated forms mainly the mineral goethite ($Fe_2O_3.H_2O$)	Yellow number of shades	(i) Prepared from natural earth. (ii) In laboratory prepared by precipitating a mixture of soluble ferrous sulphate and alum (or aluminium sulphate) with lime or potash. The yellow product is a mixture of ferric and aluminium hydroxide and gypsum.	—	Paintings

11.	White lead	Basic carbonate of Lead ($PbCO_3$)	White	(i) Occurs in nature as the mineral cerussite (ii) Mainly synthetic compound is used (iii) Get blackened in due course	—	Important pigment
12.	Zinc white (*safeda*)	Zinc oxide (ZnO)	White	Does not turn black	—	Paintings
13.	Talc	Hydrated magnesium silicate $[Mg_3Si_4O_{10}(OH]_2$	White	Found as a soft stone (or soap stone)	India, US, Itlay, and other countries	In paint
14.	Titanium white	Titanium dioxide (TiO_2)	White	Prinicipal ore—ilmentie	—	Have very good hiding power

Table 3.3 Classification of natural dyes based on botanical names

S.No.	Botanical name	Family	Common name	Pattern	Dye bearing parts	Colour
1.	*Acacia catechu* Willd.	Mimosaceae	Khair, Cutch Tree	Tree	Heartwood	Red, brown
2.	*Allium cepa* Linn.	Liliaceae	Piyaz	Bulbous biennial	Skin	Orange
3.	*Butea monosperma* (Lam.) Taub.	Fabaceae	Palas, Flame of the Forest	Tree	Flowers	Yellow-orange
4.	*Carthamus tinctorius* Linn.	Asteraceae	The Safflower, False Saffron	Herb	Flowers	Red
5.	*Haematoxylon campechianum* Linn.	Fabaceae	Logwood, Patang	Tree	Heartwood	Black
6.	*Memecylon umbellatum* Burm.f.	Melastomataceae	Iron Wood Tree, Bijasal	Tree	Leaves	Yellow
7.	*Pterocarpus marsupium* Roxb.	Fabaceae	Bijasal	Tree	Bark	Red, brown
8.	*Semecarpus anacardium* Linn. f.	Anacardiaceae	Bhilawa, Marking Nut Tree	Tree	Bark, fruits	Black
9.	*Tagetes erecta* Linn.	Asteraceae	Genda	Herb	Flowers	Yellow

(2) Classification Based on Common Name

It is also possible to classify natural dyes according to the common names assigned to different plants and animals (insects). It is very easy to remember the common or popular names rather than their complicated scientific names. Some of the dye-bearing plant species usually called by their common names are given in Table 3.4.

Table 3.4 Classification of natrual dyes based on common names

S.No.	Common name	Botanical name	Family	Pattern	Dye bearing parts	Colour
1.	Kathal, Jackfruit	*Artocarpus hetrophyllus* Lam.	Moraceae	Tree	Wood, roots, fruits	Yellow
2.	Annatto	*Bixa orellana* Linn.	Bixaceae	Tree	Seeds	Orange, red
3.	Palas, Flame of the Forest	*Butea monosperma* (Lam.) Taub.	Fabaceae	Tree	Flowers	Yellow-orange
4.	Sappan Wood	*Caesalpinia sappan* Linn.	Fabaceae	Tree	Wood	Red

Contd...

5.	Safflower, False Saffron	*Carthamus tinctorius* Linn.	Asteraceae	Herb	Flower	Yellow
6.	Haldi, turmeric	*Curcuma longa* Linn.	Zingiberaceae	Herb	Rhizome	Yellow
7.	Amla	*Emblica officinalis* Gaertn.	Euphorbiaceae	Tree	Fruits, barks	Blue-black
8.	Logwood, patang	*Haematoxylon campechianum* Linn.	Fabaceae	Tree	Heartwood	Black
9.	Mehndi, henna	*Lawsonia inermis* Linn.	Lythraceae	Shrub	Leaves	Yellow-orange, red
10.	Kamla tree, Sindur	*Mallotus philippensis* Muell. Arg.	Euphorbiaceae	Tree	Fruits	Orange
11.	Aam, Mango	*Mangifera indica* Linn.	Anacardiaceae	Tree	Bark	Yellow
12.	Al, Ach	*Morinda citrifolia* Linn.	Rubiaceae	Tree	Root bark	Red, yellow
13.	Raktachandan	*Petrocarpus santalinus* Linn.f.	Fabaceae	Tree	Wood	Red
14.	Sagwan, Teak	*Tectona grandis* Linn.f.	Verbenaceae	Tree	Leaves, bark, roots	Reddish, yellowish brown, yellow

(3) Classification Based on Traditional Colours

Natural dyes can also be classified on the basis of five traditional colours, namely, black, blue, red, and yellow. The different colours may have beneficial effects on the humans. Blue colour may be beneficial in reducing the blood pressure, green to fight depression, red in energizing, yellow for stimulating happiness, white for regeneration, and other benefits (Nature Energies, 2014). Following the principle of natural cycle, natural dyes treat our body and its organs and represent artistic activities to find compatibility in diversity. A plausible classification of some dyes on the basis of colour obtained from some plant species is given in Table 3.5.

Table 3.5 Classification of natural dyes based on traditional colours

S.No.	*Colour*	*Name of plant species*	*Family*	*Popular name*	*Dye bearing parts*	*Pattern*
1.	Black	*Anacardium occidentale* Linn.	Anacardiaceae	Cashew or cashewnut	Nut (shell)	Tree
	Black	*Anogeissus latifolia* (Roxb.) Wall.ex Bedd.	Combretaceae	Dhaura	Leaves	Tree
2.	Blue	*Indigofera tinctoria* Linn.	Fabaceae	Nil	Leaves, whole Plant	Shrub
	Blue	*Strobilanthes cusia* (Nees) Imlay	Acanthaceae	Assam indigo, the rum	Twigs	Shrub

Contd...

3.	Red	*Rubia cordifolia* Linn.	Rubiaceae	Indian madder, Manjith	Roots	Climber
	Red	*Rubia tinctorum* Linn.	Rubiaceae	European madder, alizari	Roots	Herb
4.	Yellow	*Artocarpous heterophyllus* Lam.	Moraceae	Kathal, jackfruit	Wood, roots	Tree
	Yellow	*Berberis aristata* DC.	Berberidaceae	Kilmor	Bark, roots	Shrub
	Yellow	*Reseda luteola* Linn.	Resedaceae	Weld; dyer's rocket	Whole plant	Herb

(4) Classification Based on Chemical Classes

Plants are the greatest organic chemists on the earth, and they synthesize the organic compounds having complex structures with a very fine precision. The naturally occurring compounds are optically active, while those prepared in laboratory are normally a racemic mixture. Most of the natural dyes were and are being isolated from plants and animals (insects). The chemical constituents present in these dyes belong to different classes of organic compounds, namely, alkaloids, anthocyanins, betalins, carotenoids, chlorophyll, pyran colourants (chromenes), flavonoids, indigoids, quinones, tannins, and others, as given in Table 3.6. Some important dyes based on different chemical classes are given below.

(a) Quinone-based dyes

(i) Anthraquinone dyes: The major plant-based anthraquinone dyes were obtained from the roots of *Rubia cordifolia, Rubia tinctorum,* and *Rubia peregrina.*

- Alizarin and munjistin were characterized as the prominent chemical compounds of these dyes along with other constituents (Thomson, 1971, Ch. 2).

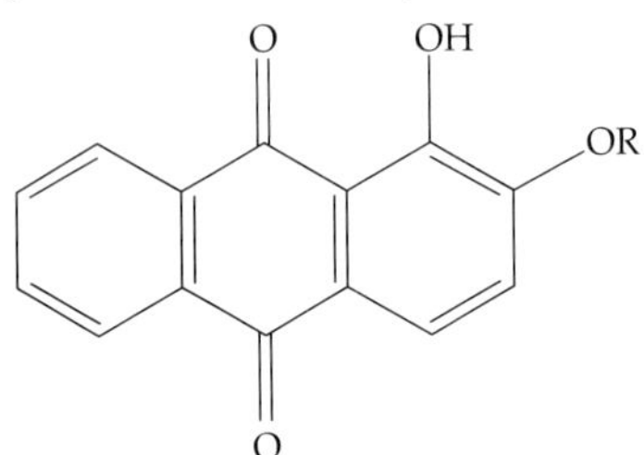

Alizarin — R: H
Ruberythric acid — R: Primeverose

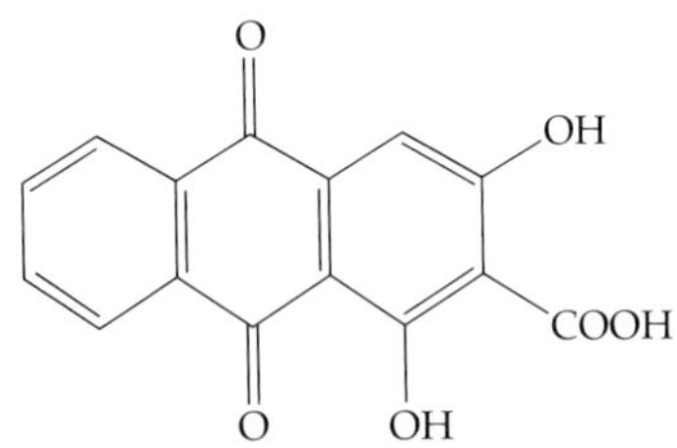

Munjistin

- Lac dye is one of the oldest dyes of the world. It is produced by a tiny lac insect *Laccifer lacca* and is isolated as a by-product while preparing shellac from sticklac. The lac dye was found to be a mixture of five closely related anthraquinone derivatives and were named as laccaic acids A, B, C, E, and laccaic acid D (Thomson, 1971, Ch. 2).

Table 3.6 Classification of natural dyes based on chemical classes

S.No.	*Chemical classes*	*Active principles*	*Name of the plant/insect species*	*Family*	*Pattern*	*Dye bearing parts*	*Colour*
1.	Alkaloids	Berberine	*Berberis aristata* DC.	Berberidaceae	Shrub	Bark, roots	Yellow
2.	Anthraquinoids	Alizarin	*Rubia cordifolia* Linn.	Rubiaceae	Climber	Roots	Scarlet-Red, Orange
		Laccaic acids A, B, C, E and laccaic acid D	*Kerria lacca* Kerr. syn. *Laccifer lacca*	Lacciferidae	Insect	Exudation of the lac insect	Red, Purple, Violet
		Carminic acid	*Dactylopius coccus* Costa	Coccidae	Insect	Insect	Red
		Kermesic acid	*Kermes vermilio* Planchon	Kermesidae	Insect	Insect	Scarlet
3.	Carotenoids	Bixin	*Bixa orelanna* Linn.	Bixaceae	Tree	seeds	Orange, red
4.	Chlorophylls	Chlorophyll a Chlorophyll b	All organisms capable of photosynthesis, Living plants alage, photosynthetic bacteria	Numerous Families	Numerous plants	Plants, alage, bacteria	Green
5.	Pyran colourants (Chromenes)	Rottlerin and isorottlerin	*Mallotus philippensis* Muell. Arg.	Euphorbiaceae	Tree	Seeds	Orange
		Haematin, Haematoxylin	*Haematoxylon campechianum* Linn.	Fabaceae	Tree	Heartwood	Black
6.	Flavonoids	Luteolin	*Reseda luteola* Linn.	Resedaceae	Herb	Whole plant	Yellow
7.	Indigoids	Indigotin	*Indigofera tinctoria* Linn.	Fabaceae	Shrub	Whole Plant	Blue
		6,6′-dibromoindigotin	*Murex trunculus L.*	Muricidae	Snail	Extract from insect	Purple
8.	Naphthoquinones	Lawsone	*Lawsonia inermis* Linn.	Lythraceae	Shrub	Leaves	Yellow-orange, red
9.	Tannins	Chebulic acid, ellagic acid, gallic acid and other phenolics	*Terminalia chebula* Retz.	Combretaceae	Tree	Fruits	Black

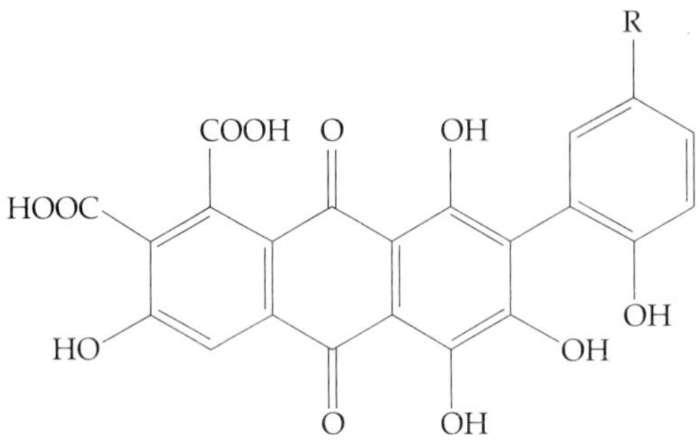

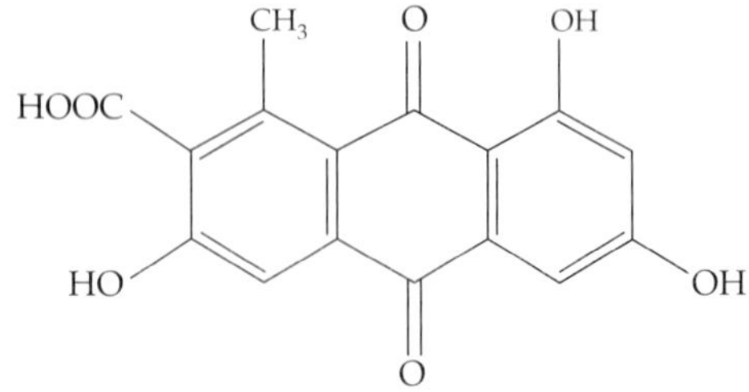

Laccaic acid D

	R
Laccaic acid A:	$CH_2CH_2NHCOCH_3$
Laccaic acid B:	CH_2CH_2OH
Laccaic acid C:	$CH_2CH(NH_2)COOH$
Laccaic acid E:	$CH_2CH_2NH_2$

- The other prominent animal-based anthraquinone dyes were isolated from the insects, namely *Kermes ilicis* and *Dactylopius coccus,* wherein kermesic acid and carminic acid were found to be the main chemical constituents, respectively (Thomson, 1971, Ch. 2).

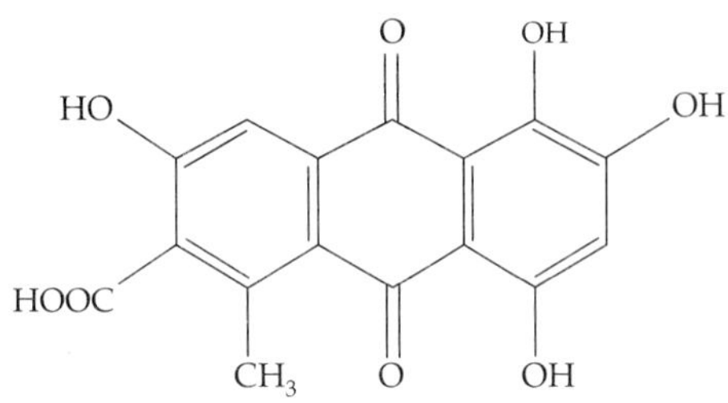

Kermesic acid

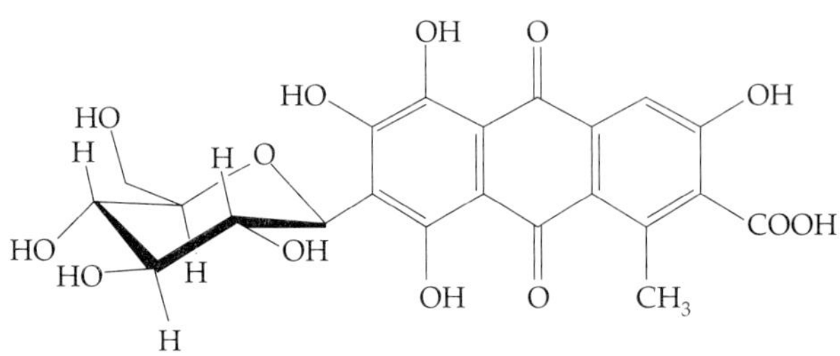

Carminic acid

(ii) Naphthaquinone-based dyes

- The powder of the leaves of *Lawsonia inerims* (mehndi) was and is being used as a dye. Lawsone was characterized as the main constituent of the dye.
- The husks of walnuts, *Juglans cinerea* (butter nuts), *Juglans regia* (Persian walnuts), and *Juglans nigra* (black walnuts) were used for the isolation of dye. Juglone was characterized as one of the active constituents of the dye.

- The reddish dye was isolated from the roots of *Anchusa tinctoria* or *Alkanna tinctoria*. The main active compound present is alkannin. Another napthoquinone dye was isolated from the roots of *Lithospermum erythrorhizon*. The active constituent was characterized as shikonin (Thomson, 1971, Ch. 5B).

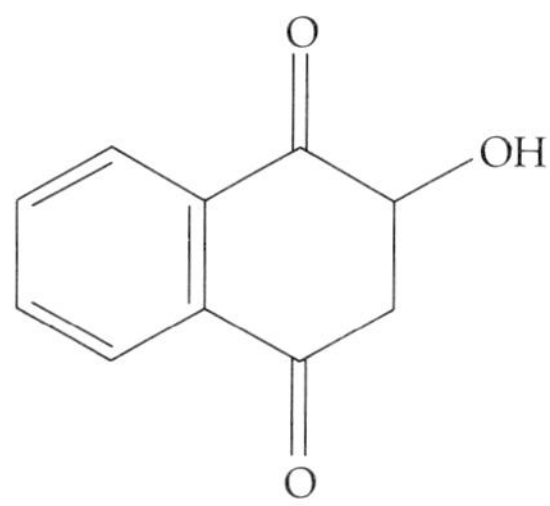

Lawsone

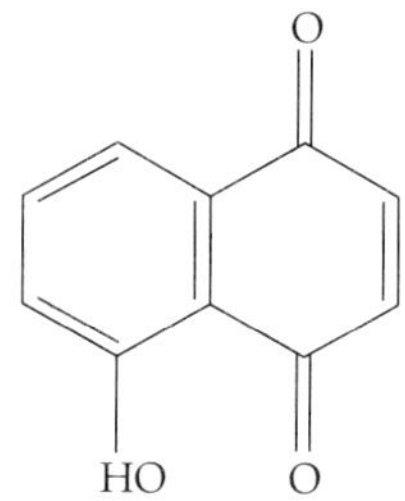

Juglone

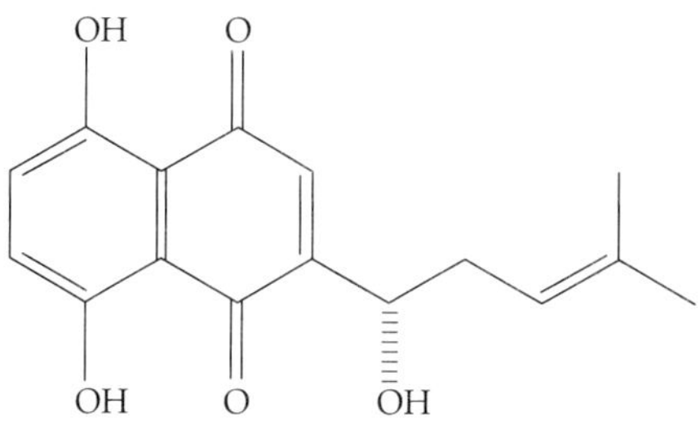

Alkannin

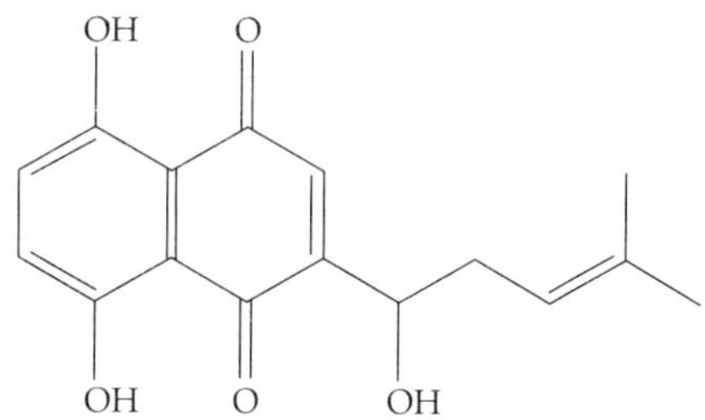

Shikonin

(b) Flavonoid-based dyes

(i) Flavones

- The yellow dye was isolated from *Reseda luteola* (whole plant). Luteolin and its glycosides were found to be the main constituents (Ref. 134–135, Ch 5A) of the dye.

(ii) Flavonols

- The flavonols containing dyes are mainly based on quercetin, kaempferol and other related compounds.

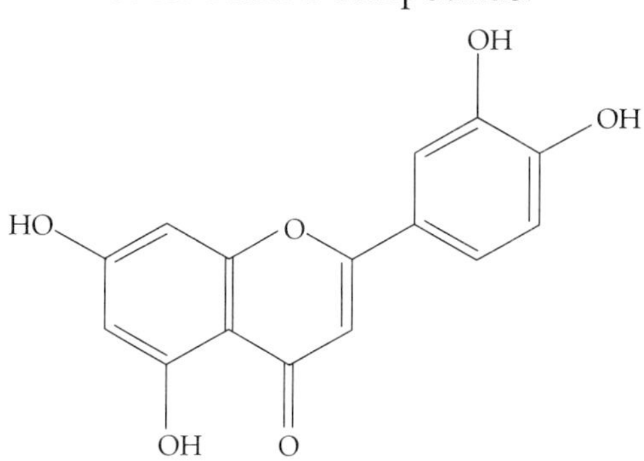

Luteolin

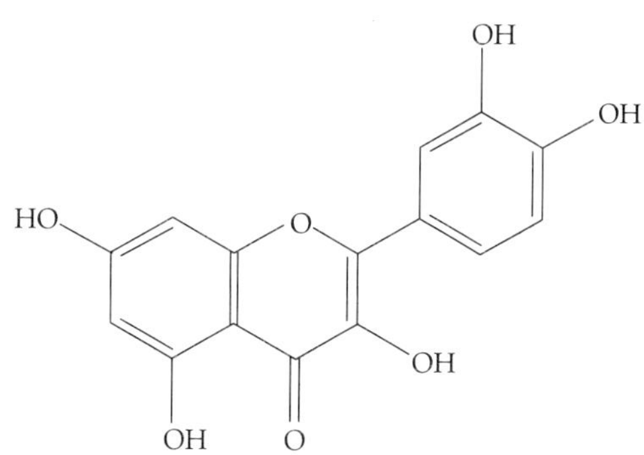

Quercetin

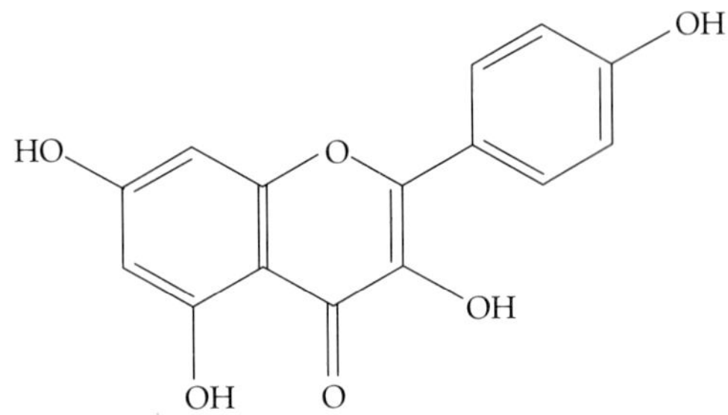

Kaempferol

(c) Pyran Colourants (Chromenes)

(i) The wood of *Haematoxylon campechianum* (logwood) was used for the extraction of the dye. Haematein and haematoxylin were characterized as the chief chemical compounds (Ref. 80-81, Ch. 5A).

(ii) *Caesalpinia sappan* (Brazil wood) wood was also used for the isolation of the dye. Brazilein and Brazilin were identified as the main compounds of the dye (Ref. 49-51, Ch. 5A).

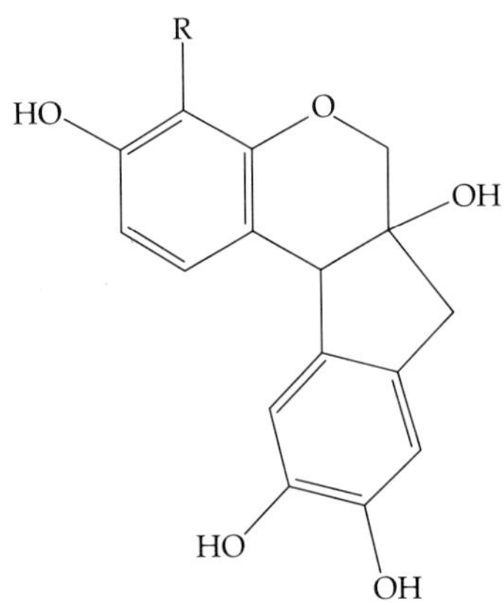

	R
Haematoxylin	OH
Brazilin	H

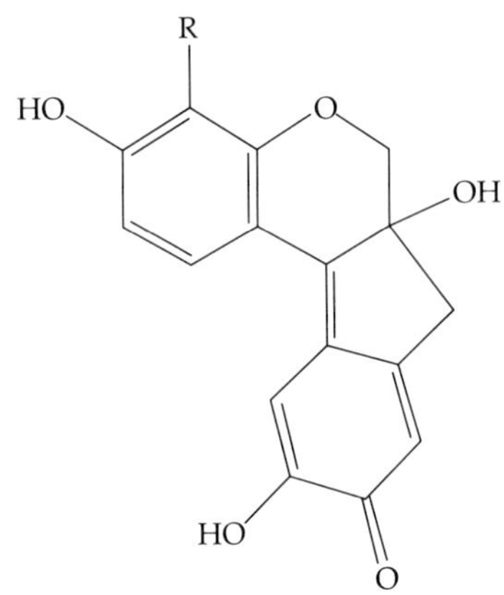

	R
Haematein	OH
Brazilein	H

(iii) Kamala dye was isolated from the ripe fruits of *Mallotus philippensis*. Rottlerin was found to be its chief chemical constituent (Ref. 94-95, Ch. 5A).

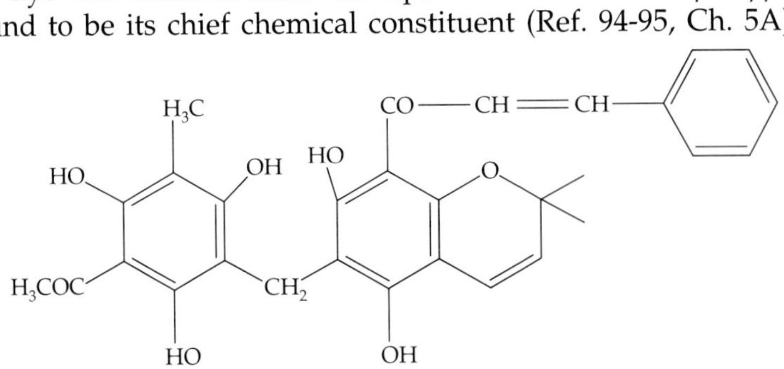

Rottlerin

(d) Anthocyanins

(i) Colourful flowers contain anthocyanins, namely, pelargonin, cyanin, delphin, and other compounds, which impart colour to these flowers. Anthocyanins are pH sensitive, as a result of which these cannot be used as commercial dyes (Finar, 1996, Ch. 5B).

(ii) Carajurin and dracorhodin containing quinoidal form of anthocyanidin were used for dyeing in the past.

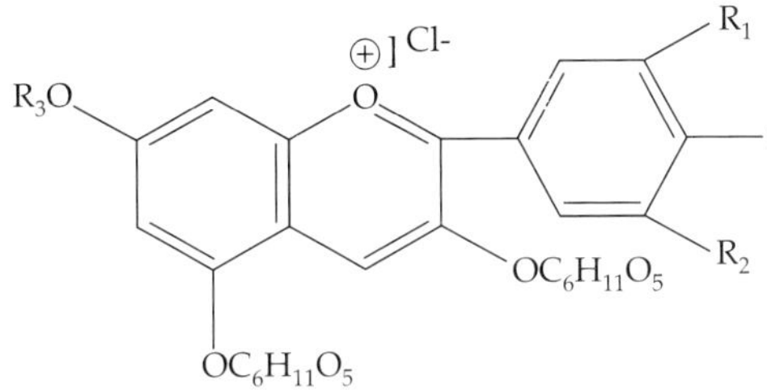

	R_1	R_2	R_3
Pelargonin chloride	H	H	H
Cyanin chloride	OH	H	H
Delphin chloride	OH	OH	H
Peonin chloride	OCH_3	H	H
Malvin chloride	OCH_3	OCH_3	H
Hirusutin chloride	OCH_3	OCH_3	OCH_3

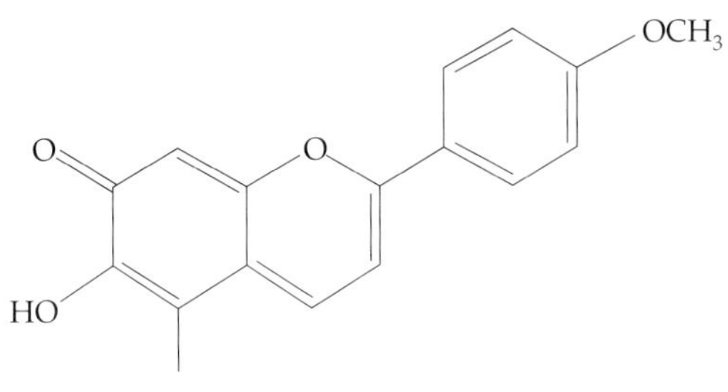

Carajurin

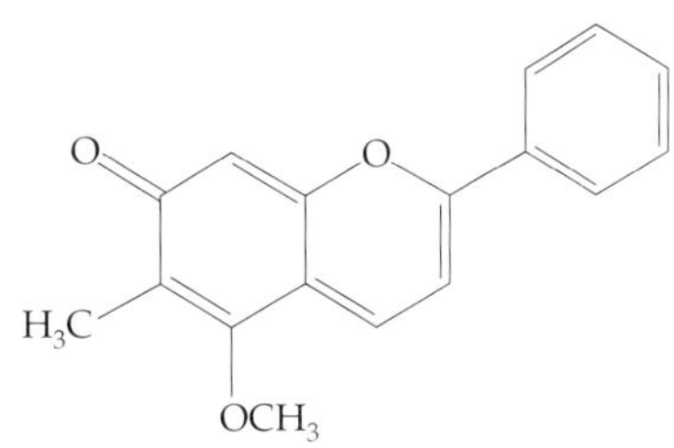

Dracorhodin

(e) Indigoid Dyes

(i) The blue dye indigo was isolated from *Indigofera tinctoria* and *Isatis tinctoria* and contained indigotin as the main constituent (Ref. 82, Ch. 5A).

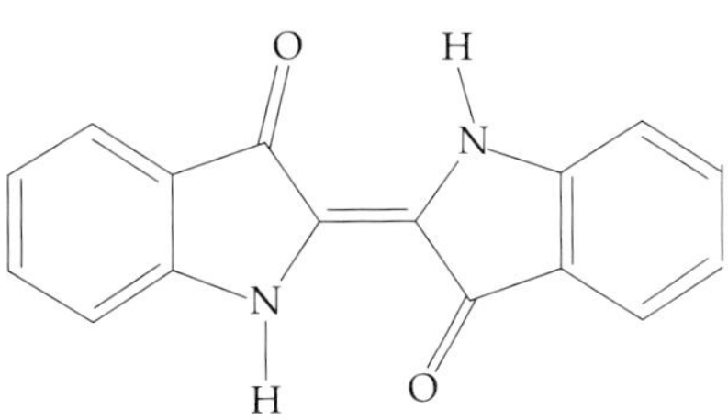

Indigotin

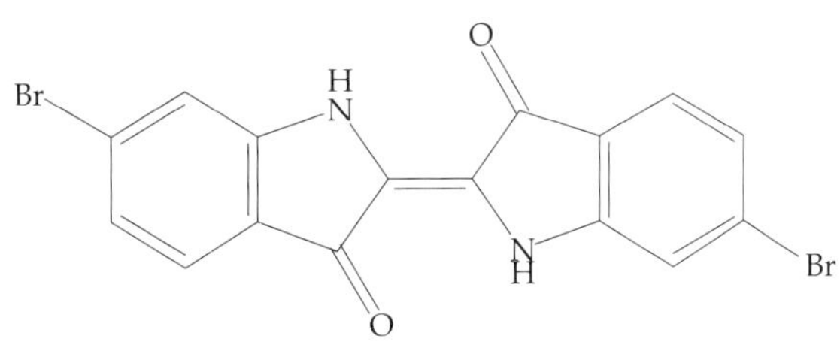

6, 6′-dibromoindigotin

(ii) The Tyrian purple dye was isolated from *Murex brandaris* and *Murex trunculus*, which were found to contain 6,6′-dibromoindigotin as the major component (Ref. 195-196, Ch. 5A).

(f) Carotenoids

(i) A red dye known as annatto was isolated from the seeds of *Bixa orellana*. Bixin was characterized as the main constituent of the dye. It dyes wool, cotton, and silk directly to give an orange-red shade and is also used to colour food stuffs (Ref. 43, Ch. 5A).

(ii) Saffron was isolated from the pistils and stigma of the flowers of *Crocus sativus*. Its major constituent was found to be crocin. It is one of the most expensive dyes of the ancient times (Ref. 60-62, Ch. 5A).

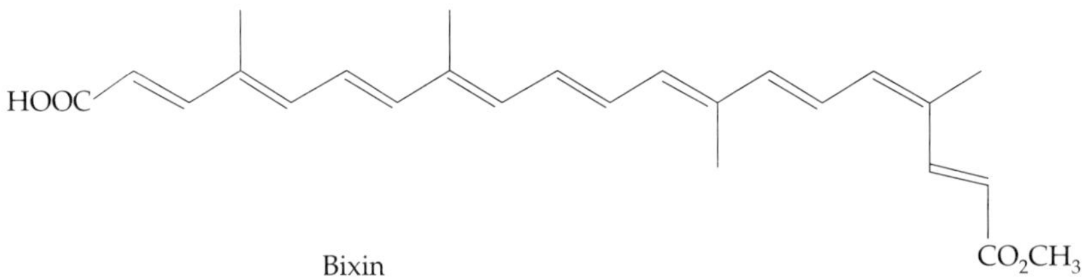

Bixin

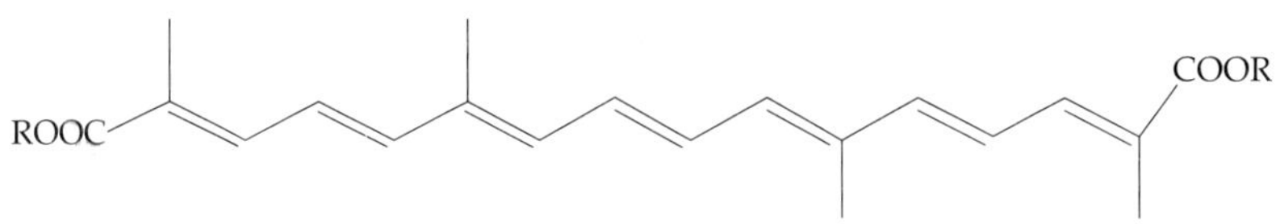

Crocetin: R = H
Crocin: R = Gentiobiose

(g) Chlorophylls

Chlorophylls are most widely occurring pigments in nature and are responsible for the green colour of plants. Chlorophyll-a and chlorophyll-b occur widely in nature (Finar, 1996, Ch. 5B).

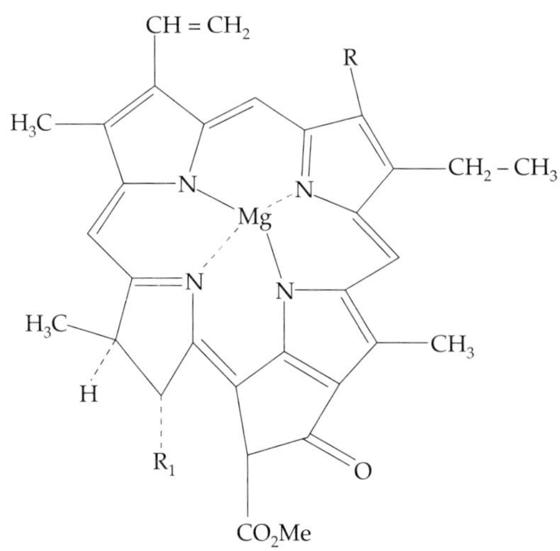

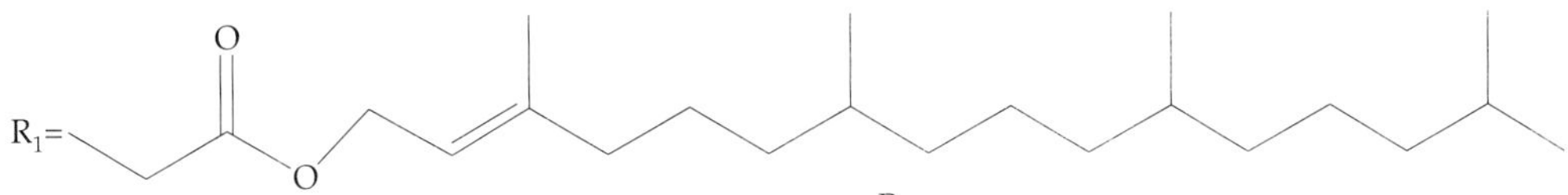

	R
Chlorophyll a	CH_3
Chlorophyll b	CHO

(5) Classification Based on Application Classes

"Treatise on Permanent Colours" was published by Bancroft about 170 years ago. According to the publication, natural dyes were classified into two groups, namely, substantive dyes and adjective dyes. Later on, this classification was replaced by another equivalent classification, according to which the dyes are known as direct dyes and mordant dyes. The dyes (colourants) that impart colour on different fibres directly without the assistance of any other chemical compound were classified as substantive dyes or direct dyes. The dyes that can only be fixed on fibres with the help of mordants (inorganic metallic salts or tannins) were put under the group of adjective dyes or mordant dyes. Natural dyes were also classified as monogenetic dyes and polygenetic dyes by Hummel. A monogenetic dye imparts only one colour, irrespective of the mordant used, while a polygenetic dye produces different colours on the fibre, depending on the mordant used (Gulrajani and Gupta, Sekar, 1999).

Generally, natural dyes are soluble in water and can be fixed on the fibre, either directly or with the help of a mordant. A few dyes are not soluble in water due to the abscence of the hydroxyl groups (–OH) in them. The ancient dye indigo is an example of this type of dye, as it is not soluble in water. The dye can be made soluble by a small chemical reaction. The reduction of two ketonic groups of indigo converts them to hydroxyl, followed by its sodium salt formation, which is soluble in water and fit for application to the fibre. The dye is converted back to its original form by air oxidation. These dyes are known as vat dyes.

The mode of application of a dye depends on the nature of both the fibre and the dye. The fibres can be put in two groups, namely, natural fibres and synthetic fibres. Natural fibres are of two types, viz., vegetable fibres and animal fibres. Vegetable fibres are cellulosic (cotton, linen, flax, hemp, and jute), while the animals fibres consist of proteins (wool, silk, leather, and fur). Synthetic fibres are high-polymer fibres built up of various types of monomers, e.g., polyamides, polyesters, and poly acrylnitrile. Natural dyes possess diverse chemical structures belonging to different class of organic compounds, but from the point of view of their application, a dye molecule may be bound to the fibre in four ways: (i) covalent bonds; (ii) hydrogen bonds; (ii) ionic bonds; (iv) van der Waals forces. The chemical nature of the fibre will determine the nature of the attachment of a given dye molecule.

Natural dyes based on their application can be classified as given below.

(a) Substantive Dyes or Direct Dyes or Permanent Dyes

A dye that is soluble in water and can be taken up directly by the material to be dyed is known as a direct or substantive dye. Turmeric, containing curcumin as the principal colouring component, is the most important example of the direct dye. This dye has affinity for natural fibres. The dyed fabric has poor colour-fastness properties, which may be improved by using mordants. Annatto, palas flowers, onion skins, walnuts husks, tea, stag horn moss, and some lichens are a few examples of this group. Although natural dyes can be used for dyeing different fibres directly, yet the use of mordant enhances the fastness properties of the dyed fabric. Dyers give preference to direct dyes over mordant dyes.

(b) Adjective Dyes or Mordant Dyes

Majority of vegetable dyes are "mordant" dyes. These dyes cannot be fixed permanently on the fibre or any other material without the help of a mordant. Most of the natural dyes belong to this class. Alum, ferrous sulphate, potassium dichromate, copper sulphate, stannous chloride, tanninc acid, and tannins are some of the examples. Each metallic salt used as a mordant produces a different range of Colours for each dye. Some of the widely used excellent sources of tannin are tea, oak galls, pomegranate fruit rinds, and myrobalan powder.

(c) Vat Dyes

The magnificent indigo blue is an example of vat dye and a very few are known to occur in nature. The crude vat dyes are insoluble in water and are made soluble by action either with chemicals or bacteria. Dyeing is done by immersing the fibre in the solution and afterwards exposing it in the air, thereby fixing the dyes on the fibre. The official name of indigo is indigotin. It is one of the oldest known dyes of the world. Indigotin was obtained from *Indigofera tinctoria* (indigo) and *Isatis tinctoria* (woad).

(d) Acid Dyes

Silk fibroin and wool keratin are composed of polypeptide chains, produced by the condensation of α-amino acids. These protein fibres contain carboxylic acid(-COOH) and amino ($-NH_2$) groups, and the acidic dyes are applied to silk and wool in the acidic medium. Natural dyes containing carboxylic acids groups are covered under this category. On the contrary, synthetic dyes contain sulphonic acid group in place of carboxylic acid group. The water-soluble anionic dyes, which could be applied on silk, wool, and other fibres in neutral to acidic dye baths, are known as acid dyes. The interaction between the anionic groups in the dyes and the cationic groups in the fibre leads to salt formation, resulting in the attachment of the dye molecules to the fibres, besides other factors. A very costly and ancient dye, saffron, isolated from the pistils of *Crocus sativus,* containing two carboxylic acid groups, is an example of an acid dye. Mordanting the dyed material with the metallic mordants or its treatment with tannic acid and tartar emetic enhances its colour-fastness properties.

(e) Basic Dyes

Basic dyes are covered under the category of water-soluble cationic dyes and the same on ionization produce coloured cations capable of forming an electrovalent bond with the carboxylic acid group of amino acids present in silk and wool. The yellow-coloured natural dye containing an alkaloid berberine, isolated from the roots of *Berberis aristata,* is an example of the basic dye. These dyes can be used to dye protein fibres, like silk and wool, by using neutral or mildly acidic conditions. These dyes can also be used for cotton dyeing already mordanted with tannic acid and tartar emetic or other metallic mordants.

(f) Disperse Dyes

The concept of disperse dyeing is a recent one as compared to the use of ancient natural dyes. Therefore, there is no mention of disperse dyes in the classification of natural dyes. Disperse dyes can be defined as the water-insoluble organic dyes, dispersed by suitable reagents before application to synthetic fibres. These were originally developed for the dyeing of cellulose acetate. The dyes are finely ground with a dispersing agent and sold as a paste or a powder after spray drying. The very fine particle size provides a large surface area, which helps dissolution to allow uptake by the fibre. These dyes possess low molecular weight, with poor solubility and presence of weak solubilizing groups. The presence of a hydroxyl or a weak basic group (-NH_2) give some solubility to the dye molecules. The dyes containing lawsone, flavones, and anthraquinones may probably be classified as disperse dyes. Disperse dyes are mainly used for dyeing of polyester, nylon, cellulose triacetates, and acrylic fibres. These dyes can also be used for dyeing of silk and wool. The post-mordanting method is suitable for fixing disperse dye having two adjacent hydroxyl groups, as the dye already present on the fibre readily complexes with metallic mordants, e.g., alizarin.

ARTIFICIAL OR SYNTHETIC DYES

The use of natural dyes represented about 6000 years of experimentation, accidental discovery, and evolving improvements. These were used for colouring textiles, food, leather, pottery, walls of houses, and for other uses. All of this was lost when the dye industry turned exclusively to the use of artificial or synthetic dyes. Not only did we lose the technology of natural dyeing and colours obtained from natural dyes, but also the contact with another home art. After the discovery of synthetic dye in 1856, the use of natural dyes started decreasing and reduced to minimum in the beginning of the twentieth century. The reason being that artificial dyes can be synthesized in large numbers and huge quantity and impart numerous bright shades. It is easy to synthesize an azo dye in the laboratory on an industrial scale. Numerous azo dyes were synthesized for colouring textiles with bright shades. Finally, with the synthetic indigo blue substitute in production in 1900, the use of natural dyes reduced to minimum and gradually wiped out from the scene. Artificial or synthetic dyes have their own merits (superiority in the speed of dyeing, printing, and fastness of colours) and thus succeeded in defeating natural dyes. The dyes that are synthesized in the laboratory using aromatic compounds originally obtained from coal tar are known as coal tar or artificial or synthetic dyes. About 7000 synthetic dyes are known today, which belong to a number of classes, namely nitro, nitroso, azo, diarylmethane, triarylmehtane, xanthen, acridine, quinoline, azine, vat, anthraquinoid, phthalocyanine, and many more. Synthetic dyes cause environmental pollution during their production and subsequent use.

DISCOVERY OF SYNTHETIC DYES

Natural dyes were used for colouring textiles, food, and in all other items until 1856. The breakthrough in the dye industry was made at that time when William Henry Perkin while trying to synthesize quinine, as suggested by his professor Augusts von Hofmann, by oxidizing two molecules of allyltoluidine with potassium dichromate, ended up in obtaining a dirty reddish-brown precipitate. He made a number of attempts to achieve this, but could not succeed. He continued to perform further experiments using aniline and its sulphate and potassium dichromate as an oxidizing agent. In this experiment, aniline reacted with toluidine impurities in it to produce a black-coloured precipitate instead of quinine. It was an unsuccessful experiment for the preparation of quinine, but Perkin while cleaning the flask with methylated spirit obtained a purple-coloured solution, which on dyeing trials proved to be an effective dye for silk and other textiles. This led to the chance discovery of first synthetic dye by Perkin in 1856. A patent entitled "Producing a new colouring matter for the dyeing with a lilac or purple colour stuffs of silk, cotton, wool, or other materials"[(sic) Patent No. 1984, AD 1856)] was filed by him on August 26, 1856, which was sealed on February 20, 1857. Thereafter, the first commercial dye called aniline purple or Perkin's mauve was produced in December, 1857. The dye was used under the name mauve, and later

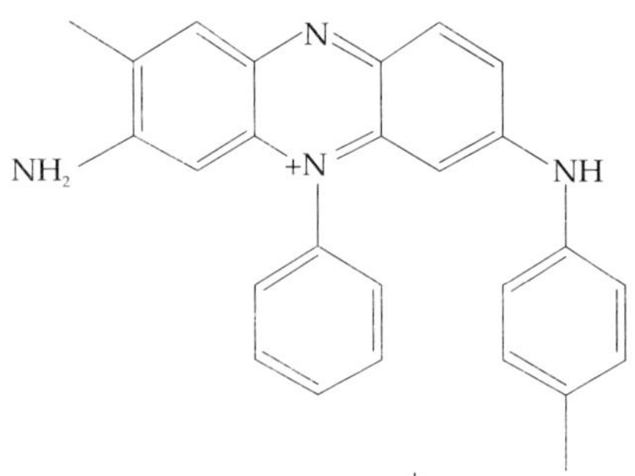

Mauveine A $C_{26}H_{23}\overset{+}{N}_4\bar{X}$

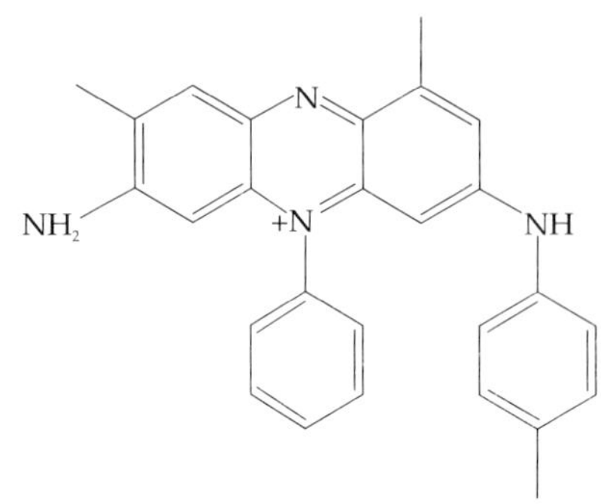

Mauveine B $C_{27}H_{25}\overset{+}{N}_4\bar{X}$

$\bar{X} = SO_4^{-2}, CH_3COO^-$

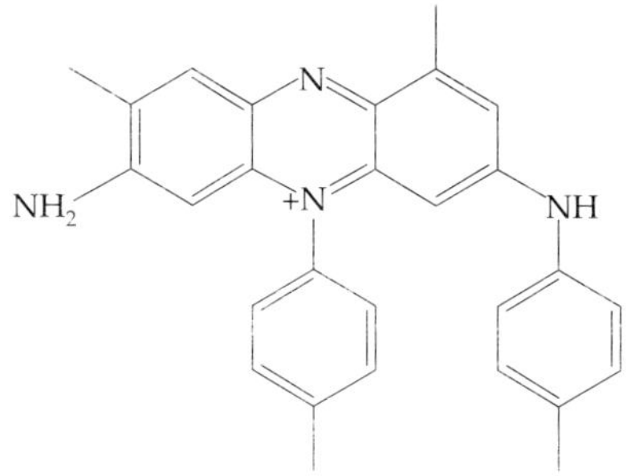

Mauveine C

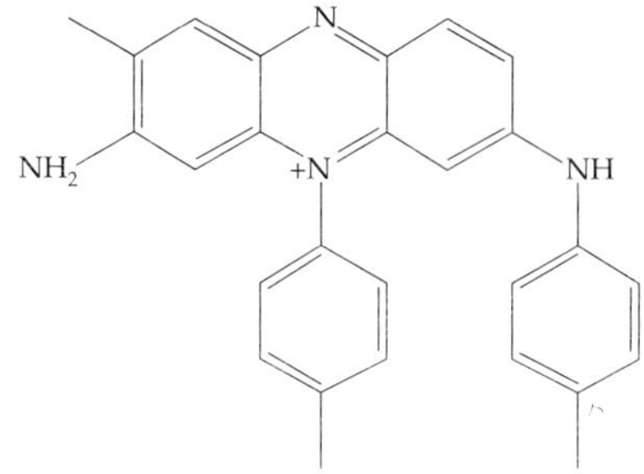

Mauveine B_2

came to be known as aniline purple. This dye was also known as mauveine and found to be costly in comparison to other synthetic dyes discovered and introduced in the market (Filarowski, 2011). Because of the uniqueness of the dye, Queen Victoria in 1862 visited the Royal exhibition in a mauve silk gown. Mauve had a short existence and fell out of fashion in the later part of 1860s due to the discovery of new synthetic colours, but not before making Perkin's fortune and giving birth to the synthetic dye chemical industry. The original sample of mauveine synthesized by Perkin is located in the Science Museum, London. Alizarin was synthesized by Perkin in 1869. The discovery of mauveine and subsequent synthetic dyes like alizarin, indigo, congo red, magenta (fuchsin), methyl violet, methyl blue, malachite green, azo dyes and others in the later part of the nineteenth century revolutionized the whole dye industry in the world (Druding, 1982). The structure of mauveine ($C_{26}H_{23}N_4^+X^-$) was not known with certainty until 1994. It was found to be a mixture of four related compounds, namely mauveine A, mauveine B, mauveine $B_{2,}$ and mauveine C, detected with the help of latest spectroscopic techniques (Meth-Cohn and Smith, 1994; Seixas de Melo et al., 2007).

AZO DYES

Azo dyes form multipurpose class of dyes and about 7,00,000 tonnes of these dyes are manufactured annually (Zollinger, 1987). Their production has increased considerably since then. An aromatic system joined to the azo group forms a chromophore, and the common auxochromes are NH_2, NR_2, and OH. Azo dyes are usually prepared by direct coupling between a diazonium salt and a phenol or amine. These dyes are classified according to the number of azo groups in the molecule: monoazo, disazo, trisazo, etc., and are also further sub-classified as acid (anionic) dyes, basic (cationic) dyes, etc. Chrysoidine (cationic dye) is an example of monoazo dye, which can be prepared by coupling benzene diazonium chloride with m-phenylenediamine.

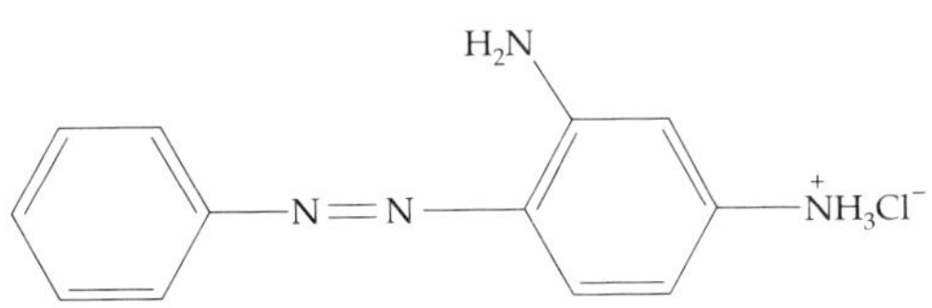

Chrysoidine

Congo red is an example of disazo dye and is prepared by coupling tetrazotised benzidine with two molecules of naphthionic acid (Finar, 1996).

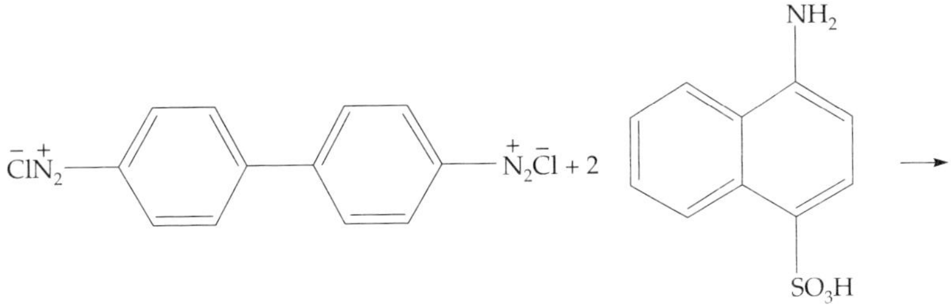

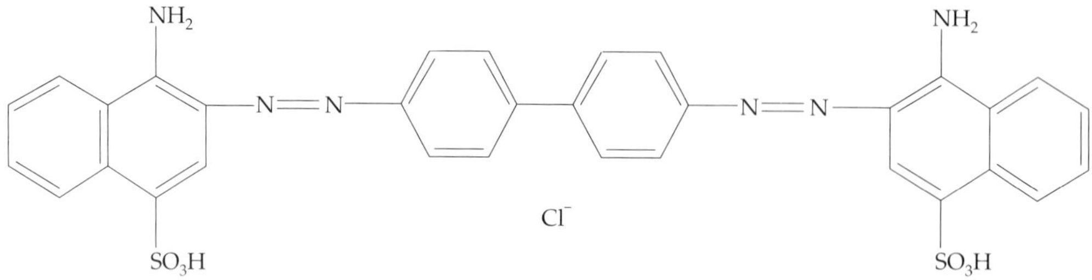

Congo red

Huge quantity of azo dyes is used for dyeing textiles, and about 10% of the dyestuffs remain unused and are thus released into rivers as toxic effluents (Hildenbrand et al., 1999). The effluent released from the dyes industries affects soil fertility, water resources, aquatic organisms, and ecosystems soundness (Puvaneswari et al., 2006; Khan and Jain, 1995).

These chemicals pass through food chain and finally reach humans and cause various health problems like cramps, hypertension, renal damage, sporadic fever, etc. Benzidine (an aromatic amine), a known human urinary bladder carcinogen and tumorigenic (Haley, 1975), is used for preparing a number of dyes that are widely used in textile dyeing, colour paper printing, and leather industries.

Azo dyes are popular because these are cheap and can be easily produced in laboratory or on a commercial scale in appropriate purity. The chemicals required for their syntheses are also easily available. The dyes are reasonably stable and the dyed fabrics have very good light and washing-fastness properties and are also resistant to microbial attack. About 2000 azo dyes are synthesized in laboratory. All these dyes, including other synthetic dyes, cause environmental pollution during their production and subsequent use, which is thus a matter of concern. The azo dyes (~150), are listed under the German legislation, which does not ban dyes as such, but certain amines which may be released by the degradation of one or more azo groups in dyes due to all or any of the following reasons (Singh et al., 2002):

(i) Prolonged storage
(ii) Exposure to intense light
(iii) Due to initiation of degradation process in the dye on contact with the spin.

The MAK Committee constituted by Germany recommended a ban on the use of 20 aromatic amines, as listed in Table 3.7 (Shenai, 1997). The ban explicitly stated that no articles of dresses (textiles and leather shoes) and bed linen can be put in trade if these have been dyed with azo dyes that release any one of the 20 named aromatic amines. It also mentioned that the articles of daily use by the people also not be dyed with specific azo dyestuffs, which are known or suspected to release harmful amines. The updated list of banned aromatic amines as per the European Union Directive is given in Table 3.8 (Directive 2002; Puntener and Page, 2004).

Table 3.7 List of forbidden amines

S.No.	Name of aromatic amine
1.	o-Aminoazotoluene
2.	4-Aminobiphenyl*
3.	2-Amino-4-nitrotoluene
4.	Benzidine
5.	p-Chloranline*
6.	4-Chloro-o-toluidine
7.	4,4′-Diamino-diphenyl methane*
8.	3,3′-Dichlorobenzidine*
9.	3,3′-Dimethylbenzidine
10.	3,3′-Dimethoxybenzidine (o-Dianisidine)
11.	3,3′-Dimethyl-4,4′ diaminodiphylmethane
12.	4-Methoxy-m-phenylenediamine
13.	2-Methoxy-5-methyl aniline*
14.	4-Methyl-1,3-phenylenediamine
15.	4,4′ -Methylenebis-(2-Chloroaniline)*
16.	2-Napthylamine*
17.	4,4′-Oxydianiline*
18.	4,4′-Thiodianiline*
19.	o-Toluidine
20.	2,4,5-Trimethylaniline*

*Not included in Indian ban

Table 3.8 Aromatic amnies as per EU directive 2002/61/EC

S.No.	Name of aromatic amine
1.	2-Aminotoluene
2.	2-Amino-4-nitroltoluene
3.	4-Aminoazobenzene
4.	4-Amino-2′, 3-dimenthylazobenzene
5.	4-Aminodiphenyl
6.	Benzidine
7.	4-Chloroaniline
8.	4-Chloro-o-toluidine
9.	4-Cresidine
10.	2,4-Diaminoanisole
11.	2,4,-Diaminotoluene

Contd...

12.	3,3′ -Dichlorobenzidine
13.	3,3′ -Dimethoxybenzidine
14.	3,3′ -Dimethylybenzidine
15.	3,3′ -Dimethyl-4,4′ diaminodiphenylmethane
16.	4,4′ -Diaminodiphenylmethane
17.	4,4′ -Methylene-bis-(2-chloroaniline)
18.	2 -Methoxyaniline
19.	2-Napthylamine
20.	4,4′ -Oxydianiline
21.	4,4′ -Thiodianiline
22.	2,4,5-Trimethylaniline

The Indian ban

After taking into consideration the German ban, the Ministry of Environment and Forests, on the recommendation of the Ministry of Textiles, Government of India, New Delhi, issued a notification [S.O. 292 (E)], dated March 29, 1996 (Prohibition and handling of azo dyes in the Gazette of India Extraordinary Part-II, Section 3, subsection II, dated April 8,1996), banning the use of 74 dyes, including four bases, due to their carcinogenic nature. The Indian banned dyes consisted of nine dyes based on aromatic amines listed in the German MAK list, including benzidine. It also included 21 dyes based on o-anisidine and p-aminoazobenzene having carcinogenic properties but not included in the MAK list. The Ministry of Environment and Forest also issued another notification [(S.O. 243 (E), dated March 26, 1997], containing all but four amines, namely, 4-chloro-o-toluidine, 2-amino-4-nitrotoluene, dianisidine, and toluidine, which have carcinogenic properties, listed in the schedule to earlier notification (Singh et al., 2002). As a result of this, the use of azo dyes that release toxic aromatic amines have been banned in the USA, the European countries, Indian subcontinent and other countries of the world (Ahlstrcom et al., 2005). Because of the health hazard of synthetic azo dyes, there is again a great demand to revive the age-old art of dyeing with natural dyes as these are non-toxic and non-carcinogenic in comparison to synthetic azo dyes (Gupta et al., 2004).

The merits and demerits of synthetic and natural dyes are discussed below.

(a) Merits of the synthetic dyes

(i) These dyes can be produced in laboratory and on industrial scale from the developed processes using readily available chemicals.

(ii) Large number of the synthetic dyes can be produced, and more than one dye can be prepared for a specific colour.

(iii) The dyes can be produced in considerable amount, depending upon the market demand.

(iv) These dyes produce bright shades, with very good colour-fastness properties.

(v) As synthetic dyes are used in pure form, the quantity required for dyeing a textile is less in comparison to natural dyes.

Despite their merits, synthetic dyes cause environmental pollution during their production and subsequent use, which is thus a matter of concern. Moreover, a number of azo dyes are harmful to humans as they break down to aromatic amines in the dyed fabrics in the presence of sunlight or due to other causes. The synthetic dyes have their own limitations and the demerits, which are described below.

(b) Demerits of the synthetic dyes

(i) Synthetic dyes have a coal tar or petrochemical base, and many of these chemicals are hazardous (carcinogenic) from the health point of view.

(ii) Production of the dyes involves reactions at high temperature and pressure using primary chemicals isolated from petroleum derivatives.

(iii) Petrochemicals are limited and irreplaceable.

(iv) The toxic or environmentally unfriendly by-products formed during the manufacturing process are discharged into air and water, causing air and water pollution. Unused dyes from the industrial dyeing units turn the rivers purple, red, and sickly green.

(v) A large number of azo dyes are based on benzidine and other similar aromatic amines. The use of azo dues, which release aromatic amines (22) on exposure to sunlight and due to other reasons are not in use in a number of countries of the world.

As synthetic dyes cause environmental pollution during their production and subsequent use, the safest option may be the use of natural dyes, which are not only eco-friendly but also prevent the environmental pollution. Plants are the greatest organic chemist of the earth. In the nature's laboratory, they carry out complex organic synthesis with ease to manufacture complicated chemical compounds with an amazing degree of accuracy, structures, and characteristics. A large number of compounds containing different functional groups and unsaturation are coloured and can be used as dyes either alone or in combination with other chemical constituents.

(c) Merits of natural dyes

(i) Natural dyes are obtained from renewable sources, viz., plants and animals (insects), and also from non-renewable minerals.

(ii) Natural dyes are eco-friendly as these are obtained from naturally occurring materials and are not synthesized in laboratory using carcinogenic chemicals, which cause water and air pollution. For example, one tonne of synthetic indigo generates considerable amount of effluents. Plants also have other tangible and non-tangible benefits.

(iii) Naturally dyed clothes do not cause any health hazard. A number of vegetable dyes are obtained from medicinal plants with proven value for human health; therefore, textiles dyed with these dyes may be beneficial for health.

(iv) There is no disposal problem as the natural dyes are biodegradeable and the quantity of mordants used for fixing the dye may be optimized to avoid water pollution by their use.

(v) Natural dyes provide a wide variety of peculiar shades, which have no matching in the synthetic range. Different shades produced using these dyes are pleasing to eyes, warm, and soothing.

(vi) As natural dyes are mainly obtained from plants, therefore a large number of plants are required to be grown for the collection of dye material and their subsequent use in the isolation of dye on a commercial scale. This will help in the utilization of waste land and employment generation. Therefore, the use of natural dyes will help in environment restoration.

(vii) A number of shades can be developed from a dye using different mordants.

(viii) Different shades similar to synthetic dyes may be produced using natural dyes.

(ix) Excellent fastness properties can be achieved using appropriate mordants.

(x) Practically, no or very mild chemical reactions are involved in the preparation of natural dyes.

(xi) It is pleasant to work with eco-friendly natural dyes, which create a rainbow of colours.

(xii) These dyes are usually non-allergic.

(xiii) Medicinally important vegetable dyes like turmeric, saffron, annatto, and others were and are being used for food colouring. It is difficult to replace them with synthetic dyes.

(xiv) Very common utensils and apparatus are required for dyeing, which are available in the rural areas also.

(d) Demerits of natural dyes

1. The collected natural dye material after air drying and pulverization is usually used for the extraction of dye by using universal solvent water. As the extracted dye solution contains a number of chemical compounds, which may or may not be contributing towards the dyeing properties of solution, the percentage of colouring principals in the dye is not the same as the dye yield. Therefore, a huge amount of the material is required for the isolation of the dye needed for dyeing textiles for large population. There may be variation in the colour yield (active principals) due to following reasons:

(a) The plants may not be of the same age.

(b) There may be geographical variation of the collected plant material.

(c) There may be variation in the time of collection of the plant material.

2. As most of the natural dyes are mordant dyes, the dyeing process with these dyes is slightly lengthy.

3. The other technical disadvantages of natural dyes, due to lack of exhaustive research work in this area, are given below:

(a) There are about 45,500 plant species known in India but only a small number of species have been screened with reference to their dyeing quality and chemical constituents. As a result of this, a limited number of suitable dye-yielding plant species are available for their use as natural dyes.

(b) Due to a lack of research work, natural dyes have not been standardized.

(c) It is difficult to dye synthetic fibres, while wool and natural silk can be dyed easily. Pretreatment and longer time are required for the dyeing of linen and cotton.

(d) The fastness properties of the dyed fabrics are on the lower side.

(e) The blending of natural dyes needs expertise.

4. Absence of the large-scale strategic cultivation of dye-yielding plants does not ensure supply of raw materials, which makes it difficult to manufacture large quantities of natural dyes.
5. Shade consistency in natural dyes is a major problem. Thus, it is difficult to execute big supply orders. Supercritical extraction and computer matching may overcome this problem.
6. Availability of raw materials is a basic problem as it takes time for growing plants.
7. There is no guarantee of the local consumption of natural dyes; therefore, entrepreneurs are reluctant to produce them.
8. The garments manufactured using natural dyes are costlier and thus will be used mainly by rich people.

FUTURE OF NATURAL DYES

1. There is an international demand for natural dyes or vegetables dyes. Domestic demand for natural dyes may also increase due to their beneficial nature.
2. In India, natural dyes may be used in carpet industry and also in dyeing woollen clothes and silk sarees.
3. Natural dyes may find potential uses in the areas like foods, drugs, cosmetics, and paints, besides textiles.
4. Natural dyes cannot be substituted to synthetic dyes for such a large population on the earth. These have their own market, and any expansion in their market will not harm the synthetic dyes market. Researches on natural dyes can overcome the demerits of the natural dyes.

RESEARCH NEEDS FOR NATURAL DYES

1. Chemical screening of plant materials obtained from forests, including herbs, shrubs, weeds, lichens, vegetable wastes, and other natural wastes, needs to be done in order to explore additional sources of natural dyes.

2. After the discovery of synthetic dyes, not much research has been carried out in the field of natural dyes, leading to a knowledge gap, as a result of which, these are not available in the standardized form. Standardization of the method of extraction, purification of dyes, agronomical practices and simple chemical modification need to be done to augment the use of the natural dyes. Simpler methods with good dye uptake, which can impart fast colours on dyeing, are also required to be developed.
3. Efficient cultivation techniques for certain staple vegetable colouring matters should be developed.
4. Natural disperse dyes should be developed for synthetic fibres
5. Standardization of parameters for fermentation is essential.
6. A process for the preservation of the extract in liquid or desiccated form should be developed. This is essential to reduce the transportation cost of raw materials and also for making the dye extract available in the usable form.
7. Most of the dyers have expertise with synthetic dyes and there is a severe shortage of trained natural dyers. The technology of natural dyeing is required to be developed.
8. There are not many books on the technology of natural dyeing that can provide the scientific basis on the subject. The scientific literature on natural dyes should be developed.

EXPORT OF DYES FROM INDIA

There are about 325 members of the Dye Stuffs Manufactures Association of India (DMAI), who are engaged in manufacturing dyes for textile, leather, paper, plastics, ink, paints, food and cosmetics, soaps, and detergents. DMAI controls around 7.5% of the $25 billion global market. The dye industry in India is export-oriented as more than 50% of its turnover is exported to major countries like the US, Europe, the Far East and Southeast, and is growing annually at the rate of 20%. The largest user of dye is the textile industry, followed by ink and paint industries. As per DMAI, the Indian dye export was Rs. 7000 crore in 2007, which was expected to touch Rs. 12,000 crore by 2011 (Mehta, 2007) and the further increase was expected in the same proportion.

References

1. Agarwal, D.P. and Tiwari, R. (1989). Mineral pigments in India. Paper presented at the First National Semmar on Natural Dyes, Jaipur, pp. 27-32.
2. Ahlstrcom, I.H., Eskilsson C.S., Bjcorklund, E. (2004). Determination of banned azo dyes in consumer goods. *Trends in Analytical Chemistry,* 24(1), 49-56.
3. Directive 2002/61/EC of the EP and of the EC of July 19, 2002.
4. Druding, S.C. (1982). Dye History from 2600 BC to 20th Century. (http.//www.Straw.com.sig/dyehist.html)

5. Filarowski Andrew. (2011) Perkin's Mauve: The history of the chemistry. *Resonance,* 16(2), 1140-1145.
6. Finar, I.L. (1973) (Reprinted 1996). *Organic Chemistry,* Vol. 1, p. 885, Longman Singapore Publishers (Pvt) Ltd, Singapore
7. Gupta, D., Gulrajani, M.L., Kumari, S. (2004). Light fastness of naturally occurring anthraquinones dyes in nylon. *Colouration Technology,* 120(5), 205-212.
8. Gulrajani, M.L. and Gupta, D. (Editors). *Natural dyes and their application* to textiles, Department of Textile Technology, Indian Institute of Technology, New Delhi.
9. Haley, T.J. (1975). Benzidine revised: A review of the literature and its congeners. *Clin. Toxicol,* 8, 13.
10. Hancock, M. and Boxworth, A. (1997). Potential for Colourants from plants sources in England and Wales. STO 106. *Arable crops and horticulture division,* Cambridge CB2 8NN.
11. Hildenbrand, S., Schmahl F.W., Wodarz, R., Kimmel, R., Dartsch, P.C. (1999). Azo dyes and carcinogenic aromatic amines in cell culture. *Int Arch Occup Environ Hlth,* 72, 52
12. Khan, I.T. and Jain, V. (1955). Effects of textile industry waste water on growth and some biochemical parameters of *Triticum aestiumvar. Raj* 3077. *J Environ. Poll;* 2, 50.
13. Mehta, J. (2007). Presdential address of 57th annual general meeting. *The Textile Magazine* 48(9), 92–94.
14. Menth-Cohn, O. and Smith, M. (1994). What did W.H. Perkin actually make when he oxidized aniline to obtain mauveine? *J. Chem. Soc. Perkin* 1, 5-7
15. Nagendrappa, Gopalpur. (2011). Chemistry triggered the first civildisobedience movement in India, indigo in Indian Independence: Role of a broken thermometer. *Resonance,* 16(12), 1168-1174.

15a. Nature Energies, Colour Therapy Resources and Equipment (2014). http://www.natures-energies.com/natural-therapies/Colour-light-therapy-chromotherapy-resources-and-equipment.

15b. Paul, R., Jayesh, M., Naik, S.R. (1996). Textile Dyes and Printer, October 23, 16-24.

16. Puntener, A. and Page, C. 2004). European ban or certain azo dyes. *TFL Leather Technology Ltd. (Quality and Environment),* 1-4.
17. Puvaneswari, N., Muthukrishnan, J., Gunasekaran. (2006). Toxicity assessment and microbial degradation of azo dyes. *Indian Journal of experimental Biology,* 44, 618-626.

17a. Sekar, N. (1999). Natural Dyes-Prospects and Future. *Covention on Natural Dyes,* Indian Institute of Technology, New Delhi. p. 35.

18. Sexixas de Melo, J., Takato, S., Melo, M.J., Parola, A.J. (2007). Revisting Perkin's dye(s): The spectroscopy and photophysics of two new mauveine compounds (B_2 and C.) *Chem. Commun;* 2624-26.

19. Shenai, V.A. (1997). Eco-friendly chemicals and ban on certain azo dyes. *Colourage*, 10, 50-64.
20. Singh, K., Singh, S., Butola, B.S. (2002). The German Ban: A realistic appraisal. *Colourage*, Vol. 1, 43-47.
21. Srivastava, M.G. (1989). Natural dyes: An overview. Paper presented in the *National Seminar on Natural Dyes*, held at Jaipur on October 20-21, pp. 1-6.
22. Tewari, D.N. (1994). *Tropical Forest Produce*. International Book Distributors, Rajpur Road, Dehradum (India), pp. 317-358.
23. Zollinger, H. (1987).Colour chemistry – synthesis, properties and applications of organic dyes and pigments (VCH, New York), 92.

4 Sources of Natural Dyes

PART A: Sources of Natural Dyes

Plants are essential for the survival of humans and animals on the earth. A forest is a natural renewable resource, which has to be managed on the basis of silvicultural principles. Forest products are divided into two groups, namely, wood and non-wood forest products (NWFPs). Wood includes timber and fuelwood, and all other products obtained from forests are covered under NWFPs, which include mainly edible plants, fodder and forage plants, medicinal plants, tanning and natural dye materials, and other products. Natural dyes are important forest products. The use of natural dyes began with the dawn of civilization. These were obtained from vegetables (plants), animals (insects), and mineral sources until 1856. Plants constituted the major renewable sources of vegetable dyes. The dyes were obtained from different parts of plants, namely, bark, flowers, fruits, leaves, roots, seeds, and wood of a number of plant species. About a thousand of different natural dye-yielding plant sources were recorded by the ancient writers. A number of plant species, of primary importance imparting fast colours, were used in dyeing of different fabrics and for other uses. Efforts have been made by different researchers (Siva, 2007; Gaur, 2008; Kumar and Sinha, 2004; Farooqui, 1993; Tewari, 1994; Saxena, 1989; Bharathan and Daniel, 2006; Gokhle et al., 2004; Mahanta and Tewari, 2005) and others to compile the list of dyeing yielding plant species. The different lists of the above-cited authors contain about 50 to 100 plant species having a number of common species. After taking into consideration the available literature, it was decided to give a comprehensive list of the dye yielding plant species which may serve as a ready reckoner, and such a list in not available at present. The list of plant species is being given in Table 4.1 containing the names of plant species, popular name, dye bearing parts, family, pattern, vital colour and references. The plants list consists of 392 dye-yielding plant species which may be used as a dye source, depending on the requirement. The vegetable dye yielding sources according to families of the plants, plant parts, and colours obtained from the dyes of these plants species in concise form are given below.

DYE SOURCES ACCORDING TO FAMILIES

Of the total number of the plant species of Table 4.1, 54 belong to the family Fabaceae, 21 to Asteraceae, and 17 to Euphorbiaceae, 15 to Rubiaceae, and the remaining to 107 other families. The list of families (21) containing at least five plant species is given in Table 4.A.

Table 4.A Families of dye-yielding plants of Table 4.1 containing at least five plant species

S. No.	*Family*	*Number of plant species*
1.	Amaranthaceae	6
2.	Acanthaceae	6
3.	Asteraceae	21
4.	Anacardiaceae	12
5.	Boraginaceae	09
6.	Berberidaceae	7
7.	Combretaceae	7
8.	Euphorbiaceae	17
9.	Fabaceae	54
10.	Clusiaceae	6
11.	Lamiaceae	6
12.	Mimosaceae	9
13.	Malvaceae	5
14.	Moraceae	10
15.	Meliaceae	5
16.	Polygonaceae	10
17.	Rubiaceae	15
18.	Rutaceae	5
19.	Rosaceae	8
20.	Symplocaceae	5
21.	Zingiberaceae	8

Dye Sources According to Plant Parts

The vegetable dyes obtained from different parts of plants of Table 4.1, namely bark, flowers, fruits, roots, seeds, wood, and even from whole plant, are given in Table 4.B and according to different colours in Table 4.C.

Dye Sources According to Colours

The number of plant species of Table 4.1 containing the dyes which impart different colours are being given in Table 4.C.

Table 4.B Dye-yielding plant species of Table 4.1 according to plant parts

S.No.	Plant part	Number of plant species	Table No
1.	Bark	101	4.6
2.	Flowers	64	4.7
3.	Fruits	57	4.8
4.	Leaves	82	4.9
5.	Roots	58	4.10
6.	Seeds	18	4.11
7.	Wood	24	4.12
8.	Whole plant	18	4.13

Table 4.C Dye-yielding plant species of Table 4.1 according to colours

S.No.	Color	Number of plant species	Table No
1.	Black	48	4.14
2.	Blue	31	4.15
3.	Brown	75	4.16
4.	Green	18	4.17
5.	Grey	6	4.18
6.	Orange	18	4.19
7.	Pink	11	4.20
8.	Purple	13	4.21
9.	Red	65	4.22
10.	Yellow	147	4.23

A glance at Table 4.A indicates that the maximum number of dye-yielding plants belong to the Family Fabaceae, followed by Acanthaceae, Euphorbiaceae, Rubiaceae, Anacardiaceae, and other Families. It is evident from Table 4.B that bark of different plant species (101) is the major source of vegetable dyes, followed by leaves, flowers, roots, fruits, wood, seeds, and whole plant. Table 4.C indicates that yellow colour can be obtained using the dyes of maximum number of plants species (147), followed by brown, red, black, blue, green, orange, purple, pink, and grey. A number of plant species impart more than one colour. Some of the colours imparted have been indicated in the tables but different colours and shades can be obtained by using different mordants or their appropriate combination. The efforts have been made to indicate the references. Lists of plant species as per the dye-yielding parts and according to different colours are given from Table 4.5 to Table 4.13 and Table 4.14 to Table 4.23, respectively.

Table 4.1 Dye bearing plant species

S.No.	*Name of plant species*	*Family*	*Popular name*	*Dye bearing parts*	*Pattern*	*Vital colour*	*References*
1.	*Abies spectabilis* (D.Don) Spach	Pinaceae	Indian Silver- Fir, Fir	Cones	Tr	Purple or violet	1
2.	*Abrus precatorius* Linn.	Fabaceae	Crab's Eye, Rati	Seeds coat	Cl	Black	2
3.	*Acacia catechu* Willd.	Mimosaceae	Khair, Cutch Tree	Heartwood	Tr	Red, brown	3, 2, 4
4.	*Acacia chundra* Willd.	Mimosaceae	Red Cutch	Wood	Tr	Grey, black	5
5.	*Acacia dealbata* Link	Mimosaceae	Silver Wattle	Bark	Tr	Brown, black	1
6.	*Acacia farnesiana* Willd.	Mimosaceae	Vilayati Kikar	Bark, seeds, pods	Tr	Black, brown	6
7.	*Acacia leucophloea* Willd.	Mimosaceae	Safed Kikar	Bark, leaves	Tr	Red	6
8.	*Acacia nilotica* Delile subsp. *indica* (Benth.) Brenan	Mimosaceae	Babul	Fruits rinds, Gum, leaves, bark	Tr	Black to dark brown	2, 6
9.	*Acacia pennata* Willd.	Mimosaceae	Aila	Bark	Sh	Brown	5
10.	*Acacia sinuata* Willd. (Lour.) Merrill	Mimosaceae	Ritha	Bark	Sh	Brown	6, 7
11.	*Acalypha indica* Linn.	Euphorbiaceae	Indian Acalypha	Leaves	H	Yellow	8
12.	*Acanthopanax trifoliatum* (Linn.) Voss	Araliaceae	Shiah-ryngkhwari	Fruits	Sh	Black	1
13.	*Achyranthes aspera* Linn.	Amaranthaceae	Latjira, Prickly- Chaff Flower	Whole plant	H	Black brown	2
14.	*Achyranthes bidentata* Blume	Amaranthaceae	Apamarga	Whole plant	H	Black brown	2
15.	*Actaea spictata* Linn.	Ranunculaceae	Baneberry, Grapewort	Berries	H	Black	1, 7
16.	*Adenanthera pavonina* Linn.	Fabaceae	Rakta kambal, Red Wood	Wood	Tr	Red	5, 9

17.	*Adhatoda zeylanica* Medic.	Acanthaceae	Vasaka, Arusa	Leaves	Sh	Yellow-green	2
18.	*Adina cordifolia* (Roxb.) Hook.f.ex Brandis	Rubiaceae	Haldu	Wood	Tr	Yellow	7
19.	*Aegle marmelos*(Linn.) Correa ex Roxb.	Rutaceae	Bel	Rinds of unripe fruits	Tr	Yellow	2, 10, 6, 4
20.	*Aesculus indica* Hook.	Hippocastanaceae	Pangar, Himalayan Chest Nuts	Bark	Tr	Brown	2
21.	*Ageratum conyzoides* Linn.	Asteraceae	Goat Weed	Whole plant	H	Light green	11
22.	*Agrimonia eupatoria* Linn.	Rosaceae	Agrimony	Leaves, roots	H	Yellow	2
23.	*Ailanthus triphysa* (Dennst.) Alston	Simaroubaceae	Guggul Dhup	Leaves	Tr	Black	1
24.	*Albizia amara* Boiv.	Fabaceae	Krishnasirisha	Leaves, pods	Tr	Brown grey, olive	8
25.	*Albizia lebbeck* Benth.	Fabaceae	Siris, Kokko	Bark	Tr	Brown	5, 12
26.	*Albizia odoratissima* (L.f.) Benth.	Fabaceae, Mimosaceae	Black Siris	Bark	Tr	Brown	6
27.	*Althea rosea* Linn.	Malvaceae	Hollyhock	Flowers	H	Red	13, 1, 9, 4
28.	*Aleurities moluccana* (Linn.)Willd.	Euphorbiaceae	Jungli Akrot	Roots	Tr	Brown	6
29.	*Alkanna tinctoria*(L.) Tausch	Boraginaceae	Dyer's Bugloss	Roots	H	Purple	14, 4
30.	*Allium cepa* Linn.	Liliaceae	Piyaz	Skin	Bulbous biennial	Orange	15, 16, 4
31.	*Alnus nepalensis* D.Don	Betulaceae	Uitis, Indian Alder	Bark	Tr	Red	2, 6
32.	*Alnus nitida* Endl.	Betulaceae	West Himalayan Alder, Seril	Bark	Tr	Brown-grey	7, 4

Contd...

S.No.	Name of plant species	Family	Popular name	Dye bearing parts	Pattern	Vital colour	References
33.	*Alpinia flabellata* Ridley	Zingiberaceae	Gingerlily	Leaves	H	Yellow	17
34.	*Alpinia galanga*(Linn.) Willd.	Zingiberaceae	Barakulanjar	Roots	H	Brown	9
35.	*Amaranthus tricolor* Linn.	Amaranthaceae	Garden Amaranth	Inflorescences, leaves	H	Red-violet	18, 4
36.	*Ampelocissus latifolia* (Roxb.) Planch.	Vitaceae	Panibel	Roots	Cl	Black	2
37.	*Anacardium occidentale* Linn.	Ancardiaceae	Cashew or The Cashew Nuts	Nuts (shells)	Tr	Black	7,9
38.	*Annona reticulata* Linn.	Annonaceae	Raamaaphal	Unripe fruits	Tr	Black	19, 6
39.	*Annona squamosa* Linn.	Annonaceae	Sharifa	Fruits	Sh	Yellow	2
40.	*Anogeissus latifolia* (Roxb.) Wall.ex Bedd.	Combretaceae	Dhaura	Leaves	Tr	Black	2, 6
41.	*Anthemis tinctoria* Linn.	Asteraceae	Dyer's Chamomile	Flowers	H	Yellow	4, 19a
42.	*Aporusa dioica* (Roxb.) Muell-Arg.syn.*A. roxburghii Baill*	Euphorbiaceae	Kokra in Bengali	Leaves	Tr	Black	20
43.	*Ardisia solanacea* Roxb. syn. *Ardisia humilis* C.B. Clarke	Myrsinaceae	Bisi	Fruits	Sh	Yellow	5, 4
44.	*Areca catechu* Linn.	Arecaceae	Supari	Nuts	Tr	Brown	6, 21
45.	*Arnebia benthamii* (Wall.ex G.Don) John.	Boraginaceae	Kashmiri kahzaban	Roots	H	Red, purple	2, 1
46.	*Arnebia densiflora* (Nordm.) Ledeb.	Boraginaceae	Ratanjot	Roots	H	Red	4, 22
47.	*Arnebia euchroma* (Royle)Johnston	Boraginaceae	Ratanjot	Roots	H	Red	23, 24, 4

48.	*Arnebia nobilis* Reichb.f.	Boranginaceae	Ratanjot	Roots	H	Purple	5
49.	*Aronia melanocarpa* (Michx.) Elliott	Rosaceae	Black choke Berry	Fruits	Sh	Blue	4, 25
50.	*Artemisia japonica* Thunb.	Asteraceae	Kunja	Leaves	H	Brown	2
51.	*Artemisia nilagirica* (C.B Clarke) Pamp.	Asteraceae	Indian WormWood	Leaves	H	Brown	2
52.	*Artocarpus communis* J.R. & G.Forst.	Moraceae	The Bread Fruits	Leaves	Tr	Pink, brown	8
53.	*Artocarpus heterophyllus* Lam .	Moraceae	Kathal, Jack Fruits	Wood, Roots,Fruits	Tr	Yellow	2, 26, 6, 4
54.	*Artocarpus lakoocha* Roxb.	Moraceae	Monkey Jack	Fruits, Wood, Roots	Tr	Yellow	2
55.	*Averrhoa carambola* Linn.	Averrhoaceae	Kamrak, Star Fruits	Whole Plant	Tr	Yellow-brown	2
56.	*Azadirachta indica* A.Juss.syn. *Melia azadirachta* Linn.	Meliaceae	Nim, Neem Tree	Bark	Tr	Red	27
57.	*Baccaurea ramiflora* Lour.	Euphorbiaceae	Kataphal	Bark, Leaves, Fruits	Tr	Black	20
58.	*Bacopa monnieri* (Linn.)Pannell	Scrophulariaceae	Brahmi	Whole Plant	H	Green	8
59.	*Baptisia australis* (L.) R.Br.	Fabaceae	Blue Wild Indigo	Flowers	H	Yellowish	28, 4
60.	*Barleria prionitis* Linn.	Acanthaceae	Kala Bansa	Flowers	Sh	Yellow	7, 1
61.	*Basella rubra* Linn.	Basellaceae	Red Vine-Spinach	Fruits	H	Violet	7
62.	*Bauhinia racemosa* Lam.	Fabaceae	Kachnal	Bark	Tr	Light green	2
63.	*Bauhinia tomentosa* Linn.	Fabaceae	Bell Bauhina	Leaves	Sh	Yellow	1,5

Contd...

S.No.	Name of plant species	Family	Popular name	Dye bearing parts	Pattern	Vital colour	References
64.	*Bauhinia variegata* Linn.	Fabaceae	Kachnar, Buddhisht Bauhinia	Flowers	Tr	Purple, violet-pink	2, 6, 4
65.	*Benthamidia capitata* (Wall.) Hara	Cornaceae	Bamora	Fruits	Tr	Red	2
66.	*Berberis aristata* DC.	Berberidaceae	Kilmor	Bark, roots	Sh	Yellow	2
67.	*Berberis asiatica* Roxb. ex. DC.	Berberidaceae	Kilmor	Bark, roots	Sh	Yellow	2
68.	*Berberis chitria* Lindl.	Berberidaceae	Chetar	Bark, roots	Sh	Yellow	2
69.	*Berberis nepalensis* (DC.) Spreng.	Berberidaceae	Nepal Barberry	Roots	Sh	Yellow	6
70.	*Berberis vulgaris* Linn.	Berberidaceae	Barberry	Bark, roots	Sh	Yellowish, yellow	2, 29, 4
71.	*Beta vulgaris* Linn.	Amaranthaceae	Common Beet	Roots	H	Red	30
72.	*Bidens bipinnata* Linn.	Asteraceae	Lumb	Leaves	H	Yellow	31, 32, 4
73.	*Bischofia javanica* Blume	Euphorbiaceae	Bishop Wood	Bark, seeds	Tr	Black	2
74.	*Bixa orellana* Linn.	Bixaceae	Annatto	Seeds	Tr	Orange, red	5, 33, 6, 4
75.	*Bougainvillea glabra* Choisy	Nyctaginaceae	Paperflower	Flowers with ivory white bracts	Vine Sh	Yellow	9
76.	*Brassica nigra* (Linn.) Kuch	Brassicaceae	Black Mustard	Seeds	H	Yellow	8
77.	*Bridelia stipularis* Blume	Euphorbiaceae	Khaji	Fruits	Cl	Black	5, 1
78.	*Brugmansia suaveolens* (Humb & Bonpl. ex.Willd.) Bercht. & J.Presl.	Solanaceae	Angel-Star	Leaves	Sh	Green	2
79.	*Butea monosperma*(Lam.) Taub.	Fabaceae	Palas,Flame of the Forest	Flowers	Tr	Yellow-orange	2, 34, 6, 4

80.	*Butea superba* Roxb.	Fabaceae	Palas Lata	Roots	H	Red	6
81.	*Caesalpinia coriaria* (Jacq.) Willd.	Fabaceae	Devi Divi Plant	Pods, seeds	Tr	Black	5
82.	*Caesalpinia pulcherrima* Sw.	Fabaceae	Peacock Flower	Flowers	Sh	Red	5
83.	*Caesalpinia sappan* Linn.	Fabaceae	Sappan Wood	Wood	Tr	Red	1, 35, 6, 4
84.	*Caesulia axillaris* Roxb.	Asteraceae	Bangra	Whole plant	H	Yellow	4, 36
85.	*Calendula officinalis* Linn.	Asteraceae	Calendula	Inflorescences	H	Orange	37, 4
86.	*Capsicum annuum* Linn. syn. *C. frutescens* C.B. Clarke	Solanaceae	Lalmirich, Chilli	Fruits	H	Red	38, 7, 4
87.	*Careya arborea* Roxb.	Barringtoniaceae	Kumbi	Bark	Tr	Yellow	2, 6
88.	*Carthamus oxyancantha* Bieb.	Asteraceae	Wild Safflower	Flowers	H	Yellow	5
89.	*Carthamus tinctorius* Linn.	Asteraceae	The Safflower, False Saffron	Flowers	H	Red	1, 39, 6, 4
90.	*Cassia auriculata* Linn.	Fabaceae	Tanner's Cassia	Flowers	Sh	Yellow	40, 1, 6, 4
91.	*Cassia fistula* Linn.	Fabaceae	Amaltas,Golden-Shower	Fruits, bark	Tr	Brown	2, 6
92.	*Cassia obtusifolia* Linn.	Fabaceae	Banarh, Cakonda	Pods, seeds	H	Yellow	5
93.	*Cassia occidentalis* Linn.	Fabaceae	Kasonda, Coffee-Senna	Pods, seeds	H	Brown, yellow	5
94.	*Cassia tora* Linn.	Fabaceae	Panwar, Foetid Cassia	Seeds	H	Red, blue, brown	6, 41
95.	*Cassytha filiformis* Linn.	Lauraceae	Love-Vine	Stem	H	Brown	1
96.	*Casuarina equisetifolia* Linn.	Casuarinaceae	Jangli saru, Casuarina	Bark	Tr	Brown	42, 6

Contd...

S.No.	Name of plant species	Family	Popular name	Dye bearing parts	Pattern	Vital colour	References
97.	*Celtis australis* Linn.	Ulmaceae	European Nettle Tree	Bark	Tr	Yellow	2
98.	*Ceriops tagal* (Perr.) C.B.Robins	Rhizophoraceae	Goran, Guttia	Bark	Mangrove	Brown	42, 9
99.	*Cheiranthus cheiri* Linn.	Brassicaceae	Wall Flower	Leaves, petal	H	Yellow	4, 43
100.	*Chlorophora tinctoria* (L.) Benth. & Hook.f.	Moraceae	Old Fustic	Wood	Tr	Yellow	44, 4
101.	*Chloroxylon swietenia* DC.	Flindersiaceae	Indian SatinWood Tree	Bark	Tree	Yellow	45, 6
102.	*Chrozophora tinctoria*(Linn.) Rafin.	Euphorbiaceae	Dyer's Litmus	Petals, fruits, sap	H	Red, blue	9, 4
103.	*Chukrasia tabularis* A.Juss.var.*tabularis*	Meliaceae	Indian Mahogany Tree	Flowers, leaves	Tr	Red, yellow	9, 7
104.	*Cinnamomum camphora* (Linn.) Presl	Lauraceae	Camphor Tree, Kapur	Bark	Tr	Brown	46
105.	*Cinnamomum tamala* (Buch.-Ham.) Nees & Eberm.	Lauraceae	Tejpat, Indian Cassia Lignea	Leaves	Tr	Brown	2
106.	*Citrus medica* Linn.	Rutaceae	Citron	Bark	Tr	Black	1, 7
107.	*Citrus sinensis* (Linn.) Osbeck	Rutaceae	Malta	Peal	Tr	Yellow	47
108.	*Cladonia verticillata* (Hoffm.) Schaer	Cladoniaceae	Cup Lichen	Whole plant	Lichen	Yellow-red	2
109.	*Clitoria mariana* Linn.	Fabaceae	Sweta, Butterfly-Pea	Flowers	Climber, Undershrub	Blue, violet	20
110.	*Clitoria ternatea* Linn.	Fabaceae	Aparajit, Conchflower	Flowers, seeds	Cl	Blue	4, 48, 1
111.	*Cocos nucifera* Linn.	Arecaceae	CocoNuts Palm	Nuts	Tr	Green	5
112.	*Colquhounia coccinea* Wallich	Lamiaceae	Himalayan Mint Plant	Whole plant	Sh	Black	20
113.	*Combretum pilosum* Roxb.ex G.Don	Combretaceae	Bhoree Loth	Leaves	Sh	Blue, black	7

114.	*Commelina benghalensis* Linn.	Commelinaceae	Benghal Dayflower	Juice of flowers	Serpentine Leaves miner	Blue	9
115.	*Convallaria majalis* Linn.	Lilaceae	May Lily	Leaves	H	Orange	1
116.	*Crocus sativus* Linn.	Iridaceae	Kesar, Saffron	Flowers	H	Yellow	49, 50, 6, 4
117.	*Cordia dichotoma* Forster. f.	Boraginaceae	Lasura, Indian Cherry	Roots, leaves, fruits	Tr	Yellow, red, green	1, 7, 5, 6
118.	*Coreopsis tinctoria* Nutst.	Asteraceae	Calliopsis	Flowers	H	Yellowish	4, 51
119.	*Coriaria nepalensis* Wall.	Coriariaceae	Makola	Wood	Tr or Sh	Red	6, 4
120.	*Corylus avellana* Linn.	Betulaceae	Hazel	Flowers	Tr	Yellow	8
121.	*Corylus jacquemotii* Decne	Betulaceae	Indian Hazel Tree, Kabasi	Fruits rinds	Tr	Camel	2
122.	*Coscinium fenestratum* Clolebr.	Menispermaceae	Tree -Turmeric, Jharihuldi	Seeds, bark, wood	Sh	Red	1
123.	*Cotinus coggygria* Scop.	Anacardiaceae	Young Fustic	Wood, leaves	Sh	Yellowish, orange	52
124.	*Crotalaria striata* DC.	Fabaceae	Can janka	Seeds	Sh	Brown	5
125.	*Cudrania cochinchinensis* (Lour.) Kudo et Masam.	Moraceae	Pulikant	Leaves, seeds	Sh	Orange, yellow	20
126.	*Cuminum cyminum* Linn.	Apiaceae	Cumin Seeds	Seeds	H	Bright yellow	8, 52a
127.	*Cupressus torulosa* Don var. *corneyana* (Knight &Perry) Carr.	Cupressaceae	Surai	Leaves	Tr	Green	2
128.	*Curcuma amada* Roxb.	Zingiberaceae	Amada, Amba Haldi	Rhizomes	H	Yellow	5

Contd...

S.No.	Name of plant species	Family	Popular name	Dye bearing parts	Pattern	Vital colour	References
129.	*Curcuma aromatica* Salisb.	Zingiberaceae	Jangli Haldi, Wild Turmeric	Rhizomes	H	Yellow	2
130.	*Curcuma longa* Linn. syn *C. domestica* Val.	Zingiberaceae	Haldi, Turmeric	Rhizomes	H	Yellow	2, 53, 54, 6, 4
131.	*Curcuma zedoaria* (Christm.) Rosc.	Zingiberaceae	Zedoary, Kachura	Rhizomes	H	Yellow	9, 7
132.	*Cyamopsis tetragonoloba* (Linn.) Tanb.	Fabaceae	Guar	Pods, seeds	H	Yellow	5
133.	*Cynometra ramiflora* Linn.	Fabaceae	Cynometra	Roots	Tr	Purple	6
134.	*Cyperus scariosus* R. Br.	Cyperaceae	Nagarmotha	Roots	H	Brown	2
135.	*Daphne papyracea* Wall.ex. Steud. emend.	Thymelaeaceae	Satpura	Bark, fruits	Sh	Red	2, 20, 55
136.	*Datisca cannabina* Linn.	Datiscaceae	Akalbir, Bujr-Bhang	Roots	H	Yellowish-red	2, 6, 4
137.	*Datura metel* Linn.	Solanaceae	Thorn Apple	Leaves	H	Grey, brown	8, 55a
138.	*Delonix regia* Rafin.	Fabaceae	Gul Mohar	Flowers	Tr	Deep crimson	56
139.	*Delphinium grandiflorum* L.	Ranunculaceae	Sieberian Larkspur	Flowers	H	Blue, grey	56a
140.	*Delphinium zalil* Aitch &Hemsl.	Ranunculaceae	Guljali, Larkspur	Flowers	H	Yellow	5, 6, 4
141.	*Desmodium multiflorum* DC.	Fabaceae	Bhatmase Phool	Flowers	Sh	Purple	20
142.	*Dioscorea bulbifera* Linn.	Dioscoreaceae	Potatoyam, Air Potato	Tubers	Sh	Pale colour	1
143.	*Diospyros malabarica* Desr. syn. *D. peregrina* (Gartn.) Gurke	Ebenaceae	Gaub Persimmon, Gab	Unripe fruits	Tr	Brown	2, 6, 4

144.	*Diospyros melanoxylon* Roxb.	Ebenaceae	Bidi Patta	Fruits	Tr	Brown, black	7
145.	*Dipterocarpus turbinatus* Gaertn.f.	Dipterocarpaceae	Common Gurjan Tree	Twigs, bark	Tr	Yellow, brown	1, 7
146.	*Eclipta alba* (Linn.) Hassk.	Asteraceae	Bhangra	Whole plant	H	Green, grey	57
147.	*Elaeodendron glaucum* (Rottb.) pers.	Celasteraceae	Chouri	Bark	Tr	Red	1
148.	*Emblica officinalis* Gaertn.syn *Phyllanthus emblica* Linn.	Euphorbiaceae	Amla	Fruits, bark	Tr	Blue-black	2, 58, 6
149.	*Embellia ribes* Burm.f.	Myrsinaceae	Barberang, False Pepper	Fruits	H	Red	59
150.	*Engèlhardtia spicata* Lesch.ex. Blume	Juglandaceae	Corcorshing, Mahwa	Bark	Tr	Dark brown, Dark blue	20, 2
151.	*Entada pursaetha* DC.	Mimosaceae	Ghilla	Young pods	Cl	Dark blue	20
152.	*Ervatamia coronaria* Stap.f.	Apocynaceae	Chandni	Seeds pulp	Sh	Red	58a
153.	*Erythrina suberosa* Roxb.	Fabaceae	Madar	Flowers, bark	Tr	Dark brown	2
154.	*Erythrina variegata* Linn.	Fabaceae	Indian Coral (Tree)	Flower	Tr	Red	5
155.	*Eucalyptus hybrid* (mainly *E. umbellata*Domin syn.*E. tereticornis* Sm.)	Myrtaceae	Safeda	Bark, leaves	Tr	Yellow, brown, black	41
156.	*Euonymus tigens* Wall.	Celastraceae	Dog Wood, Prick Wood	Bark	Tr	Yellow	2
157.	*Eupatorium glandulosum* Michex.	Asteraceae	Stickly Snake Roots	Whole plant	Sh	Green	2, 60

Contd...

S.No	Name of plant species	Family	Popular name	Dye bearing parts	Pattern	Vital colour	References
158.	*Euphorbia hirta* Linn.	Euphorbiaceae	Dudhi	Whole plant	Tr	Yellow	8, 60a
159.	*Euphorbia tirucalli* Linn.	Euphorbiaceae	Konpal	Wood	Sh	Red	1
160.	*Euterpe edulis* Mart.	Arecaceae	Jucara Palm	Fruits	Tr	Red	61
161.	*Euterpe oleracea* Mart.	Arecaceae	Acai Palm	Fruits	Tr	Red	61
162.	*Everniastrum cirrhatum* (Fr.)Hale ex Sipman	Parmeliaceae	Jhau, Charata	Whole plant	Lichen	Red, brown	2
163.	*Excoecaria agallocha* Linn.	Euphorbiaceae	Gengwa	Bark	Tr	Brown, black	5
164.	*Fagopyrum esculentum* Moench	Polygonaceae	Kotu,Common Buckwheat	Grain	H	Yellow	62
165.	*Ficus altissima* Blume	Moraceae	Gadgubar	Stem, roots	Tr	Yellow	20
166.	*Ficus benghalensis* Linn.	Moraceae	The Banynan	Leaves, bark, fruits	Tr	Orange, brown	62a
167.	*Ficus religiosa* Linn.	Moraceae	Pipal	Bark, leaves	Tr	Red, brown	6, 63
168.	*Flemingia macrophylla* (Willd.)Merr.	Fabaceae	Bara, Solpan Waras	Leaves, bark	Sh	Black, yellow	20, 6, 1
169.	*Flemingia semialata* Roxb.	Fabaceae	Winged-Stalk Flemingia, Ban Chola	Pods	Sh	Yellow	7
170.	*Galium aparine* Linn.	Rubiaceae	Catch Weed	Roots	H	Purple	1, 7
171.	*Galium rotundifolium* Linn.	Rubiaceae	Caglio a foglie rotonde	Roots	Herb	Yellow, brown	1, 4
172.	*Galium verum* Linn.	Rubiaceae	Cheese Rennet	Roots	H	Yellow, red	1
173.	*Garcinia mangostana* Linn.	Guttiferae	Mangosteen	Fruits	Tr	Black	1
174.	*Garcinia morella* Desr.	Clusiaceae	Indian Gamboge, Tamal	Bark (exudate)	Tr	Yellow	5
175.	*Garcinia xanthochymus* Hook. f.	Clusiaceae	Dephal, Vika	Unripe fruits	Tr	Yellow	64, 6, 4

176.	*Gardenia jasminoides* Ellis=*G.augusta* (Linn.) Merrill	Rubiaceae	Gandhraj	Fruits	Tr/Sh	Yellow	64a, 4
177.	*Garuga gamlie* King	Burseraceae	Pani-Amora	Bark	Tr	Brownish black	20
178.	*Garuga pinnata* Roxb.	Burseraceae	Chogar	Leaves	Tr	Red	65, 6, 4
179.	*Genipa americana* Linn	Rubiaceae	Jagua	Fruits	Tr	Blue	66
180.	*Genista tinctoria* Linn	Fabaceae	Dyer's Brown	Flowers	Sh	Yellow	67, 4
181.	*Gentiana pneumonanthe* Linn	Gentiananceae	Gentiana	Flowers	H	Blue	68, 4
182.	*Geranium nepalense* Sweet	Geraniaceae	Bhanda	Roots	H	Red	2, 6, 4
183.	*Geranium wallichianum* D. Don	Geraniaceae	Laljhari	Roots	H	Red-brown	2
184.	*Girardinia heterophylla* Decne	Urticaceae	Himalayan Nettle	Fibres	H	Blue	69
185.	*Gossypium herbaceum* Linn.	Malvaceae	Uppam Cotton	Flowers	Sh	Yellow	70, 6, 4
186.	*Grevillea robusta* A. Cunn.	Proteaceae	Silver-Oak	Flowers	Tree	Yellow	2
187.	*Grewia optiva* Drumm.	Tiliaceae	Bhimal	Seeds	Tr	Brown, black	41
188.	*Grewia subinaequalis* DC.	Tiliaceae	Phalsa	Fruits	Tree	Yellow-orange	2
189.	*Gymnema tingens* Spreng.	Asclepiadaceae	Gurmar	Leaves	Cl	Blue	5
190.	*Haematoxylon campechianum* Linn.	Fabaceae	LogWood, Patang	Heartwood	Tr	Black	71, 72
191.	*Hedera helix* Linn.	Araliaceae	Common Nepal Ivy, Lablab	Bark	Cl	Yellow	73, 4

Contd...

S.No.	Name of plant species	Family	Popular name	Dye bearing parts	Pattern	Vital colour	References
192.	*Hedychium spicatum* Buch.-Ham.	Zingiberaceae	Bagaldu	Rhizomes	H	Yellow	2
193.	*Hedyotis capitellata* Wall ex G.Don	Rubiaceae	Tou hua er cao	Leaves	H or Sh	Green	7
194.	*Helianthus annus* Linn.	Asteraceae	Sunflower	Flowers	H	Yellow	73a
195.	*Heliotropium indicum* Linn.	Boraginaceae	Hatta-Juri	Leaves	H	Black	7
196.	*Heliotropium strigosum* Willd.	Boraginaceae	Chitiphul	Leaves	H	Black	1
197.	*Hibiscus rosa-sinensis* Linn.	Malvaceae	Jabaphool	Flowers	Sh	Purple	74, 75
198.	*Hibiscus sabdariffa* Linn.	Malvaceae	Lalambari	Flowers	Sh	Yellow	76, 4
199.	*Hippophae salicifolia* D.Don	Elaeagnaceae	Chuk	Fruits	Sh	Yellow	2
200.	*Hypericum oblongifolium* Choisy	Hypericaceae	Basant	Flowers	H	Yellow	2
201.	*Hypericum perforatum* Linn.	Hypericaceae	St. Jhon's Wort	Wood	H	Green	76a
202.	*Hyptis suaveolens* Poit.	Lamiaceae	Mint Weed,Wilayati	Whole plant	H	Yellow	77, 4
203.	*Illicium griffithii* Hook f.&Thomson	Illiciaceae	Lissi	Seeds, bark, leaves	Shrub, Tree	Yellow (traditional dye)	20, 78
204.	*Impatiens balsamina* Linn	Balsaminaceae	Gul Mehndi	Flowers, leaves	H	Red, brown orange	79, 9, 4
205.	*Indigofera arrecta* Hochst.	Fabaceae	Bengal Indigo	Leaves	Sh	Blue	5
206.	*Indigofera articulata* Gouan	Fabaceae	Surat Indigo	Leaves	Sh	Blue	5

207.	*Indigofera aspalathoides* Vahl ex DC.	Fabaceae	Wiry Indigo	Leaves	Under shrub	Blue, black	1
208.	*Indigofera atropurpurea* Buch.- Ham.	Fabaceae	Kala-Sakina	Flowers	Sh	Purple	2
209.	*Indigofera heterantha* Brandis	Fabaceae	Sakina	Flowers	Sh	Pink-purple	2
210.	*Indigofera hirusta* Linn.	Fabaceae	Hairy Indigo	Leaves	H	Blue	5, 1
211.	*Indigofera tinctoria* Linn.	Fabaceae	Nil, Indian Indigo	Leaves, whole plant	Sh	Blue	80, 6, 4
212.	*Ipomoea asarifolia* R.et Schult.	Convolvulaceae	Sweet Potato	Grains	H	Pink	4, 81
213.	*Ipomoea batatas* (Linn.)Lam.	Convolvulaceae	Sweet Potato	Flowers	H	Purple	82, 4
214.	*Isatis tinctoria* Linn.	Cruciferae	Dyer's Woad	Leaves	H	Blue	83, 84
215.	*Ixora coccinea* Linn.	Rubiaceae	Jungle Flame, Ixora	Flowers	Sh, Tr	Grey, brown	8, 84a
216.	*Jasminum humile* Linn.	Oleaceae	Peeli Chameli	Roots	Sh	Yellow	6
217.	*Jatropha curcas* Linn.	Euphorbiaceae	Safed arand	Bark, leaves, flowers	Tr	Blue	1, 85, 4
218.	*Juglans regia* Linn.	Juglandaceae	Akhrot	Bark, fruits, leaves	Tr	Camel, brown, black, red	86, 2, 20, 4
219.	*Kandelia candel* (Linn.) Druce	Rhizophoraceae	Goria	Bark	Sh, Tree	Red, brown	6
220.	*Kigelia pinnata* DC.	Bignoniaceae	Common Sausage Tree	Wood, roots	Tr	Greyish brown	87
221.	*Kirganelia reticulata* (Poir.) Baill.	Euphorbiaceae	Panjuli	Bark, roots	Sh	Red	1
222.	*Lagerstroemia parviflora* Roxb.	Lythraceae	Dhaura	Bark	Sh	Black	2, 6

Contd...

S.No.	Name of plant species	Family	Popular name	Dye bearing parts	Pattern	Vital colour	References
223.	*Lannea coromandelica* (Houtt.) Merrill	Anacardiaceae	Jhingan	Bark, resin	Tr	Yellow, brown	2
224.	*Lannea microcarpa* Engl. & K.Krause	Anacardiaceae	Microcarpa	Fruits, bark	Tr	Yellow, red -brown	88, 4
225.	*Lantana camara* Linn.	Verbenaceae	Wild Sage	Leaves	Sh	Green	89
226.	*Lawsonia inermis* Linn.	Lythraceae	Mehndi, Henna	Leaves	Sh	Yellow-orange, red	90, 91, 6, 4, 2
227.	*Leucas aspera* Spreng.	Lamiaceae	Chota Halkusa	Whole plant	H	Yellow	8, 91a
228.	*Ligustrum vulgare* Linn.	Oleaceae	European Privet	Fruits	Sh	Blue, yellow	1, 9
229.	*Lithospermum arvense* Linn.	Boraginaceae	Corn Gromwell	Roots	H	Red	7
230.	*Lonicera periclymenum* Linn.	Caprifoliaceae	Woodbine,Honey Suckle	Flowers	Sh	Yellow	7
231.	*Lotus corniculatus* Linn.	Fabaceae	Birdsfoot Trefoil	Flowers, leaves	H	Orange, yellow	92, 4, 96
232.	*Lychnis chalcedonica* Linn.	Caryophyllaceae	Red Robin	Flowers	H	Red	7
233.	*Lycopus europaeus* Linn.	Lamiaceae	Gipsywort	Fruits	H	Green	7
234.	*Maclura pomifera* Schneid.	Moraceae	Osage- orange	Wood, bark, roots	Tr	Yellow	93, 4
235.	*Madhuca longifolia* (Koenig) Macb.	Sapotaceae	South Indian Mahua	Bark	Tr	Yellow-brown	2, 6
236.	*Mahonia aquifolium* (Pursh) Nutst.	Berberidaceae	Holly Leaved Barberry	Fruits juice	Sh	Red	94, 4
237.	*Mahonia borealis* Takeda	Berberidaceae	Pualeng	Bark, roots, sap	Sh	Yellow	2
238.	*Mallotus philippensis* Muell.Arg.	Euphorbiaceae	Kamala Tree, Sindur	Fruits	Tr	Orange	95, 6, 4

239.	*Malphigia glabra* Linn.	Malpighiaceae	Barbados Cherry	Flowers, fruits, bark	Sh	Yellow, blue, coffee tint	1, 5
240.	*Mammea longifolia* Planch. & Triana	Clusiaceae	Nagkesar	Flower buds	Tr	Red	5
241.	*Mangifera indica* Linn.	Anacardiaceae	Aam, The Mango	Bark	Tr	Yellow	96, 97, 6, 4
242.	*Manilkara hexandra* (Roxb.) Dubard syn. *Mimusops hexandra* Roxb.	Sapotaceae	Khirni	Bark	Tr	Red, brown	7
243.	*Marsdenia tinctoria* R.Br.	Asclepiadaceae	Manri, Kali lara	Leaves	Cl	Blue	96
244.	*Matricaria chamomilla* Linn.	Asteraceae	Babuna	Flowers	H	Blue	5
245.	*Melastoma malabathricum* Linn.	Melastomataceae	Phutuka	Fruits leaves, roots,	Sh	Purple, black, pink	1, 20, 98, 6, 4
246.	*Memecylon umbellatum* Burm.f.	Melastomataceae	Iron Wood Tree	Leaves	Sh,Tree	Yellow	99, 7, 6, 4
247.	*Mentha piperita* Linn. emend-Huds.	Lamiaceae	Peppermint	Whole plant	H	Yellow, green	8, 99a
248.	*Mesua ferrea* Linn.	Clusiaceae	Nagkesar, Mesua	Wood, flower buds	Tr	Yellow	7
249.	*Michelia champaca* Linn.	Magnoliaceae	Champak	Flowers	Tr	Yellow	7
250.	*Michelia montana* Blume	Magnoliaceae	Pan sopa	Bark, flowers	Tr	Yellow	20
251.	*Miliusa roxburghiana* Hook. f .& Thoms.	Annonaceae	Chhag-loti	Bark, ripe fruits	Tree	Red	20
252.	*Mimusops elengi* Linn.	Sapotaceae	BulletWood	Bark	Tr	Brown	100, 6, 4
253.	*Mirabilis jalapa* Linn.	Nyctaginaceae	Gulabans	Flowers	H	Pink-red	2
254.	*Morinda angustifolia* Roxb.	Rubiaceae	Asho	Roots	Sh,Tree	Yellow	7, 101, 4

Contd...

S.No.	Name of plant species	Family	Popular name	Dye bearing parts	Pattern	Vital colour	References
255.	*Morinda citrifolia* Linn.	Rubiaceae	Al, Ach	Roots, bark	Tr	Red, yellow	102, 6
256.	*Morinda umbellata* Linn.Hook.f.	Rubiaceae	Vomit Vine	Roots, bark	Sh	Yellow	6
257.	*Moringa oleifera* Lam.	Moringaceae	Drumstick Tree	Wood	Tr	Blue	6, 103
258.	*Murraya koenigii* (Linn.) Spreng.	Rutaceae	Kurry patta	Leaves, fruits	Sh,Tr	Grey, brown	8, 103a
259.	*Myrica esculenta* Buch.-Ham.	Myricaceae	Kaiphal, Box Mrytle	Bark	Tr	Yellow	2, 7, 6
260.	*Myrtillocactus geometrizans* (Mart.) Console	Cactaceae	Blue Candle	Fruits	Shrubby Cactus	Red	104, 4
261.	*Nardostachys jatamansi* DC.	Valerianaceae	Jatamansi, Spikenard	Inflorescences	H	Red	2, 4
262.	*Nelumbo nucifera* Gaertn.	Nymphaeaceae	Sacred Lotus, Kamal	Leaves	Aquatic H	Sandal yellow	8, 4
263.	*Nerium indicum* Mill.	Apocyanaceae	Indian Oleander, Kaner	Roots, bark	Sh	Black	20
264.	*Nyctanthes arbortristis* Linn.	Oleaceae	Night Jasmine,Harsinghar	Flowers	Sh	Orange, yellow	105, 2, 6
265.	*Nymphaea alba* Linn.	Nymphaeaceae	European White Nilofar, Water -Lily	Rhizomess	Aquatic H	Brown	1, 9, 7
266.	*Oldenlandia umbellata* Linn. syn. *Hedyotis umbellata* (Linn.) Lam.	Rubiaceae	Indian Madder, Chay-Roots	Roots	Woody annual	Red	7, 106, 107
267.	*Oroxylum indicum* Vent.	Bignoniaceae	Ullu	Bark, fruits	Tr	Black, mordant in dyeing & tanning	2

268.	*Osbeckia stellata* Buch.- Ham ex Kew Gawler	Melastomataceae	Osbeckia	Fruits	Sh	Brown	2
269.	*Oxalis corniculata* Linn.	Oxalidiaceae	Indian Sorrel	Leaves	H	Blue	1
270.	*Papaver rhoeas* Linn.	Papaveraceae	Corn Poppy,Lalpost	Petals	H	Red	1
271.	*Parthenium hysterophorous* Linn.	Asteraceae	Congress Grass	Whole plant	H	Green	108
272.	*Peganum harmala* Linn.	Zygophyllaceae	Harmal	Seeds	H	Brown to black, yellow, red	7
273.	*Peltophorum pterocarpum* Backer ex K. Heyne	Caesalpiniaceae	Copper Pods	Wood, leaves, bark	Tr	Yellowish, brown, black	1
274.	*Perilla frutescens* (Linn.) Britton	Lamiaceae	Bhanjira	Leaves	H	Purple, black	109, 1, 4
275.	*Peristrophe bicalyculata* Nees	Acanthaceae	Atrilal	Leaves	H	Blue	109a
276.	*Peristrophe bivalvis* Merr.	Acanthaceae	Bet-rang	Twigs	H	Yellow orange to deep red orange	5
277.	*Pfaffia iresinoides* (Kunth.) Spreng.	Amaranthaceae	Brazil Gingseng	Roots	H	Yellow	110
278.	*Phaius tankervilliae* Blume	Orchidaceae	Tipui	Leaves, scape	H	Indigo	20
279.	*Phaseolus mungo* Linn.	Fabaceae	Urd	Seeds cover	H	Brown, yellow	8
280.	*Phlogacanthus thyrsiflorus* Nees	Acanthaceae	Banheka	Flower	Sh	Orange-yellow	2
281.	*Phragmites communis* Trin.	Poaceae	Common Reed, Dila	Flowers	Reed	Green	5

Contd...

S.No.	Name of plant species	Family	Popular name	Dye bearing parts	Pattern	Vital colour	References
282.	*Phytolacca dioica* Linn.	Phytolaccaceae	Ombu	Leaves	Sh	Yellow	7
283.	*Phyllanthus fraternus* Webster	Euphorbiaceae	Jungle Amli	Leaves, stem	H	Black	96
284.	*Pinus palustris* Mill.	Pinaceae	LongLeaves Pine	Flowers, needles	Tr	Olive, brown	8
285.	*Pinus roxburghii* Sarg.	Pinaceae	Chir Pine, Chir	Bark, needles	Tr	Black, brown	111, 6
286.	*Pinus wallichiana* A.B. Jackson	Pinaceae	Kal, Blue Pine	Bark	Tr	Black	2,6
287.	*Piper longum* Linn.	Piperaceae	Pipli, Indian Long Pepper	Dry fruits	Cl	Grey	8,111a
288.	*Pisonia grandis* R.Br.	Nyctaginaceae	Lettuce Tree	Leaves	Tr	Yellow	96
289.	*Pistacia integerrima* Stewart. ex. Brandis	Anacardiaceae	Kakra	Galls of leaves and petioles	Tr	Yellow	1, 2, 112
290.	*Pistacia khinjuk* Stocks	Anacardiaceae	Bombay Mastic	Stem galls	Tr	Brown	2
291.	*Pithecellobium dulce* Benth.	Fabaceae	Jangle Jalebi, Quamachil	Bark	Tr	Yellow	5
292.	*Plecospermum spinosum* Trec.	Moraceae	Gumbenfong	Bark, wood	Sh or small tree	Yellow	113
293.	*Plumbago indica* Linn.	Plumbaginaceae	Chitra	Roots	Shrubby perennial	Orange	5
294.	*Polygonum hydropiper* Linn.	Polygonaceae	Water Peeper	Whole plant	H	Blue, black, yellow	20
295.	*Polygonum tortuosum* D.Don.	Polygonaceae	Niala	Flowers, bark	Sh	Yellow	114
296.	*Pongamia pinnata* Pierre	Fabaceae	Karanj	Leaves, bark, seeds cover	Tr	Olive brown	8, 114a
297.	*Populus deltoides* Marsh.	Salicaceae	Poplar	Bark	Tr.	Yellow, brown	115
298.	*Prinsepia utilis* Royle	Rosaceae	Bekkra	Fruits	Sh	Blue	2

299.	*Prunus cerasoides* D.Don	Rosaceae	Himalayan Wild Cherry, Paddam	Fruits	Tr	Yellowish	2
300.	*Prunus cerasus* Linn.	Rosaceae	Sour Cherry, Gilas	Fruits	Tr	Pink	116
301.	*Prunus persica* Batsch.	Rosaceae	Peach, Aru	Leaves, roots bark	Tr	Brown	9
302.	*Psidium guajava* Linn.	Myrtaceae	Amrud	Bark	Tr	Black-brown	2
303.	*Pterocarpus dalbergiodes* Roxb.	Fabaceae	Padauk	Wood, bark	Tr	Brown, Pink	8
304.	*Pterocarpus indicus* Willd.non Baker	Fabaceae	Malay Padauk	Wood	Tr	Red	5
305.	*Pterocarpus marsupium* Roxb.	Fabaceae	Bijasal, Indian Kino Tree	Bark	Tr	Red, Brown	6,
306.	*Pterocarpus santalinus* Linn. f.	Fabaceae	Red Sandal Wood,Raktachandan	Wood	Tr	Red	1, 6, 117, 4
307.	*Punica granatum* Linn.	Punicaceae	Anar,Pomegranate	Rinds, flowers	Tr	Yellowish brown, light red	2, 118, 4
308.	*Pyrus pashia* Buch. Ham.ex D.Don	Rosaceae	Mehal	Fruits	Tr	Black	20
309.	*Quercus infectoria* Olivier	Fagaceae	Majuphal, Gall Oak	Galls	Tr	Yellow, dark grey	9, 119, 4
310.	*Quercus robur* Linn.	Fagaceae	English Oak	Bark	Tr	Yellow	8, 119a
311.	*Rauvolfia serpentina*(L.) Benth.ex Kurz.	Apocynaceae	Sarpagandha, Rauvolfia	Roots	Sh	Orange pinkish	96, 8
312.	*Rauvolfia tetraphylla* Linn.	Apocynaceae	Barachandrika	Fruits	Sh	Black	96
313.	*Reseda luteola* Linn.	Resedaceae	Weld,Dyer's Rocket	Whole plant	H	Deep yellow	120, 121, 4
314.	*Rheum emodi* Wall.ex Meissn.	Polygonaceae	Dolu, Himalayan Rhubarb	Rhizomes	H	Reddish, yellow	122, 123, 4
315.	*Rheum moorcroftianum* Royle	Polygonaceae	Luan guo da huang	Roots	H	Yellow	2

Contd...

S.No.	Name of plant species	Family	Popular name	Dye bearing parts	Pattern	Vital colour	References
316.	*Rheum webbianum* Royle	Polygonaceae	Archa	Roots	H	Yellow	2
317.	*Rhizophora mucronata* Lam.	Rhizophoraceae	True Mangrove	Bark	Tr	Chocolate	124, 6, 4
318.	*Rhododendron arboreum* Sm.	Ericaceae	Burans, Rose-Tree	Flowers	Tr	Red	2, 125
319.	*Rhododendron lepidotum* Wall. ex G.Don	Ericaceae	Talisfur,Thyampa	Leaves, flowers	Sh	Pink-red	2
320.	*Rhus coriaria* Linn.	Anacardiaceae	Sicilian Sumac	Leaves, twigs	Sh	Yellow	7
321.	*Rhus cotinus* Linn.	Anacardiaceae	Royal Purple	Leaves, twigs	Sh	Yellow	7
322.	*Rhus paniculata* Wall. ex. Hook.f.	Anacardiaceae	Benwe, Khaung-bin	Bark	Tr	Yellow	96
323.	*Rimelia reticulata* (Taylor) Hale & Fletcher	Parmeliaceae	Charata	Whole plant	Lichenous	Orange-yellow	2
324.	*Rivina humilis* Linn.	Phytolaccaceae	Blood-Berry	Fruits	H	Red	7
325.	*Rubia cordifolia* Linn.	Rubiaceae	Indian Madder, Manjit	Roots	Cl	Scarlet-red, orange	2, 126, 127, 6, 4
326.	*Rubia sikkimensis* Kurz.	Rubiaceae	Naga-Madder	Roots	Cl	Scarlet-red, orange	7
327.	*Rubia tinctorum* Linn.	Rubiaceae	European Madder, Alizari	Roots	H	Red	128, 4
328.	*Rubus fruticosus* Linn. var. *discolor*.	Rosaceae	Alish,Akhi	Fruits	Sh	Brown, red	9, 7
329.	*Rumex dentatus* Linn.	Polygonaceae	Ambavah, Jungli Palak	Roots	Deep Rootsed Annual	Red	1
330.	*Rumex acetosa* Linn.	Polygonaceae	Garden Soorel	Roots	H	Red	5
331.	*Rumex hastatus* D. Don	Polygonaceae	Amlora	Roots, bark	H	Yellow, green	2

332.	*Rumex nepalensis* Spreng.	Polygonaceae	Pahari Palang, Palak	Roots	H	Yellow, green	2
333.	*Salix caprea* Linn.	Salicaceae	Goat Willow, The Sallow	Bark	Tr	Pink	129, 4
334.	*Sanguinaria canadensis* Linn.	Papaveraceae	Blood Roots	Roots	H	Red	8
335.	*Sapium baccatum* Roxb.	Euphorbiaceae	Seleng, Mousedeer's Rubber-Tree	Leaves	Tr	Blackish green	20
336.	*Sapium indicum* Willd.	Euphorbiaceae	Hurna	Leaves	Tr	Greenish	96
337.	*Saraca asoca* (Roxb.) De Wilde	Fabaceae	Ashoka	Leaves, bark	Tr	Yellow, brown	129a
338.	*Sarothamnus scoparius* (Linn.)Wimm. ex Koch.	Fabaceae	Broom	Flowers	Sh	Yellow	130, 4
339.	*Schleichera oleosa*(Lour.)Oken.	Sapindaceae	Kusum, The Lac Tree	Flowers	Tr	Greenish, yellow	7
340.	*Securinega virosa* (Roxb.ex Willd.) Pax &Hoffm.	Euphorbiaceae	Patala	Bark	Sh	Black	7
341.	*Semecarpus anacardium* Linn.f.	Anacardiaceae	Bhilawa The Marking Nuts Tree,	Bark, fruits	Tr	Black	131, 2, 4
342.	*Serratula tinctoria* Linn.	Asteraceae	Saw Wort	Whole plant juice except roots	H	Yellow	120, 4
343.	*Sesbania bispinosa* W.F. White	Fabaceae	Dhaincha	Pods, seeds	H	Yellow	5
344.	*Shorea robusta* Roxb.ex Gaertn.f.	Dipterocarpaceae	Sal	Bark	Tr	Red, black	132, 133, 6, 4
345.	*Smilax glabra* Roxb.	Liliaceae	Barichob-Chini	Berries	Cl	Blue	5
346.	*Solanum indicum* Linn.	Solanaceae	Barhanta, Poison-Berry	Half ripe fruits	Sh	Black, yellow	20

Contd...

S.No.	Name of plant species	Family	Popular name	Dye bearing parts	Pattern	Vital colour	References
347.	*Solidago canadensis* Linn.	Asteraceae	Goldenrod	Whole Plant	Sh	Golden yellow	19
348.	*Sophora japonica* Linn.	Fabaceae	Chinese Scholar Tree	Flowers	Tr	Yellow	5
349.	*Sophora mollis* Grah. ex Baker	Fabaceae	The Himalayan Laburnum	Roots, flowers	Sh	Brown, yellow	2
350.	*Sorghum caudatum* (Hack.) Stapf.	Poaceae	Caudatum	Leaf sheaths	Stout Grass	Yellow orange	134
351.	*Soymida febrifuga* A. Juss.	Meliaceae	Indian RedWood, Rohun	Bark	Tr	Brown	7
352.	*Spathodea campanulata* Beauv.	Bignoniaceae	African Tulip Tree	Leaves	Tr	Olive brown	134a
353.	*Spinacia oleracea* Linn.	Amaranthaceae	Palak, Garden Spinach	Leaves	H	Green	5
354.	*Spondias pinnata* (Linn.f.) Kurz	Anacardiaceae	Wild Mango	Bark	Tr	Black	7
355.	*Strobilanthes cusia* (Nees) Imlay	Acanthaceae	Assam Indigo, The Rum	Twigs	Sh	Blue	5
356.	*Strychnos nux-vomica* Linn.	Loganiaceae	Snake Wood, Nux-Vomica	Seeds	Tr	Brown-black	7
357.	*Symplocos paniculata* Miq.	Symplocaceae	Sweet Leaves, Ludh	Bark, leaves	Sh	Yellow	2, 20, 135
358.	*Symplocos racemosa* Roxb.	Symplocaceae	Lodh	Leaves, bark	Tr	Yellow	136, 6, 4
359.	*Symplocos ramosissima* Wall.ex G.Don	Symplocaceae	Lodh	Leaves, bark	Sh,Tr	Yellow	2
360.	*Symplocos laurina* Wall.ex Rehd.&E.H.Wils	Symplocaceae	Bholia, Chungan	Leaves	Tr	Yellow	5
361.	*Symplocos sumunita* Buch-Ham.ex D.Don	Symplocaceae	Lodh, Pathani Lodh	Leaves, bark	Tr	Yellow	5
362.	*Syzygium cuminii*(Linn.) Skeels	Myrtaceae	Jaman	Bark, fruits	Tr	Black, purple	2, 20, 137, 6

363.	*Tagetes erecta* Linn.	Asteraceae	Genda, Azetac or African Marigold	Flowers	H	Yellow	2, 138, 139, 140, 4
364.	*Tagetes minuta* Linn.	Asteraceae	Jangli Genda, Stinking-Roger	Flowers, whole plant	H	Yellow	141
365.	*Tagetes patula* Linn.	Asteraceae	French Marigold, Genda	Flowers	H	Yellow	142, 4
366.	*Tamarindus indica* Linn.	Fabacaeae	Imli, Amli	Leaves	Tr	Reddish-yellow	2
367.	*Taxus baccata* Linn.	Taxaceae	Thuner, Common Yew	Bark	Tr	Red	2
368.	*Tectona grandis* Linn.f.	Verbenaceae	Sagwan, Teak	Leaves, bark, roots	Tr	Reddish, yellowish brown, yellow	2, 143, 6, 4
369.	*Tephrosia candida* DC.	Fabaceae	Lashtia	Leaves	Sh	Dark blue	20
370.	*Tephrosia purpura* (Linn.) Pers.	Fabaceae	Wild Indigo, Dhamasia	Leaves	H	Green	8, 143a
371.	*Terminalia arjuna* (Roxb.) Wight &Arn.	Combretaceae	Arjuna	Wood, bark	Tr	Brown	143 b
372.	*Terminalia bellerica* Roxb. (Gaertn.)	Combretaceae	Bahera	Fruits	Tr	Blue	2, 6, 144
373.	*Terminalia catappa* Linn.	Combretaceae	Indian Almond-Tree, Deshi-Badam	Bark, leaves	Tr	Pinkish red, light brownish, yellow	145, 146, 4
374.	*Terminalia chebula* Retz.	Combretaceae	Harra, Myrobalan	Fruits, bark	Tr	Yellow, brown	147, 148, 2, 6, 4
375.	*Terminalia alata* Heyne ex Roth	Combretaceae	Asan, Laurel	Bark, fruits	Tr	Brown	6
376.	*Thespesia populnea* Soland. and. ex Correa	Malvaceae	Parsipu, Portia Tree	Bark, wood, leaves, flowers	Tr	Orange, yellow, brown, green	7, 8
377.	*Toddalia asiatica* (L.) Lam.	Rutaceae	Forest Peeper, Jangli-Kalimirch	Roots, bark	Sh	Yellow	1

Contd...

S.No.	Name of plant species	Family	Popular name	Dye bearing parts	Pattern	Vital colour	References
378.	*Toona ciliata* M. Roemer	Meliaceae	Tun, Red Cedar	Flowers, sawdust	Tr	Red, pink	6, 4
379.	*Toona hexandra* (Wall. ex Roxb.) M. Roem. ssp. *hexandra* var.*gamblei* (DC.) Almedia &Almedia	Meliaceae	Toon	Flowers, seeds	Tr	Yellow-brown	2
380.	*Toona serrata* (Royle) M.Roem.	Meliaceae	Pahari-Toon	Flowers, seeds	Tr	Yellow-brown	2
381.	*Urtica dioica* Linn.	Urticaceae	Bichhu booti	Roots	H	Brown-black	2
382.	*Usnia verticillata* (Hoffm.) Schaer.	Usneaceae	Chulau	Whole plant	L	Red brown	2
383.	*Ventilago denticulata* Willd.	Rhamnaceae	Pitti	Bark, roots	Sh	Red	6, 149
384.	*Ventilago madraspatna* Gaertn.	Rhamnaceae	Raktavalli	Bark, roots	Cl	Red	149
385.	*Vitex altissima* Linn.f.	Verbenaceae	Milla	Wood, bark	Tr	Yellow	7
386.	*Wedelia chinensis* Merrill	Asteraceae	Bhanra	Roots, leaves	H	Black, blue black	8, 149a
387.	*Woodfrodia fruitscosa* Kurz.	Lythraceae	Dawi, Fire-Flame Bush	Flowers, leaves	Sh	Pink-red	150, 2, 6, 4
388.	*Wrightia tinctoria* R.Br.	Apocyanaceae	Pala Indigo-Plant, Indrajau	Leaves	Tr	Blue	5
389.	*Wrightia tomentosa* Roem.&Schult.	Apocynaceae	Dudhi	Yellow milky juice	Tr	Yellow	7, 4
390.	*Zingiber officinale* Rosc.	Zingiberaceae	Adrak, Ginger	Rhizomes	H	Brown	151
391.	*Ziziphus jujuba* Mill.	Rhamnaceae	Pitni-ber	Fruits	Tr	Reddish pink	9, 152
392.	*Zizyphus mauritiana* Lam.	Rhamnaceae	Indian Jujuba, Ber	Leaves	Tr	Red-pink	6

Table 4.2 Dye bearing plant species (Family: Asteraceae)

S.No.	*Name of plant species*	*Family*	*Popular name*	*Dye bearing parts*	*Pattern*	*Vital colour*	*References*
1.	*Ageratum conyzoides* Linn.	Asteraceae	Goat Weed	Whole plant	H	Light-green	11
2.	*Anthemis tinctoria* Linn.	Asteraceae	Dyer's Chamomile	Flowers	H	Yellow	4, 19a
3.	*Artemisia japonica* Thunb.	Asteraceae	Kunja	Leaves	H	Brown	2
4.	*Artemisia nilagirica* (C.B Clarke) Pamp.	Asteraceae	Indian Wormwood	Leaves	H	Brown	2
5.	*Bidens bipinnata* Linn.	Asteraceae	Lumb	Leaves	H	Yellow	31, 32, 4
6.	*Caesulia axillaris* Roxb.	Asteraceae	Bangra	Whole plant	H	Yellow	4, 36
7.	*Calendula officinalis* Linn.	Asteraceae	Calendula	Inflorescence	H	Orange	37, 4
8.	*Carthamus oxyancantha* Bieb.	Asteraceae	Wild Safflowers	Flowers	H	Yellow	5
9.	*Carthamus tinctorius* Linn.	Asteraceae	The Safflowers, False Saffron	Flowers	H	Red	1, 6, 39, 4
10.	*Coreopsis tinctoria* Nutt.	Asteraceae	Calliopsis	Flowers	H	Yellowish	4, 51
11.	*Eclipta alba* (Linn.) Hassk.	Asteraceae	Bhangra	Whole plant	H	Green, grey	57
12.	*Eupatorium glandulosum* Michex.	Asteraceae	Stickly Snake Root	Whole Plant	Sh	Green	2, 60
13.	*Helianthus annuus* Linn.	Asteraceae	Sunflowers	Flowers	H	Yellow	73 a
14.	*Matricaria chamomilla* Linn.	Asteraceae	Babuna	Flowers	H	Blue	5
15.	*Parthenium hysterophorous* Linn.	Asteraceae	Congress Grass	Whole plant	H	Green	108
16.	*Serratula tinctoria* Linn.	Asteraceae	Saw Wort	Whole plant juice except roots	H	Yellows	120, 4
17.	*Solidago canadensis* Linn.	Asteraceae	Goldenrod	Whole plant	Sh	Golden yellow	19
18.	*Tagetes erecta* Linn.	Asteraceae	Genda, Azetac or African Marigold	Flowers	H	Yellow	21, 38, 139, 140, 4
19.	*Tagetes minuta* Linn.	Asteraceae	Jangli Genda, Stinking-Roger	Flowers, whole plant	H	Yellow	141
20.	*Tagetes patula* Linn.	Asteraceae	French Marigold, Genda	Flowers	H	Yellow	142
21.	*Wedelia chinensis* Merrill	Asteraceae	Bhanra	Roots, leaves	H	Black, blue black	149a

Table 4.3 Dye bearing plant species (Family: Euphorbiaceae)

S.No.	Name of plant species	Family	Popular name	Dye bearing parts	Pattern	Vital colour	References
1.	*Acalypha indica* Linn.	Euphorbiaceae	Indian Acalypha	Leaves	H	Yellow	8
2.	*Aleurities moluccana* (Linn.)Willd.	Euphorbiaceae	Jungli Akrot	Root	Tr	Brown	6
3.	*Aporusa dioica* (Roxb.) Muell-Arg. syn. *A. roxburghii* Baill	Euphorbiaceae	Kokra in Bengali	Leaves	Tr	Black	20
4.	*Baccaurea ramiflora* Lour.	Euphorbiaceae	Kataphal	Bark, leaves and fruits	Tr	Black	20
5.	*Bischofia javanica* Blume	Euphorbiaceae	Bishop Wood	Bark, seeds	Tr	Black	2
6.	*Bridelia stipularis* Blume	Euphorbiaceae	Khaji	Fruits	Cl	Black	5,1
7.	*Chrozophora tinctoria* (Linn) Rafin.	Euphorbiaceae	Dyer's Litmus	Herb	H	Red, blue	9
8.	*Emblica officinalis* Gaertn.syn. *Phyllanthus emblica* Linn.	Euphorbiaceae	Amla	Fruits, barks	Tr	Blue, black	2, 58, 6
9.	*Euphorbia hirta* Linn.	Euphorbiaceae	Dudhi	Whole plant	Tr	Yellow	8, 60a
10.	*Euphorbia tirucalli* Linn	Euphorbiaceae	Konpal	wood	Sh	Red	1
11.	*Excoecaria agallocha* Linn.	Euphorbiaceae	Gengwa	Bark	Tr	Brown, black	5
12.	*Jatropha curcas* Linn.	Euphorbiaceae	Safedarand	Roots	Tr	Blue	1, 85, 4
13.	*Kirganelia reticulata* (Poir.) Baill.	Euphorbiaceae	Panjuli	Woods, roots	Sh	Red	1
14.	*Mallotus philippensis* Muell.Arg.	Euphorbiaceae	Kamala Tree, Sindur	Bark roots	Tr	Orange	95, 6, 4
15.	*Sapium baccatum* Roxb.	Euphorbiaceae	Seleng, Mousedeer's Rubber-Tree	Roots	Tr	Blackish green	20
16.	*Sapium indicum* Willd.	Euphorbiaceae	Hurna	Leaves	Tr	Greenish	96
17.	*Securinega virosa* (Roxb.ex Willd.) Pax &Hoffm.	Euphorbiaceae	Patala	Roots	Sh	Black	7

Table 4.4 Dye bearing plant species (Family: Fabaceae)

S.No.	*Name of plant species*	*Family*	*Popular name*	*Dye bearing parts*	*Pattern*	*Vital colour*	*References*
1.	*Abrus precatorius* Linn.	Fabaceae	Crab's Eye, Rati	Seeds coats	Cl	Black	2
2.	*Adenanthera pavonina* Linn.	Fabaceae	Rakta kambal, Red Wood	Wood	Tr	Red	5, 9
3.	*Albizia amara* Boiv.	Fabaceae	Krishnasirisha	Leaves, pod	Tr	Brown grey, olive	8
4.	*Albizia lebbeck* Benth.	Fabaceae	Siris, Kokko	Bark	Tr	Brown	5, 12
5.	*Albizia odoratissima* (L.f.) Benth.	Mimosaceae	Black Siris	Bark	Tr	Brown	6
6.	*Baptisia australis* (L.) R.Br.	Fabaceae	Blue Wild Indigo	Flowers	H	Yellowish	28, 4
7.	*Bauhinia racemosa* Lam.	Fabaceae	Kachnal	Bark	Tr	Light-green	2
8.	*Bauhinia tomentosa* Linn.	Fabaceae	Bell Bauhina	Leaves	Sh	Yellow	1, 5
9.	*Bauhinia variegata* Linn.	Fabaceae	Kachnar, Buddhisht Bauhinia	Flowers	Tr	Purple, violet-pink	2, 4
10.	*Butea monosperma*(Lam.) Taub.	Fabaceae	Palas,Flame of the Forest	Flowers	Tr	Yellow-orange	2, 4
11.	*Butea superba* Roxb.	Fabaceae	Palas lata	Roots	H	Red	6
12.	*Caesalpinia coriaria* (Jacq.) Willd.	Fabaceae	Devi Divi Plant	Pods, seeds	Tr	Black	5
13.	*Caesalpinia pulcherrima* Sw.	Fabaceae	Peacock Flowers	Flowers	Sh	Red	5
14.	*Caesalpinia sappan* Linn.	Fabaceae	Sappan Wood	Wood	Tr	Red	1, 6, 35, 4
15.	*Cassia auriculata* Linn.	Fabaceae	Tanner's Cassia	Flowers	Sh	Yellow	1, 6, 40, 4
16.	*Cassia fistula* Linn.	Fabaceae	Amaltas,Golden-Shower	Fruits, bark	Tr	Brown	2, 6
17.	*Cassia obtusifolia* Linn.	Fabaceae	Banarh, Cakonda	Pods, seeds	H	Yellow	5
18.	*Cassia occidentalis* Linn.	Fabaceae	Kasonda, Coffee-senna	Pods, seeds	H	Brown, yellow	5
19.	*Cassia tora* Linn.	Fabaceae	Panwar, Foetid Cassia	Seeds	H	Red, blue, brown	6, 41

Contd...

S.No.	Name of plant species	Family	Popular name	Dye bearing parts	Pattern	Vital colour	References
20.	*Clitoria mariana* Linn.	Fabaceae	Sweta, Butterfly-Pea	Flowers	Climber, Undershrub	Blue, violet	20
21.	*Clitoria ternatea* Linn.	Fabaceae	Aparajit, Conch Flowers	Flowers, seeds	Cl	Blue	4, 48, 1
22.	*Crotalaria striata* DC.	Fabaceae	Can janka	Seeds	Sh	Brown	5
23.	*Cyamopsis tetragonoloba* (Linn.) Tanb.	Fabaceae	Guar	Pods, seeds	H	Yellow	5
24.	*Cynometra ramiflora* Linn.	Fabaceae	Cynometra	Roots	Tr	Purple	6
25.	*Delonix regia* Rafin.	Fabaceae	Gul Mohar	Flowers	Tr	Deep crimson	56
26.	*Desmodium multiflorum* DC.	Fabaceae	Bhatmase Phool	Flowers	Sh	Purple	20
27.	*Erythrina suberosa* Roxb.	Fabaceae	Madar	Flowers, bark	Tr	Dark, brown	2
28.	*Erythrina variegata* Linn.	Fabaceae	Indian Coral (Tree)	Flowers	Tr	Red	5
29.	*Flemingia macrophylla* (Willd.)Merr.	Fabaceae	Varas	Leaves, bark	Sh	Black, yellow	20, 6, 1
30.	*Flemingia semialata* Roxb.	Fabaceae	Winged-Stalk Flemingia, Ban Chola	Pods	Sh	Yellow	7
31.	*Haematoxylon campechianum* Linn.	Fabaceae	Logwood, Patang	Heartwood	Tr	Black	71, 72
32.	*Genista tinctoria* Linn	Fabaceae	Dyer's Brown	Flowers	Sh	Yellow	67
33.	*Indigofera arrecta* Hochst.	Fabaceae	Bengal Indigo	Leaves	Sh	Blue	5
34.	*Indigofera articulata* Gouan	Fabaceae	Surat Indigo	Leaves	Sh	Blue	5
35.	*Indigofera aspalathoides* Vahl ex.DC	Fabaceae	Wiry Indigo	Leaves	Under Sh	Blue, black	1
36.	*Indigofera atropurpurea* Buch- Ham.	Fabaceae	Kala-Sakina	Flowers	Sh	Purple	2
37.	*Indigofera heterantha* Brandis	Fabaceae	Sakina	Flowers	Sh	Pink-purple	2

38.	*Indigofera hirusta* Linn.	Fabaceae	Hairy Indigo	Leaves	H	Blue	5, 1
39.	*Indigofera tinctoria* Linn.	Fabaceae	Nil, Indian Indigo	Leaves, whole plant		Blue	80, 6, 4
40.	*Lotus corniculatus* Linn.	Fabaceae	Birdsfoot Trefoil	Flowers, leaves	H	Orange, yellow	92, 4, 96
41.	*Phaseolus mungo* Linn.	Fabaceae	Urd	Seeds covers	H	Brown, yellow	8
42.	*Pithecellobium dulce* Benth.	Fabaceae	Jangle Jalebi, Quamachil	Bark	Tr	Yellow	5
43.	*Pongamia pinnata* Pierre	Fabaceae	Karanj	Leaves, bark, seeds covers	Tr	Olive brown	8, 114a
44.	*Pterocarpus dalbergiodes* Roxb.	Fabaceae, Papilionaceae	Padauk	Wood, bark	Tr	Brown, pink	8
45.	*Pterocarpus indicus* Willd. non Baker	Fabaceae	Malay Padauk	Wood	Tr	Red	5
46.	*Pterocarpus marsupium* Roxb.	Fabaceae	Bijasal, Indian Kino Tree	Bark	Tr	Red, brown	6
47.	*Pterocarpus santalinus* Linn. f.	Fabaceae	Red Sandal wood, Raktachandan	Wood	Tr	Red	1, 6, 117, 4
48.	*Saraca asoca* (Roxb.)de Wilde	Fabaceae	Ashoka	Leaves, bark	Tr	Yellow, brown	129a
49.	*Sarothamnus scoparius* (Linn.)Wimm. ex Koch.	Fabaceae	Broom	Flowers	Sh	Yellow	130
50.	*Sesbania bispinosa* W.F. White	Fabaceae	Dhaincha	Pods, seeds	H	Yellow	5
51.	*Sophora japonica* Linn.	Fabaceae	Chinese Scholar Tree	Flowers	Tr	Yellow	5
52.	*Sophora mollis* Grah. ex Baker	Fabaceae	The Himalayan Laburnum	Roots, flowers	Sh	Brown, yellow	2
53.	*Tephrosia candida* DC.	Fabaceae	Lashtia	Leaves	Sh	Dark blue	20
54.	*Tephrosia purpura* Pers. Baker	Fabaceae	Wild Indigo, Dhamasia	Leaves	H	Green	8, 143a

Table 4.5 Dye bearing plant species (Family: Rubiaceae)

S.No.	Name of plant species	Family	Popular name	Dye bearing parts	Pattern	Vital colour	References
1.	*Adina cordifolia* (Roxb.) Hook.f.ex Brandis	Rubiaceae	Haldu	Wood	Tr	Yellow	7
2.	*Galium aparine* Linn.	Rubiaceae	Catch Weed	Roots	H	Purple	1, 7
3.	*Galium rotundifolium* Linn.	Rubiaceae	Caglio a foglie rotonde	Roots	Herb	Yellow, brown	1
4.	*Galium verum* Linn.	Rubiaceae	Cheese Rennet	Roots	H	Yellow, red	1, 4
5.	*Gardenia jasminoides* Ellis=*G. augusta* (Linn.) Merrill	Rubiaceae	Gandhraj	Fruits	Tr/Sh	Yellow	64a, 4
6.	*Genipa americana* Linn	Rubiaceae	Jagua	Fruits	Tr	Blue	66
7.	*Hedyotis capitellata* Wall ex G.Don	Rubiaceae	Tou Hua er cao	Leaves	Herb or Sh	Green	7
8.	*Ixora coccinea* Linn.	Rubiaceae	Jungle Flame, Ixora	Flower	Sh, Tree	Grey, brown	84a
9.	*Morinda angustifolia* Roxb.	Rubiaceae	Asho	Roots	Sh, Tree	Yellow	7, 101, 4
10.	*Morinda citrifolia* Linn.	Rubiaceae	Al, Ach	Root bark	Tr	Red, yellow	102, 6
11.	*Morinda umbellata* Linn. Hook.f.	Rubiaceae	Vomit Vine	Root bark	Sh	Yellow	6
12.	*Oldenlandia umbellata* Linn. syn. *Hedyotis umbellata* (Linn.) Lam.	Rubiaceae	Indian Madder, Chay-Root	Roots	Woody Annual	Red	7, 106, 107
13.	*Rubia cordifolia* Linn.	Rubiaceae	Indian Madder, Manjit	Roots	Cl	Scarlet-red, orange	2, 126, 127, 6, 4
14.	*Rubia sikkimensis* Kurz.	Rubiaceae	Naga-Madder	Roots	Cl	Scarlet-red, orange	7
15.	*Rubia tinctorum* Linn.	Rubiaceae	European Madder Alizari	Roots	H	Red	128, 4

Table 4.6 Plant species bearing bark dyes

S.No.	Name of plant species	Family	Popular name	Dye bearing parts	Pattern	Vital colour	References
1.	*Acacia dealbata* Link	Mimosaceae	Silver Wattle	Bark	Tr	Brown, black	1
2.	*Acacia farnesiana* Willd.	Mimosaceae	Vilayati Kikar	Bark, seeds, pods	Tr	Black, brown	6
3.	*Acacia leucophloea* Willd.	Mimosaceae	Safed Kikar	Bark, leaves	Tr	Red	6
4.	*Acacia nilotica* Delile subsp. *indica*(Benth.) Brenan	Mimosaceae	Babul	Bark, fruits, rind, gum, leaves	Tr	Black to dark brown	2, 6
5.	*Acacia pennata* Willd.	Mimosaceae	Aila	Bark	Sh	Brown	5
6.	*Acacia sinuata* Willd.(Lour.) Merrill	Mimosaceae	Ritha	Bark	Sh	Brown	6, 7
7.	*Aesculus indica* Hook.	Hippocastanaceae	Pangar, Himalayan Chestnut	Bark	Tr	Brown	2
8.	*Albizia lebbeck* Benth.	Fabaceae	Siris, Kokko	Bark	Tr	Brown	5, 12
9.	*Albizia odoratissima* (L.f.) Benth.	Fabaceae, Mimosaceae	Black Siris	Bark	Tr	Brown	6
10.	*Alnus nepalensis* D.Don	Betulaceae	Uitis, Indian Alder	Bark	Tr	Red	2, 6
11.	*Alnus nitida* Endl.	Betulaceae	West Himalayan Alder, Seril	Bark	Tr	Brown grey	7, 4
12.	*Baccaurea ramiflora* Lour.	Euphorbiaceae	Kataphal	Bark, leaves, fruits	Tr	Black	20
13.	*Bauhinia racemosa* Lam.	Fabaceae	Kachnal	Bark	Tr	Light green	2
14.	*Berberis aristata* DC.	Berberidaceae	Kilmor	Bark, roots	Sh	Yellow	2
15.	*Berberis asiatica* Roxb. ex. DC.	Berberidaceae	Kilmor	Bark, roots	Sh	Yellow	2
16.	*Berberis chitria* Lindl.	Berberidaceae	Chetar	Bark, roots	Sh	Yellow	2

Contd...

S.No.	Name of plant species	Family	Popular name	Dye bearing parts	Pattern	Vital colour	References
17.	*Berberis vulgaris* Linn.	Berberidaceae	Barberry	Bark, roots	Sh	Yellowish, yellow	2, 29, 4
18.	*Bischofia javanica* Blume	Euphorbiaceae	Bishop Wood	Bark, seeds	Tr	Black	2
19.	*Careya arborea* Roxb.	Barringtoniaceae	Kumbi	Bark	Tr	Yellow	2, 6
20.	*Cassia fistula* Linn.	Fabaceae	Amaltas,Golden-Shower	Bark, fruits	Tr	Brown	2, 6
21.	*Casuarina equisetifolia* Linn.	Casuarinaceae	Jangli Saru, Casuarina	Bark	Tr	Brown	6, 42
22.	*Celtis australis* Linn.	Ulmaceae	European Nettle Tree	Bark	Tr	Yellow	2
23.	*Ceriops tagal* (Perr.) C.B.Robins	Rhizophoraceae	Goran, Guttia	Bark	Mangrove	Brown	9, 42
24.	*Chloroxylon swietenia* DC.	Flindersiaceae	Indian SatinWood Tree	Bark	Tree	Yellow	6, 45, 4
25.	*Cinnamomum camphora* (Linn.) Presl	Lauraceae	Camphor Tree, Kapur	Bark	Tr	Brown	46
26.	*Citrus medica* Linn.	Rutaceae	Citron	Bark	Tr	Black	1, 7
27.	*Coscinium fenestratum* Clolebr.	Menispermaceae	Tree -Turmeric, Jharihuldi	Bark, seeds, wood	Sh	Red	1
28.	*Daphne papyracea* Wall.ex. Steud.emend.	Thymelaeaceae	Satpura	Bark, fruits	Sh	Red	2, 20, 55
29.	*Dipterocarpus turbinatus* Gaertn.f.	Dipterocarpaceae	Common Gurjan Tree	Bark, twigs	Tr	Yellow, brown	1, 7
30.	*Elaeodendron glaucum* (Rottb.) pers.	Celasteraceae	Chouri	Bark	Tr	Red	1
31.	*Emblica officinalis* Gaertn.syn *Phyllanthus emblica* Linn.	Euphorbiaceae	Amla	Bark, fruits	Tr	Blue-black	2, 6, 58
32.	*Engalhardtia spicata* Lesch.ex. Blume	Juglandaceae	Corcorshing, Mahwa	Bark	Tr	Dark brown, dark blue	20, 2
33.	*Erythrina suberosa* Roxb.	Fabaceae	Madar	Bark, flowers	Tr	Dark brown	2

34.	Eucalyptus hybrid (mainly *E.umbellata* Domin syn.*E.tereticornis* Sm.)	Myrtaceae	Safeda	Bark, leaves	Tr	Yellow, brown, black	41
35.	*Euonymus tigens* Wall.	Celastraceae	Dog Wood, Prick Wood	Bark	Tr	Yellow	2
36.	*Excoecaria agallocha* Linn.	Euphorbiaceae	Gengwa	Bark	Tr	Brown, black	5
37.	*Ficus benghalensis* Linn.	Moraceae	The Banynan	Bark, leaves, fruits	Tr	Orange, brown	62a
38.	*Ficus religiosa* Linn.	Moraceae	Pipal	Bark, leaves	Tr	Red, brown	6, 63
39.	*Flemingia macrophylla* (Willd.)Merr.	Fabaceae	Bara Saolpan, Waras	Bark, leaves	Sh	Black, yellow	20, 6, 1
40.	*Garuga gamlie* King	Burseraceae	Pani amora	Bark	Tr	Brownish black	20
41.	*Illicium griffithii* Hook f.&Thomson	Illiciaceae	Lissi	Bark, seeds, leaves	Shrub, Tree	Yellow (traditional dye)	20, 78
42.	*Jatropha curcas* Linn.	Euphorbiaceae	Safedarand	Bark, leaves	Tr	Blue	1, 85, 4
43.	*Juglans regia* Linn.	Juglandaceae	Akhrot	Bark, fruits, leaves,	Tr	Camel, brown, black, red	86, 2, 20, 4
44.	*Kandelia candel* (Linn.) Druce	Rhizophoraceae	Goria	Bark	Sh,Tree	Red, brown	6
45.	*Kirganelia reticulata* (Poir.) Baill.	Euphorbiaceae	Panjuli	Bark, roots	Sh	Red	1
46.	*Lagerstroemia parviflora* Roxb.	Lythraceae	Dhaura	Bark	Sh	Black	2, 6
47.	*Lannea coromandelica* (Houtt.) Merrill	Anacardiaceae	Jhingan	Bark, resin	Tr	Yellow-brown	2
48.	*Lannea microcarpa* Engl. & K.Krause	Anacardiaceae	Microcarpa	Bark, fruits	Tr	Yellow, red-brown	88, 4
49.	*Maclura pomifera* Schneid.	Moraceae	Osage-orange	Bark, wood, roots	Tr	Yellow	93, 4

Contd...

S.No.	Name of plant species	Family	Popular name	Dye bearing parts	Pattern	Vital colour	References
50.	*Madhuca longifolia* (Koenig) Macb.	Sapotaceae	South Indian Mahua	Bark	Tr	Yellow-brown	2, 6
51.	*Mahonia borealis* Takeda	Berberidaceae	Pualeng	Bark, roots, sap	Sh	Yellow	2
52.	*Malphigia glabra* Linn.	Malpighiaceae	Barbados Cherry	Bark, flowers, fruits,	Sh	Yellow, blue, coffee tint	1, 5
53.	*Mangifera indica* Linn.	Anacardiaceae	Aam, The Mango	Bark	Tr	Yellow	6, 96, 97
54.	*Manilkara hexandra* (Roxb.) Dubard syn. *Mimusops hexandra* Roxb.	Sapotaceae	Khirni	Bark	Tr	Red, brown	7
55.	*Michelia montana* Blume	Magnoliaceae	Pan sopa	Bark, flowers	Tr	Yellow	20
56.	*Miliusa roxburghiana* Hook. f. & Thoms.	Annonaceae	Chhag-loti	Bark, ripe fruits	Tree	Red	20
57.	*Mimusops elengi* Linn.	Sapotaceae	Bulletwood	Bark	Tr	Brown	6, 100
58.	*Morinda citrifolia* Linn.	Rubiaceae	Al, Ach	Roots bark	Tr	Red, yellow	6, 102
59.	*Morinda umbellata* Linn.	Rubiaceae	Vomit-Vine	Bark, roots	Sh	Yellow	6
60.	*Myrica esculenta* Buch.-Ham.	Myricaceae	Kaiphal, Box Mrytle	Bark	Tr	Yellow	2, 6, 7
61.	*Nerium indicum* Mill.	Apocynaceae	Indian Oleander, Kaner	Bark, roots	Sh	Black	20
62.	*Oroxylum indicum* Vent.	Bignoniaceae	Ullu	Bark, fruits	Tr	Black, mordant in dyeing & tanning	2
63.	*Peltophorum pterocarpum* Backer ex K. Heyne	Caesalpiniaceae	Copper Pods	Bark, wood, leaves	Tr	Yellowish, brown, black	1
64.	*Pinus roxburghii* Sarg.	Pinaceae	Chir Pine, Chir	Bark, needles	Tr	Black, brown	6, 111
65.	*Pinus wallichiana* A.B. Jackson	Pinaceae	Kal, Blue Pine	Bark	Tr	Black	2, 6

66.	*Pithecellobium dulce* Benth.	Fabaceae	Jangle Jalebi, Quamachil	Bark	Tr	Yellow	5
67.	*Plecospermum spinosum* Trec.	Moraceae	Gumbenfong	Bark, wood	Sh or small tree	Yellow	113
68.	*Polygonum tortuosum* D.Don.	Polygonaceae	Niala	Bark, flowers	Sh	Yellow	114
69.	*Pongamia pinnata* Pierre	Fabaceae	Karanj	Bark, leaves, seeds cover	Tr	Olive brown	8, 114a
70.	*Populus deltoides* Marsh.	Salicaceae	Poplar	Bark	Tr.	Yellow, brown	115
71.	*Prunus persica* Batsch.	Rosaceae	Peach, Aru	Roots bark, leaves	Tr	Brown	9
72.	*Psidium guajava* Linn.	Myrtaceae	Amrud	Bark	Tr	Black-brown	2
73.	*Pterocarpus dalbergiodes* Roxb.	Fabaceae, Papilionaceae	Padauk	Bark, wood	Tr	Brown, pink	8
74	*Pterocarpus marsupium* Roxb.	Fabaceae	Bijasal, Indian Kino Tree	Bark	Tr	Red, brown	6
75.	*Quercus robur* Linn.	Fagaceae	English Oak	Bark	Tr	Yellow	8, 119a
76.	*Rhizophora mucronata* Lam.	Rhizophoraceae	True Mangrove	Bark	Tr	Chocolate	6, 124, 4
77.	*Rhus paniculata* Wall. ex. Hook.f.	Anacardiaceae	Benwe, Khaung-bin	Bark	Tr	Yellow	9, 96
78.	*Rumex hastatus* D. Don	Polygonaceae	Amlora	Bark, roots	H	Yellow-green	2
79.	*Salix caprea* Linn.	Salicaceae	Goat Willow, The Sallow	Bark	Tr	Pink	129, 4
80.	*Saraca asoca* (Roxb.) De Wilde	Fabaceae	Ashoka	Bark, leaves	Tr	Yellow, brown	129a
81.	*Securinega virosa* (Roxb.ex Willd.) Pax &Hoffm.	Euphorbiaceae	Patala	Bark	Sh	Black	7
82.	*Semecarpus anacardium* Linn.f.	Anacardiaceae	Bhilawa, The Marking Nut Tree,	Bark, fruits	Tr	Black	131, 2

Contd...

S.No.	Name of plant species	Family	Popular name	Dye bearing parts	Pattern	Vital colour	References
83.	*Shorea robusta* Roxb.ex Gaertn.f.	Dipterocarpaceae	Sal	Bark	Tr	Red, black	6, 132, 133, 4
84.	*Soymida febrifuga* A. Juss.	Meliaceae	Indian RedWood, Rohun	Bark	Tr	Brown	7
85.	*Spondias pinnata* (Linn.f.) Kurz	Anacardiaceae	Wild Mango	Bark	Tr	Black	7
86.	*Symplocos paniculata* Miq .	Symplocaceae	Sweet Leaves, Ludh	Bark, leaves	Sh	Yellow	2, 20, 135
87.	*Symplocos racemosa* Roxb.	Symplocaceae	Lodh	Bark, leaves	Tr	Yellow	6, 136, 4
88.	*Symplocos ramosissima* Wall. ex G.Don	Symplocaceae	Lodh	Bark, leaves	Sh, Tr	Yellow	2
89.	*Symplocos sumunita* Buch-Ham.ex D.Don	Symplocaceae	Lodh, Pathani Lodh	Bark, leaves	Tr	Yellow	5
90.	*Syzygium cuminii* (Linn.) Skeels	Myrtaceae	Jaman	Bark, fruits	Tr	Black, purple	2, 6, 20, 137
91.	*Taxus baccata* Linn.	Taxaceae	Thuner, Common Yew	Bark	Tr	Red	2
92.	*Tectona grandis* Linn.f.	Verbenaceae	Sagwan, Teak	Bark, leaves, roots	Tr	Reddish, yellowish brown	2, 6, 143, 4
93.	*Terminalia alata Heyne ex* Roth	Combretaceae	Asan, Laurel	Bark, fruits	Tr	Brown	6
94.	*Terminalia arjuna* (Roxb.) Wight &Arn.	Combretaceae	Arjuna	Bark, wood	Tr	Brown	143b
95.	*Terminalia catappa* Linn.	Combretaceae	Indian Almond-Tree, Deshi-badam	Bark, leaves	Tr	Pinnk red, light brownish, yellow	145, 146, 4
96.	*Terminalia chebula* Retz.	Combretaceae	Harra, Myrobalan	Barks, fruits	Tr	Yellow, brown	26, 147, 148, 4

97.	*Thespesia populnea* Soland. ex Correa	Malvaceae	Parsipu, Portia Tree	Bark, wood, leaves, flowers	Tr	Orange, yellow, brown, green	7
98.	*Toddalia asiatica* (L.) Lam.	Rutaceae	Forest Peeper, Jangli-Kalimirch	Bark, roots	Sh	Yellow	1
99.	*Ventilago denticulata* Willd.	Rhamnaceae	Pitti	Bark, roots	Sh	Red	6, 149
100.	*Ventilago madraspatna* Gaertn.	Rhamnaceae	Raktavalli	Bark, roots	Cl	Red	149
101.	*Vitex altissima* Linn.f.	Verbenaceae	Milla	Bark, wood	Tr	Yellow	7

Table 4.7 Plant species bearing flowers dyes

S.No.	Name of plant species	Family	Popular name	Dye bearing parts	Pattern	Vital colour	References
1.	*Althea rosea* Linn.	Malvaceae	Hollyhock	Flowers	H	Red	13, 1, 9, 4
2.	*Anthemis tinctoria* Linn.	Asteraceae	Dyer's Chamomile	Flowers	H	Yellow	4, 19a
3.	*Baptisia australis* (L.) R.Br.	Fabaceae	Blue Wild Indigo	Flowers	H	Yellowish	28, 4
4.	*Barleria prionitis* Linn.	Acanthaceae	Kala Bansa	Flowers	Sh	Yellow	7, 1
5.	*Bauhinia variegata* Linn.	Fabaceae	Kachnar, Buddhisht Bauhinia	Flowers	Tr	Purple, violet-pink	2, 6, 4
6.	*Bougainvillea glabra* Choisy	Nyctaginaceae	Paper Flowers	Flowers with ivory white bracts	Vine Sh	Yellow	9
7.	*Butea monosperma* (Lam.)Taub.	Fabaceae	Palas, Flame of Forest	Flowers	Tr	Yellow-orange	2, 34, 6, 4
8.	*Caesalpinia pulcherrima* Sw.	Fabaceae	Peacock Flowers	Flowers	Sh	Red	5
9.	*Carthamus oxyancantha* Bieb.	Asteraceae	Wild Safflowers	Flowers	H	Yellow	5
10.	*Carthamus tinctorius* Linn.	Asteraceae	The Safflowers, False Saffron	Flowers	H	Red	1, 6, 39, 4
11.	*Cassia auriculata* Linn.	Fabaceae	Tanner's Cassia	Flowers	Sh	Yellow	1, 6, 40, 4
12.	*Chukrasia tabularis* A.Juss.var.*tabularis*	Meliaceae	Indian Mahogany Tree	Flowers, leaves	Tr	Red & yellow	9, 7
13.	*Clitoria mariana* Linn.	Fabaceae	Sweta, Butterfly-Pea.	Flowers	Climber, Undershrub	Blue, violet	20
14.	*Clitoria ternatea* Linn.	Fabaceae	Aparajit, Conch Flowers	Flowers, seeds	Cl	Blue	1, 4, 48
15.	*Commelina benghalensis* Linn.	Commelinaceae	Benghal Dayflower	Juice of flowers	Serpentine Leaves miner	Blue	9
16.	*Crocus sativa* Linn.	Iridaceae	Kesar, Saffron	Flowers	H	Yellow	6, 49, 50, 4

17.	*Coreopsis tinctoria* Nutt.	Asteraceae	Calliopsis	Flowers	H	Yellowish	4, 51
18.	*Corylus avellana* Linn.	Betulaceae	Hazel	Flowers	Tr	Yellow	8
19.	*Delonix regia* Rafin.	Fabaceae	Gul Mohar	Flowers	Tr	Deep crimson	56
20.	*Delphinium grandiflorum* L.	Ranunculaceae	Sieberian Larkspur	Flowers	H	Blue, grey	56a
21.	*Delphinium zalil* Aitch &Hemsl.	Ranunculaceae	Guljali, Larkspur	Flowers	H	Yellow	4, 5, 6
22.	*Desmodium multiflorum* DC.	Fabaceae	Bhatmase Phool	Flowers	Sh	Purple	20
23.	*Erythrina suberosa* Roxb.	Fabaceae	Madar	Flowers, bark	Tr	Dark, brown	2
24.	*Erythrina variegata* Linn.	Fabaceae	Indian coral (Tree)	Flowers	Tr	Red	5
25.	*Genista tinctoria* Linn	Fabaceae	Dyer's Brown	Flowers	Sh	Yellow	67
26.	*Gentiana pneumonanthe* Linn.	Gentiananceae	Gentiana	Flowers	H	Blue	68
27.	*Gossypium herbaceum* Linn.	Malvaceae	Uppam Cotton	Flowers	Sh	Yellow	6, 70
28.	*Grevillea robusta* A. Cunn.	Proteaceae	Silver-Oak	Flowers	Tree	Yellow	2
29.	*Helianthus annuus* Linn.	Asteraceae	Sunflowers	Flowers	H	Yellow	73a
30.	*Hibiscus rosa-sinensis* Linn.	Malvaceae	Jabaphool	Flowers	Sh	Purple	74, 75
31.	*Hibiscus sabdariffa* Linn.	Malvaceae	Lalambari	Flowers	Sh	Yellow	76
32.	*Impatiens balsamina* Linn	Balsaminaceae	Gul Mehndi	Flowers, leaves	H	Red, brown orange	79, 9
33.	*Indigofera atropurpurea* Buch- Ham.	Fabaceae	Kala-Sakina	Flowers	Sh	Purple	2
34.	*Indigofera heterantha* Brandis	Fabaceae	Sakina	Flowers	Sh	Pink-purple	2

Contd...

S.No.	*Name of plant species*	*Family*	*Popular name*	*Dye bearing parts*	*Pattern*	*Vital colour*	*References*
35.	*Ixora coccinea* Linn.	Rubiaceae	Jungle Flame, Ixora	Flowers	Sh, Tree	Grey, brown	84a
36.	*Lonicera periclymenum* Linn.	Caprifoliaceae	Woodbine, Honey Suckle	Flowers	Sh	Yellow	7
37.	*Lotus corniculatus* Linn.	Fabaceae	Birdsfoot Trefoil	Flowers, leaves	H	Orange, yellow	92, 4, 96
38.	*Lychnis chalcedonica* Linn.	Caryophyllaceae	Red Robin	Flowers	H	Red	7
39.	*Malphigia glabra* Linn.	Malpighiaceae	Barbados Cherry	Flowers, fruits, bark	Sh	Yellow, blue, coffee tint	1, 5
40.	*Mammea longifolia* Planch. & Triana	Guttiferae	Nagkesar	Flowers buds	Tr	Red	5
41.	*Matricaria chamomilla* Linn.	Asteraceae	Babuna	Flowers	H	Blue	5
42.	*Mesua ferrea* Linn.	Guttiferae	Nagkesar, Mesua	Flowers buds, wood	Tr	Yellow	7
43.	*Michelia champaca* Linn.	Magnoliaceae	Champak	Flowers	Tr	Yellow	7
44.	*Michelia montana* Blume	Magnoliaceae	Pan sopa	Flowers, bark	Tr	Yellow	20
45.	*Mirabilis jalapa* Linn.	Nyctaginaceae	Gulabans	Flowers	H	Pink-red	2
46.	*Phlogacanthus thyrsiflorus* Nees	Acanthaceae	Banheka	Flowers	Sh	Orange-yellow	2
47.	*Phragmites communis* Trin.	Gramineae	Common Reed, Dila	Flowers	Reed	Green	5
48.	*Pinus palustris* Mill.	Pinaceae	Long Leaves Pine	Flowers, needles	Tr	Olive, brown	8
49.	*Polygonum tortuosum* D.Don.	Polygonaceae	Niala	Flowers, bark	Sh	Yellow	114
50.	*Punica granatum* Linn.	Punicaceae	Anar, Pomegranate	Flowers, rind	Tr	Yellowish brown light red	2, 118, 4
51.	*Rhododendron arboreum* Sm.	Ericaceae	Burans, Rose-Tree	Flowers	Tr	Red	2, 125

52.	*Rhododendron lepidotum* Wall. ex G.Don	Ericaceae	Talisfur,Thyampa	Flowers, leaves	Sh	Pink-red	2
53.	*Sarothamnus scoparius* (Linn.)Wimm. ex Koch.	Fabaceae	Broom	Flowers	Sh	Yellow	130, 4
54.	*Schleichera oleosa* (Lour.) Oken.	Sapindaceae	Kusum, The Lac Tree	Flowers	Tr	Greenish, yellow	7
55.	*Sophora japonica* Linn.	Fabaceae	Chinese Scholar Tree	Flowers	Tr	Yellow	5
56.	*Sophora mollis* Grah. ex Baker	Fabaceae	The Himalayan Laburnum	Flowers, roots	Sh	Brown, yellow	2
57.	*Tagets erecta* Linn.	Asteraceae	Genda, Azetac or African Marigold	Flowers	H	Yellow	2, 138, 139, 140, 4
58.	*Tagetes minuta* Linn.	Asteraceae	Jangli Genda, Stinking-Roger	Flowers, whole plant	H	Yellow	141
59.	*Tagetes patula* Linn.	Asteraceae	French Marigold, Genda	Flowers	H	Yellow	142, 4
60.	*Thespesia populnea* Soland. and ex Correa	Malvaceae	Parsipu, Portia Tree	Flowers, bark, wood, leaves	Tr	Orange, yellow, brown, green	7, 8
61.	*Toona ciliata* M.Roemer	Meliaceae	Tun Red Cedar	Flowers, sawdust	Tr	Red, pink	6
62.	*Toona hexandra* (Wall. ex Roxb.) M. Roem. ssp. *hexandra* var. *gamblei* (DC.) Almedia & Almedia	Meliaceae	Toon	Flowers, seeds	Tr	Yellow-brown	2
63.	*Toona serrata* (Royle) M.Roem.	Meliaceae	Pahari-Toon	Flowers, seeds	Tr	Yellow-brown	2
64.	*Woodfrodia fruitcosa* Kurz.	Lythraceae	Dawi, Fire-Flame Bush	Flowers, leaves	Sh	Pink-red	2, 6, 150, 4

Table 4.8 Plant species bearing fruits dyes

S.No.	Name of plant species	Family	Popular name	Dye bearing parts	Pattern	Vital colour	References
1.	*Acacia nilotica* Delile subsp. *indica*(Benth.) Brenan	Mimosaceae	Babul	Fruits rind, leaves, bark	Tr	Black to dark brown	2, 6
2.	*Acanthopanax trifoliatum* (Linn.) Voss	Araliaceae	Shiah-ryngkhwari	Fruits	Sh	Black	1
3.	*Aegle marmelos* (Linn.) Correa ex Roxb.	Rutaceae	Bel	Rind of unripe fruits	Tr	Yellow	2, 6, 10, 4
4.	*Annona reticulata* Linn.	Annonaceae	Raamaaphal	Unripe fruits	Tr	Black	9, 19
5.	*Annona squamosa* Linn.	Annonaceae	Sharifa	Fruits	Sh	Yellow	2
6.	*Ardisia solanacea* Roxb. syn. *Ardisia humilis* C.B. Clarke	Myrsinaceae	Bisi	Fruits	Sh	Yellow	5
7.	*Aronia melanocarpa* (Michx.) Elliott	Rosaceae	Black Choke Berry	Fruits	Sh	Blue	4, 25
8.	*Artocarpus hetrophyllus* Lam .	Moraceae	Kathal, JackFruits	Fruits, wood, roots	Tr	Yellow	2, 6, 26, 4
9.	*Artocarpus lakoocha* Roxb.	Moraceae	Monkey Jack	Fruits, wood, roots	Tr	Yellow	2
10.	*Baccaurea ramiflora* Lour.	Euphorbiaceae	Kataphal	Fruits, bark, leaves	Tr	Black	20
11.	*Benthamidia capitata* (Wall.) Hara	Cornaceae	Bamora	Fruits	Tr	Red	2
12.	*Bridelia stipularis* Blume	Euphorbiaceae	Khaji	Fruits	Cl	Black	5, 1
13.	*Capsicum annuum* Linn syn. *C. frutescens* C.B. Clarke	Solanaceae	Lalmirich, Chilli	Fruits	H	Red	38, 7, 4
14.	*Cassia fistula* Linn.	Fabaceae	Amaltas, Golden- Shower	Fruits, bark	Tr	Brown	2, 6
15.	*Chrozophora tinctoria* (Linn) Rafin.	Euphorbiaceae	Dyer's Litmus	Fruits, petal, sap	H	Red, blue	9
16.	*Daphne papyracea* Wall.ex. Steud.emend.	Thymelaeaceae	Satpura	Fruits, bark	Sh	Red	2, 20, 55

17.	*Diospyros malabarica* Desr. syn. *D. peregrina* (Gartn.) Gurke	Ebenaceae	Gaub Persimmon, Gab	Unripe fruits	Tr	Brown	2, 6, 4
18.	*Diospyros melanoxylon* Roxb.	Ebenaceae	Bidi Patta	Fruits	Tr	Brown, black	2, 6, 58
19.	*Emblica officinalis* Gaertn. syn *Phyllanthus emblica* Linn.	Euphorbiaceae	Amla	Fruits, barks	Tr	Blue-black	2, 58
20.	*Embellia ribes* Burm. f.	Myrsinaceae	Barberang, False pepper	Fruits	H	Red	59
21.	*Euterpe edulis* Mart.	Arecaceae	Jucara Palm	Fruits	Tr	Red	61
22.	*Euterpe oleracea* Mart.	Arecaceae	Acai Palm	Fruits	Tr	Red	61
23.	*Ficus benghalensis* Linn.	Moraceae	The Banynan	Fruits, leaves, bark	Tr	Orange, brown	62a
24.	*Garcinia mangostana* Linn.	Clusiaceae	Mangosteen	Fruits	Tr	Black	1
25.	*Garcinia xanthochymus* Hook. f.	Clusiaceae	Dephal, Vika	Unripe fruits	Tr	Yellow	6, 64, 4
26.	*Gardenia jasminoides* Ellis=*G. augusta* (Linn.) Merrill	Rubiaceae	Gandhraj	Fruits	Tr/Sh	Yellow	64a, 4
27.	*Genipa americana* Linn	Rubiaceae	Jagua	Fruits	Tr	Blue	66
28.	*Grewia subinaequalis* DC.	Tiliaceae	Phalsa	Fruits	Tree	Yellow-orange	2
29.	*Hippophae salicifolia* D.Don	Elaeagnaceae	Chuk	Fruits	Sh	Yellow	2
30.	*Juglans regia* Linn.	Juglandaceae	Akhrot	Fruits, leaves, bark	Tr	Camel brown, black, red	86, 2, 20, 4
31.	*Lannea microcarpa* Engl. & K.Krause	Anacardiaceae	Microcarpa	Fruits, bark	Tr	Yellow, red-brown	88
32.	*Ligustrum vulgare* Linn.	Oleaceae	European Privet	Fruits	Sh	Blue, yellow	1, 9
33.	*Lycopus europaeus* Linn.	Lamiaceae	Gipsywort	Fruits	H	Green	7
34.	*Mahonia aquifolium* (Pursh) Nutt.	Berberidaceae	Holly leaved Barberry	Fruits juice	Sh	Red	94, 4

Contd...

S.No.	Name of plant species	Family	Popular name	Dye bearing parts	Pattern	Vital colour	References
35.	*Mallotus philippensis* Muell. Arg.	Euphorbiaceae	Kamala Tree, Sindur	Fruits	Tr	Orange	6, 95, 4
36.	*Malphigia glabra* Linn.	Malpighiaceae	Barbados Cherry	Fruits, flowers, bark	Sh	Yellow, blue, coffee tint	1, 5
37.	*Melastoma malabathricum* Linn.	Melastomataceae	Phutuka	Fruits leaves, roots	Sh	Purple, black, pink	1, 6, 20, 98, 4
38.	*Miliusa roxburghiana* Hook.f. & Thoms.	Annonaceae	Chhag-loti	Ripe fruits, bark	Tree	Red	20
39.	*Murraya koenigii* (Linn.) spreng.	Rutaceae	Kurry patta	Fruits, leaves	Sh,Tr	Grey, brown	8, 103a
40.	*Myrtillocactus geometrizans* (Mart.) Console	Cactaceae	Blue Candle	Fruits	Shrubby Cactus	Red	104, 4
41.	*Oroxylum indicum* Vent.	Bignoniaceae	Ullu	Fruits, bark	Tr	Black, mordant in dyeing & tanning	2
42.	*Osbeckia stellata* Buch.- Ham. ex Kew Gawler	Melastomataceae	Osbeckia	Fruits	Sh	Brown	2
43.	*Piper longum* Linn.	Piperaceae	Pipli, Indian Long Pepper	Dry fruits	Cl	Grey	8
44.	*Prinsepia utilis* Royle	Rosaceae	Bekkra	Fruits	Sh	Blue	2
45.	*Prunus cerasoides* D.Don	Rosaceae	Himalayan Wild Cherry, Paddam	Fruits	Tr	Yellowish	2
46.	*Prunus cerasus* Linn.	Rosaceae	Sour Cherry, Gilas	Fruits	Tr	Pink	116
47.	*Pyrus pashia Buch.* Ham.ex D.Don	Rosaceae	Mehal	Fruits	Tr	Black	20
48.	*Rauvolfia tetraphylla* Linn.	Apocynaceae	Barachandrika	Fruits	Sh	Black	96
49.	*Rivina humilis* Linn.	Phytolaccaceae	Blood-Berry	Fruits	H	Red	7
50.	*Rubus fruticosus* Linn. var. *discolor.*	Rosaceae	Alish,Akhi	Fruits	Sh	Brown, red	9, 7

51.	*Semecarpus anacardium* Linn.f.	Anacardiaceae	Bhilawa, The Marking Nut Tree	Fruits, bark	Tr	Black	131, 2, 4
52.	*Solanum indicum* Linn.	Solanaceae	Barhanta, Poison-Berry	Half ripe fruits	Sh	Black, yellow	20
53.	*Syzygium cuminii* (Linn.) Skeels	Myrtaceae	Jaman	Fruits, bark	Tr	Black, purple	2, 6, 20, 137
54.	*Terminalia bellirica* Roxb.	Combretaceae	Bahera	Fruits	Tr	Blue	2, 6, 144
55.	*Terminalia chebula* Retz.	Combretaceae	Harra, Myrobalan	Fruits, barks	Tr	Yellow, brown	2, 6, 147, 148, 4
56.	*Terminalia alata* Heyne ex Roth	Combretaceae	Asan, Laurel	Fruits, bark	Tr	Brown	6
57.	*Ziziphus jujuba* Mill.	Rhamnaceae	Pitni-Ber	Fruits	Tr	Reddish pink	9, 152

Table 4.9 Plant species bearing leaves dyes

S.No.	Name of plant species	Family	Popular name	Dye bearing parts	Pattern	Vital colour	References
1.	*Acacia leucophloea* Willd.	Mimosaceae	Safed Kikar	Leaves, bark	Tr	Red	6
2.	*Acacia nilotica* Delile subsp. *indica* (Benth.) Brenan	Mimosaceae	Babul	Leaves, fruit rind, gum, bark	Tr	Black to dark brown	2, 6
3.	*Acalypha indica* Linn.	Euphorbiaceae	Indian Acalypha	Leaves	H	Yellow	8
4.	*Adhatoda zeylanica* Medic.	Acanthaceae	Vasaka, Arusa	Leaves	Sh	Yellow-green	2
5.	*Agrimonia eupatoria* Linn.	Rosaceae	Agrimony	Leaves, roots	H	Yellow	2
6.	*Ailanthus triphysa* (Dennst.) Alston	Simaroubaceae	Guggul Dhup	Leaves	Tr	Black	1
7.	*Albizia amara* Boiv.	Fabaceae	Krishnasirisha	Leaves, pods	Tr	Brown grey, olive	8
8.	*Alpinia flabellata* Ridley	Zingiberaceae	Gingerlily	Leaaves	H	Yellow	17
9.	*Amaranthus tricolor* Linn.	Amaranthaceae	Garden Amaranth	Leaves, inflorescences	H	Red-violet	18
10.	*Anogeissus latifolia* (Roxb.) Wall. ex Bedd.	Combretaceae	Dhaura	Leaves	Tr	Black	2, 6
11.	*Aporusa dioica* (Roxb.) Muell-Arg.syn. *A. roxburghii* Baill	Euphoribiaceae	Kokra in Bengali	Leaves	Tr	Black	20
12.	*Artemisia japonica* Thunb.	Asteraceae	Kunja	Leaves	H	Brown	2
13.	*Artemisia nilagirica* (C.B Clarke) Pamp.	Asteraceae	Indian Wormwood	Leaves	H	Brown	2
14.	*Artocarpus communis* J.R. & G.Forst.	Moraceae	The Bread fruit	Leaves	Tr	Pink, brown	8
15.	*Baccaurea ramiflora* Lour.	Euphorbiaceae	Kataphal	Bark, leaves, fruits	Tr	Black	20
16.	*Bauhinia tomentosa* Linn.	Fabaceae	Bell Bauhina	Leaves	Sh	Yellow	1, 5
17.	*Bidens bipinnata* Linn.	Asteraceae	Lumb	Leaves	H	Yellow	31, 32, 4

18.	*Choiranthus cheiri* Linn.	Brassicaceae	Wall Flower	Leaves, petals	H	Yellow	4, 43
19.	*Chukrasia tabularis* A.Juss. var.*tabularis*	Meliaceae	Indian Mahogany Tree	Leaves, flowers	Tr	Red & yellow	9, 7
20.	*Cinnamomum tamala* (F.Hamilt.) Nees & Eberm.	Lauraceae	Tejpat, Indian Cassia Lignea	Leaves	Tr	Brown	2
21.	*Combretum pilosum* Roxb. ex G.Don	Combretaceae	Bhoree Loth	Leaves	Sh	Blue, black	7
22.	*Convallaria majalis* Linn.	Lilaceae	May Lily	Leaves	H	Orange	1
23.	*Cordia dichotoma* Forster. f.	Boraginaceae	Lasura, Indian Cherry	Leaves, roots, fruits	Tr	Yellow, red, green	1, 5, 6, 7
24.	*Cudrania cochinchinensis* (Lour.) Kudo et Masam.	Moraceae	Pulikant	Leaves, seeds	Sh	Orange, yellow	20
25.	*Cupressus torulosa* Don var. *corneyana* (Knight &Perry) Carr.	Cupressaceae	Surai	Leaves	Tr	Green	2
26.	*Datura metel* Linn.	Solanaceae	Thorn apple	Leaves	H	Grey, brown	8, 55a
27.	*Eucalyptus hybrid* (mainly *E.umbellata*Domin syn.*E.tereticornis* Sm.)	Myrtaceae	Safeda	Leaves, bark	Tr	Yellow, brown, black	41
28.	*Ficus benghalensis* Linn.	Moraceae	The Banynan	Leaves, bark, fruits	Tr	Orange, brown	62a
29.	*Ficus religiosa* Linn.	Moraceae	Pipal	Leaves, bark,	Tr	Red, brown	6, 63
30.	*Flemingia macrophylla* (Willd.)Merr.	Fabaceae	Bara, Solpan Waras	Leaves, bark	Sh	Black, yellow	20, 6, 1
31.	*Gymnema tingens* Spreng.	Asclepiadaceae	Gurmar	Leaves	Cl	Blue	5
32.	*Heliotropium indicum* Linn.	Boraginaceae	Hatta-juri	Leaves	H	Black	7
33.	*Heliotropium strigosum* Willd.	Boraginaceae	Chitiphul	Leaves	H	Black	1
34.	*Illicium griffithii* Hook f.&Thomson	Illiciaceae	Lissi	Leaves, seeds, bark	Shrub, Tree	Yellow (traditional dye)	20, 78

Contd...

S.No.	Name of plant species	Family	Popular name	Dye bearing parts	Pattern	Vital colour	References
35.	*Impatiens balsamina* Linn	Balsaminaceae	Gul Mehndi	Leaves, flowers	H	Red, brown orange	79, 9, 4
36.	*Indigofera arrecta* Hochst.	Fabaceae	Bengal Indigo	Leaves	Sh	Blue	5
37.	*Indigofera articulata* Gouan	Fabaceae	Surat Indigo	Leaves	Under Sh	Blue, black	5
38.	*Indigofera aspalathoides* Vahl ex DC.	Fabaceae	Surat Indigo	Leaves	Under Sh	Blue, black	1
39.	*Indigofera hirusta* Linn.	Fabaceae	Hairy Indigo	Leaves	H	Blue	5, 1
40.	*Indigofera tinctoria* Linn.	Fabaceae	Nil	Leaves, whole plant	Sh	Blue	80, 6, 4
41.	*Isatis tinctoria* Linn.	Cruciferae	Dyer's Woad	Leaves	H	Blue	83, 84
42.	*Jatropha curcas* Linn.	Euphorbiaceae	Safedarand	Leaves, bark	Tr	Blue	1, 85, 4
43.	*Juglans regia* Linn.	Juglandaceae	Akhrot	Leaves, fruits, bark	Tr	Camel, brown, black, red	86, 2, 20, 4
44.	*Lantana camara* Linn.	Verbenaceae	Wild Sage	Leaves	Sh	Green	89
45.	*Lawsonia inermis* Linn.	Lythraceae	Mehndi, Henna	Leaves	Sh	Yellow-orange, red	6, 90, 91, 4, 2
46.	*Lotus corniculatus* Linn.	Fabaceae	Birdsfoot Trefoil	Leaves, flowers	H	Orange, yellow	92, 4, 96
47.	*Marsdenia tinctoria* R.Br.	Asclepiadaceae	Manri, Kali lara	Leaves	Cl	Blue	96
48.	*Melastoma malabathricum* Linn.	Melastomataceae	Phutuka	Leaves, fruits, roots,	Sh	Purple, black, pink	1, 6, 20, 98, 4
49.	*Memecylon umbellatum* Burm.f.	Melastomataceae	Iron Wood Tree	Leaves	Sh,Tree	Yellow	6, 7, 99, 4
50.	*Nelumbo nucifera* Gaertn.	Nymphaeaceae	Sacred Lotus, Kamal	Leaves	Aquatic H	Sandal yellow	8
51.	*Peltophorum pterocarpum* Backer ex K. Heyne	Caesalpiniaceae	Copper pod	Leaves, bark, wood	Tr	Yellowish, brown, black	1
52.	*Perilla frutescens* (Linn.) Britton	Lamiaceae	Bhanjira	Leaves	H	Purple, black	109, 1, 4

53.	*Peristrophe bicalyculata* Nees	Acanthaceae	Atrilal	Leaves	H	Blue	109a
54.	*Phaius tankervilliae* Blume	Orchidaceae	Tipui	Leaves, scape	H	Indigo	20
55.	*Phytolacca dioica* Linn.	Phytolaccaceae	Ombu	Leaves	Sh	Yellow	7
56.	*Phyllanthus fraternus* Webster	Euphorbiaceae	Jungle amli	Leaves, stem	H	Black	96
57.	*Pisonia grandis* R.Br.	Nyctaginaceae	Lettuce Tree	Leaves	Tr	Yellow	96
58.	*Pistacia integerrima* Stewart. ex. Brandis	Anacardiaceae	Kakara	Galls of leaves and petioles	Tr	Yellow	1, 2, 112
59.	*Pongamia pinnata* Pierre	Fabaceae	Karanj	Leaves, bark, seeds cover	Tr	Olive brown	8, 114a
60.	*Prunus persica* Batsch.	Rosaceae	Peach, Aru	Leaves, roots bark	Tr	Brown	9
61.	*Rhododendron lepidotum* Wall. ex G.Don	Ericaceae	Talisfur, Thyampa	Leaves, flowers	Sh	Pink-red	2
62.	*Rhus coriaria* Linn.	Anacardiaceae	Sicilian Sumac	Leaves, twigs	Sh	Yellow	7
63.	*Rhus cotinus* Linn.	Anacardiaceae	Royal Purple	Leaves, twigs	Sh	Yellow	7
64.	*Sapium baccatum* Roxb.	Euphorbiaceae	Seleng, Mousedeer's Rubber-Tree	Leaves	Tr	Blackish green	20
65.	*Sapium indicum* Willd.	Euphorbiaceae	Hurna	Leaves	Tr	Greenish	96
66.	*Saraca asoca* (Roxb.) De Wilde	Fabaceae	Ashoka	Leaves, bark	Tr	Yellow, brown	129a
67.	*Sorghum caudatum* (Hack.) Stapf.	Poaceae	Caudatum	Leaves sheaths	Stout Grass	Yellow orange	134
68.	*Spathodea campanulata* Beauv.	Bignoniaceae	African Tulip Tree	Leaves	Tr	Olive brown	134a
69.	*Spinacia oleracea* Linn.	Amaranthaceae	Palak, Garden Spinach	Leaves	H	Green	5
70.	*Symplocos paniculata* Miq .	Symplocaceae	Sweet Leaves, Ludh	Leaves, bark	Sh	Yellow	2, 20, 135

Contd...

S.No.	Name of plant species	Family	Popular name	Dye bearing parts	Pattern	Vital colour	References
71.	*Symplocos ramosissima* Wall.ex G.Don	Symplocaceae	Lodh	Leaves, bark	Sh,Tr	Yellow	2
72.	*Symplocos laurina* Wall.ex Rehd.&E.H.Wils	Symplocaceae	Bholia, Chungan	Leaves	Tr	Yellow	5
73.	*Tamarindus indica* Linn.	Caesalpiniaceae	Imli, Amli	Leaves	Tr	Reddish-yellow	2
74.	*Tectona grandis* Linn.f.	Verbenaceae	Sagwan, Teak	Leaves, bark, roots	Tr	Reddish, yellow, yellowish brown	2, 6, 143, 4
75.	*Tephrosia candida* DC.	Fabaceae	Lashtia	Leaves	Sh	Dark blue	20
76.	*Tephrosia purpura* (Linn.) Pers.	Fabaceae	Wild Indigo, Dhamasia	Leaves	H	Green	8, 143a
77.	*Terminalia catappa* Linn.	Combretaceae	Indian Almond-Tree, Deshi-badam	Leaves, bark	Tr	Pinkish red, Light brownish, yellow	145, 146, 4
78.	*Thespesia populnea* Soland. and. ex Correa	Malvaceae	Parsipu, Portia Tree	Leaves, bark, wood, flowers	Tr	Orange, yellow, brown, green	7, 8
79.	*Wedelia chinensis* Merrill	Asteraceae	Bhanra	Leaves, roots	H	Black, blue black	149a
80.	*Woodfrodia fruitcosa* Kurz.	Lythraceae	Dawi, Fire-Flame Bush	Leaves, flowers,	Sh	Pink-red	2, 6, 150, 4
81.	*Wrightia tinctoria* R.Br.	Apocyanaceae	Pala Indigo-Plant, Indrajau	Leaves	Tr	Blue	5, 4
82.	*Zizyphus mauritiana* Lam.	Rhamnaceae	Indian Jujuba, Ber	Leaves	Tr	Red-pink	6

Table 4.10 Plant species bearing roots dyes

S.No.	*Name of plant species*	*Family*	*Popular name*	*Dye bearing parts*	*Pattern*	*Vital colour*	*References*
1.	*Agrimonia eupatoria* Linn.	Rosaceae	Agrimony	Roots, leaves	H	Yellow	2
2.	*Aleurities moluccana* (Linn.) Willd.	Euphorbiaceae	Jungli Akrot	Roots	Tr	Brown	6
3.	*Alkanna tinctoria* (L.) Tausch	Boraginaceae	Dyer's Bugloss	Roots	H	Purple	14, 4
4.	*Alpinia galanga* (Linn.) Willd.	Zingiberaceae	Barakulanjar	Roots	H	Brown	9
5.	*Ampelocissus latifolia* (Roxb.) Planch.	Vitaceae	Panibel	Roots	Cl	Black	2
6.	*Arnebia benthamii* (Wall.exG. Don) John.	Boraginaceae	Kashmiri kahzaban	Roots	H	Red, purple	2, 1
7.	*Arnebia densiflora* (Nordm.) Ledeb.	Boraginaceae	Ratanjot	Roots	H	Red	22, 4
8.	*Arnebia euchroma* (Royle) Johnston	Boraginaceae	Ratanjot	Roots	H	Red	23, 24, 4
9.	*Arnebia nobilis* Reichb.f.	Boranginaceae	Ratanjot	Roots	H	Purple	5
10.	*Artocarpus heterophyllus* Lam.	Moraceae	Kathal, JackFruits	Roots, wood, fruits	Tr	Yellow	2, 26, 6, 4
11.	*Artocarpus lakoocha* Roxb.	Moraceae	Monkey Jack	Roots, wood, fruits	Tr	Yellow	2
12.	*Berberis aristata* DC.	Berberidaceae	Kilmor	Roots, bark	Sh	Yellow	2
13.	*Berberis asiatica* Roxb. ex. DC.	Berberidaceae	Kilmor	Roots, bark	Sh	Yellow	2
14.	*Berberis chitria* Lindl.	Berberidaceae	Chetar	Roots, bark	Sh	Yellow	2
15.	*Berberis nepalensis* (DC.) Spreng.	Berberidaceae	Nepal barberry	Roots	Sh	Yellow	6
16.	*Berberis vulgaris* Linn.	Berberidaceae	Barberry	Roots, bark	Sh	Yellowish, yellow	2, 29, 4
17.	*Beta vulgaris* Linn.	Amaranthaceae	Common beet	Roots	H	Red	30

Contd...

S.No.	*Name of plant species*	*Family*	*Popular name*	*Dye bearing parts*	*Pattern*	*Vital colour*	*References*
18.	*Butea superba* Roxb.	Fabaceae	Palas lata	Roots	H	Red	6
19.	*Cynometra ramiflora* Linn.	Fabaceae	Cynometra	Roots	Tr	Purple	6
20.	*Cyperus scariosus* R. Br.	Cyperaceae	Nagarmotha	Roots	H	Brown	2
21.	*Datisca cannabina* Linn.	Datiscaceae	Akalbir, Bujr-Bhang	Roots	H	Yellowish-red	2, 6
22.	*Ficus altissima* Blume	Moraceae	Gadgubar	Roots, stem	Tr	Yellow	20
23.	*Galium aparine* Linn.	Rubiaceae	Catch Weed	Roots	H	Purple	1, 7
24.	*Galium rotundifolium* Linn.	Rubiaceae	Caglio a foglie rotonde	Roots	H	Yellow, brown	1
25.	*Galium verum* Linn.	Rubiaceae	Cheese Rennet	Roots	H	Yellow, red	1, 4
26.	*Geranium nepalense* Sweet	Geraniaceae	Bhanda	Roots	H	Red	2, 6
27.	*Geranium wallichianum* D. Don	Geraniaceae	Laljhari	Roots	H	Red-brown	2
28.	*Jasminum humile* Linn.	Oleaceae	Peeli Chameli	Roots	Sh	Yellow	6
29.	*Kigelia pinnata* DC.	Bignoniaceae	Common Sausage Tree	Roots, woods	Tr	Greyish brown	87
30.	*Kirganelia reticulata* (Poir.) Baill.	Euphorbiaceae	Panjuli	Roots, bark	Sh	Red	1
31.	*Lithospermum arvense* Linn.	Boraginaceae	Corn Gromwell	Roots	H	Red	7
32.	*Maclura pomifera* Schneid.	Moraceae	Osage-orange	Roots, wood, bark	Tr	Yellow	93, 4
33.	*Mahonia borealis* Takeda	Berberidaceae	Pualeng	Roots, bark, sap	Sh	Yellow	2
34.	*Morinda angustifolia* Roxb.	Rubiaceae	Asho	Roots	Sh,Tree	Yellow	7, 101, 4
35.	*Morinda citrifolia* Linn.	Rubiaceae	Al, Ach	Roots bark	Tr	Red, yellow	102, 6
36.	*Morinda umbellata* Linn. Hook.f.	Rubiaceae	Vomit vine	Roots, bark	Sh	Yellow	6
37.	*Nerium indicum* Mill.	Apocyanaceae	Indian Oleander, Kaner	Roots, bark	Sh	Black	20

38.	*Oldenlandia umbellata* Linn. syn. *Hedyotis umbellata* (Linn.) Lam.	Rubiaceae	Indian Madder, Chay-Roots	Roots	Woody annual	Red	7, 106, 107
39.	*Pfaffia iresinoides* (Kunth.) Spreng.	Amaranthaceae	Brazil Gingseng	Roots	H	Yellow	110
40.	*Plumbago indica* Linn.	Plumbaginaceae	Chitra	Roots	Shrubby perennial	Orange	5
41.	*Prunus persica* Batsch.	Rosaceae	Peach, Aru	Leaves, roots bark	Tr	Brown	9
42.	*Rauvolfia serpentina* (L.) Benth.ex Kurz.	Apocynaceae	Sarpagandha, Rauvolfia	Roots	Sh	Orange pinkish	96, 8
43.	*Rheum moorcroftianum* Royle	Polygonaceae	Luan gauo da huang	Roots	H	Yellow	2
44.	*Rheum webbianum* Royle	Polygonaceae	Archa	Roots	H	Yellow	2
45.	*Rubia cordifolia* Linn.	Rubiaceae	Indian Madder, Manjit	Roots	Cl	Scarlet-red, orange	2, 126, 127, 6, 4
46.	*Rubia sikkimensis* Kurz.	Rubiaceae	Naga-Madder	Roots	Cl	Scarlet-red, orange	7
47.	*Rubia tinctorum* Linn.	Rubiaceae	European Madder Alizari	Roots	H	Red	128, 4
48.	*Rumex dentatus* Linn.	Polygonaceae	Ambavah, Jungli palak	Roots	Deep Rootsed Annual	Red	1
49.	*Rumex acetosa* Linn.	Polygonaceae	Garden Soorel	Roots	H	Red	5
50.	*Rumex hastatus* D. Don	Polygonaceae	Amlora	Roots, bark	H	Yellow, green	2
51.	*Rumex nepalensis* Spreng.	Polygonaceae	Pahari Palang, Palak	Roots	H	Yellow-green	2
52.	*Sanguinaria canadensis* Linn.	Papaveraceae	Blood Roots	Roots	H	Red	8
53.	*Sophora mollis* Grah. ex Baker	Fabaceae	The Himalayan Laburnum	Roots, flowers	Sh	Brown, yellow	2

Contd...

S.No.	Name of plant species	Family	Popular name	Dye bearing parts	Pattern	Vital colour	References
54.	*Toddalia asiatica* (L.) Lam.	Rutaceae	Forest Peeper, Jangli-Kalimirch	Roots bark	Sh	Yellow	1
55.	*Urtica dioica* Linn.	Urticaceae	Bichhu booti	Roots	H	Brown-black	2
56.	*Ventilago denticulata* Willd.	Rhamnaceae	Pitti	Roots, bark	Sh	Red	6, 149
57.	*Ventilago madraspatna* Gaertn.	Rhamnaceae	Raktavalli	Roots, bark	Cl	Red	149
58.	*Wedelia chinensis* Merrill	Asteraceae	Bhanra	Roots, leaves	H	Black, blue black	149a

Table 4.11 Plant species bearing seeds dyes

S.No.	Name of plant species	Family	Popular name	Dye bearing parts	Pattern	Vital colour	References
1.	*Abrus precatorius* Linn.	Fabaceae	Crab's Eye, Rati	Seeds coat	Cl	Black	2
2.	*Acacia farnesiana* Willd.	Mimosaceae	Vilayati Kikar	Seeds, bark, pods	Tr	Black, brown	6
3.	*Caesalpinia coriaria* (Jacq.) Willd.	Fabaceae	Devi Divi Plant	Seeds, pods	Tr	Black	5
4.	*Cassia obtusifolia* Linn.	Fabaceae	Banarh, Cakonda	Seeds, pods	H	Yellow	5
5.	*Cassia occidentalis* Linn.	Fabaceae	Kasonda, Coffee-Senna	Seeds, pods	H	Brown, yellow	5
6.	*Cassia tora Linn.*	Fabaceae	Panwar, Foetid Cassia	Seeds	H	Red, blue, brown	6, 41
7.	*Coscinium fenestratum* Clolebr.	Menispermaceae	Tree-Turmeric, Jharihuldi	Seeds, bark, wood	Sh	Red	1
8.	*Crotalaria striata* DC.	Fabaceae	Can Janka	Seeds	Sh	Brown	5
9.	*Cudrania cochinchinensis* (Lour.) Kudo et Masam.	Moraceae	Pulikant	Seeds, leaves	Sh	Orange, yellow	20
10.	*Cuminum cyminum* Linn.	Apiaceae	Cumin Seeds	Seeds		Bright yellow	52a
11.	*Cyamopsis tetragonoloba* (Linn.) Tanb.	Fabaceae	Guar	Seeds, pods	H	Yellow	5
12.	*Illicium griffithii* Hook f.&Thomson	Illiciaceae	Lissi	Seeds, bark, leaves	Shrub, Tree	Yellow (traditional dye)	20, 78
13.	*Peganum harmala* Linn.	Zygophyllaceae	Harmal	Seeds	H	Brown to black, yellow, red	7
14.	*Phaseolus mungo* Linn.	Fabaceae	Urd	Seeds cover	H	Brown, yellow	8

Contd...

S.No.	Name of plant species	Family	Popular name	Dye bearing parts	Pattern	Vital colour	References
15.	*Pongamia pinnata* Pierre	Fabaceae	Karanj	Seeds cover, leaves, bark	Tr	Olive brown	8, 114a
16.	*Sesbania bispinosa* W.F. White	Fabaceae	Dhaincha	Seeds, pods	H	Yellow	5
17.	*Toona hexandra* (Wall. ex Roxb.) M. Roem. ssp. *hexandra* var. *gamblei* (DC.) Almedia &Almedia	Meliaceae	Toon	Seeds, flowers	Tr	Yellow-brown	2
18.	*Toona serrata* (Royle) M.Roem.	Meliaceae	Pahari-Toon	Seeds, flowers	Tr	Yellow-brown	2

Table 4.12 Plant species bearing wood dyes

S.No.	Name of plant species	Family	Popular name	Dye bearing parts	Pattern	Vital colour	References
1.	Acacia catechu Willd.	Mimosaceae	Khair, Cutch Tree	Heartwood	Tr	Red, brown	5
1a.	*Acacia chundra* Willd.	Mimosaceae	Red Cutch	Wood	Tr	Grey, black	5
2.	*Adenanthera pavonina* Linn.	Fabaceae	Rakta Kambal, Red Wood	Wood	Tr	Red	5, 9
3.	*Adina cordifolia* (Roxb.) Hook.f.ex Brandis	Rubiaceae	Haldu	Wood	Tr	Yellow	7
4.	*Artocarpus hetrophyllus* Lam.	Moraceae	Kathal, JackFruits	Wood, roots, fruits	Tr	Yellow	2, 6, 26, 4
5.	*Artocarpus lakoocha* Roxb.	Moraceae	Monkey Jack	Wood, fruits, roots	Tr	Yellow	2
6.	*Caesalpinia sappan* Linn.	Fabaceae	Sappan Wood	Wood	Tr	Red	1, 6, 35
7.	*Chlorophora tinctoria* (L.) Benth. & Hook.f.	Moraceae	Old Fustic	Wood	Tr	Yellow	44, 4
8.	*Coriaria nepalensis* Wall.	Coriariaceae	Makola	Wood	Tr or Sh	Red	6
9.	*Coscinium fenestratum* Clolebr.	Menispermaceae	Tree -Turmeric, Jharihuldi	Wood, seed, bark	Sh	Red	1
10.	*Cotinus coggygria* Scop.	Anacardiaceae	Young Fustic	Wood, leaves	Sh	Yellowish, orange	52
11.	*Euphorbia tirucalli* Linn	Euphorbiaceae	Konpal	Wood	Sh	Red	1
12.	*Haematoxylon campechianum* Linn.	Fabaceae	LogWood, Patang	Heartwood	Tr	Black	71, 72
13.	*Hypericum perforatum* Linn.	Hypericaceae	St. Jhon's Wort	Wood	H	Green	76a
14.	*Maclura pomifera* Schneid.	Moraceae	Osage-orange	Wood, bark, roots	Tr	Yellow	93, 4
15.	*Mesua ferrea* Linn.	Clusiaceae	Nagkesar, Mesua	Wood, flowers bud	Tr	Yellow	7
16.	*Moringa oleifera* Lam.	Moringaceae	Drumstick Tree	Wood	Tr	Blue	6, 103

Contd...

S.No.	Name of plant species	Family	Popular name	Dye bearing parts	Pattern	Vital colour	References
17.	*Peltophorum pterocarpum* Backer ex K. Heyne	Caesalpiniaceae	Copper Pods	Wood, leaves, bark	Tr	Yellowish, brown, black	1
18.	*Plecospermum spinosum* Trec.	Moraceae	Gumbenfong	Wood, bark	Sh or Small tree	Yellow	113
19.	*Pterocarpus dalbergiodes* Roxb.	Fabaceae, Papilionaceae	Padauk	Wood, bark	Tr	Brown, pink	8
20.	*Pterocarpus indicus* Willd. non Baker	Fabaceae	Malay Padauk	Wood	Tr	Red	5
21.	*Pterocarpus santalinus* Linn. f.	Fabaceae	Red Sandal Wood, Raktachandan	Wood	Tr	Red	1, 6, 117, 4
22.	*Terminalia arjuna* (Roxb.) Wight & Arn.	Combretaceae	Arjuna	Wood, bark	Tr	Brown	8, 143b
23.	*Thespesia populnea* Soland. and. ex Correa	Malvaceae	Parsipu, Portia Tree	Wood, bark, leaves flowers	Tr	Orange, yellow, brown, green	7
24.	*Vitex altissima* Linn.f.	Verbenaceae	Milla	Wood, bark	Tr	Yellow	7

Table 4.13 Plant species bearing whole plants dyes

S.No.	Name of plant species	Family	Popular name	Dye bearing parts	Pattern	Vital colour	References
1.	*Achyranthes aspera* Linn.	Amaranthaceae	Latjira, Prickly- Chaff Flower	Whole plant	H	Black-brown	2
2.	*Achyranthes bidentata* Blume	Amaranthaceae	Apamarga	Whole plant	H	Black-brown	2
3.	*Averrhoa carambola* Linn.	Averrhoaceae	Kamrak, Star Fruit	Whole plant	Tr	Yellow-brown	2
4.	*Bacopa monnieri* (Linn.) Pannell	Scrophulariaceae	Brahmi	Whole plant	H	Green	8
5.	*Caesulia axillaris* Roxb.	Asteraceae	Bangra	Whole plant	H	Yellow	4, 36
6.	*Cladonia verticillata* (Hoffm.) Schaer	Cladoniaceae	Cholara	Whole plant		Yellow-red	2
7.	*Colquhounia coccinea* Wallich	Lamiaceae	Himalayan Mint Plant	Whole plant	Sh	Black	20
8.	*Eupatorium glandulosum* Michex.	Asteraceae	Stickly Snake root	Whole plant	Sh	Green	2, 60
9.	*Euphorbia hirta* Linn.	Euphorbiaceae	Dudhi	Whole plant	Tr	Yellow	8, 60a
10.	*Everniastrum cirrhatum* (Fr.) Hale ex Sipman	Parmeliaceae	Jhau, Charata	Whole plant	Lichen	Red, brown	2
11.	*Leucas aspera* Spreng.	Lamiaceae	Chota Halkusa	Whole plant	H	Yellow	8, 91a
12.	*Parthenium hysterophorous* Linn.	Asteraceae	Congress Grass	Whole plant	H	Green	108
13.	*Polygonum hydropiper* Linn.	Polygonaceae	Water Peeper	Whole plant	H	Blue, black, yellow	20
14.	*Reseda luteola* Linn.	Resedaceae	Weld, Dyer's Rocket	Whole plant	H	Deep yellow	120, 121, 4
15.	*Rimelia reticulata* (Taylor) Hale & Fletcher	Parmeliaceae	Charata	Whole plant	Lichenous	Orange-yellow	2

Contd...

S.No.	Name of plant species	Family	Popular name	Dye bearing parts	Pattern	Vital colour	References
16.	*Serratula tinctoria* Linn.	Asteraceae	Saw Wort	Whole plant juice except roots	H	Yellow	120, 4
17.	*Solidago canadensis* Linn.	Asteraceae	Goldenrod	Whole plant	Sh	Golden yellow	19
18.	*Tagetes minuta* Linn.	Asteraceae	Jangli Genda, Stinking-Roger	Whole plant, flowers	H	Yellow	141

Table 4.14 Black colour imparting dye bearing plant species

S.No.	Name of plant species	Family	Popular name	Dye bearing parts	Pattern	Vital colour	References
1.	*Abrus precatorius* Linn.	Fabaceae	Crab's Eye, Rati	Seeds coats	Cl	Black	2
2.	*Acacia chundra* Willd.	Mimosaceae	Red Cutch	Wood	Tr	Black, grey	5
3.	*Acacia dealbata* Link	Mimosaceae	Silver Wattle	Bark	Tr	Black, brown	1
4.	*Acacia farnesiana* Willd.	Mimosaceae	Vilayati Kikar	Bark, seeds and pods	Tr	Black, brown	6
5.	*Acacia nilotica* Delile subsp. *indica* (Benth.) Brenan	Mimosaceae	Babul	Fruits rinds, gum, leaves, & bark	Tr	Black to dark brown	2, 6
6.	*Acanthopanax trifoliatum* (Linn.) Voss	Araliaceae	Shiah-ryngkhwari	Fruits	Sh	Black	1
7.	*Achyranthes aspera* Linn.	Amaranthaceae	Latjira, Prickly-Chaff Flower	Whole plant	H	Black-brown	2
8.	*Achyranthes bidentata* Blume	Amaranthaceae	Apamarga	Whole plant	H	Black-brown	2
9.	*Actaea spictata* Linn.	Ranunculaceae	Baneberry, Grapewort	Berries	H	Black	1, 7
10.	*Ailanthus triphysa* (Dennst.) Alston	Simaroubaceae	Guggul Dhup	Leaves	Tr	Black	1
11.	*Ampelocissus latifolia* (Roxb.) Planch.	Vitaceae	Panibel	Roots	Cl	Black	2
12.	*Anacardium occidentale* Linn.	Ancardiaceae	Cashew or The Cashewnutt	Nuts (shells)	Tr	Black	7, 9
13.	*Annona reticulata* Linn.	Annonaceae	Raamaaphal	Unripe fruits	Tr	Black	19, 6
14.	*Anogeissus latifolia* (Roxb.) Wall.ex Bedd.	Combretaceae	Dhaura	Leaves	Tr	Black	2, 6
15.	*Aporusa dioica* (Roxb.) Muell-Arg.syn.*A.roxburghii* Baill	Euphorbiaceae	Kokra in Bengali	Leaves	Tr	Black	20

Contd...

S.No.	Name of plant species	Family	Popular name	Dye bearing parts	Pattern	Vital colour	References
16.	*Baccaurea ramiflora* Lour.	Euphorbiaceae	Kataphal	Bark, leaves and fruits	Tr	Black	20
17.	*Bischofia javanica* Blume	Euphorbiaceae	Bishop Wood	Bark, seeds	Tr	Black	2
18.	*Bridelia stipularis* Blume	Euphorbiaceae	Khaji	Fruits	Cl	Black	5, 1
19.	*Caesalpinia coriaria* (Jacq.) Willd.	Fabaceae	Devi Divi Plant	Pods and seeds	Tr	Black	5
20.	*Citrus medica* Linn.	Rutaceae	Citron	Bark	Tr	Black	1, 7
21.	*Colquhounia coccinea* Wallich	Lamiaceae	Himalayan Mint Plant	Whole plant	Sh	Black	20
22.	*Combretum pilosum* Roxb.ex G.Don	Combretaceae	Bhoree Loth	Leaves	Sh	Blue, black	7
23.	*Flemingia macrophylla* (Willd.) Merr.	Fabaceae	Bara solpan waras	Leaves, bark	Sh	Black, yellow	20, 6, 1
24.	*Garcinia mangostana* Linn.	Clusiaceae	Mangosteen	Fruits	Tr	Black	1
25.	*Garuga gamlie* King	Burseraceae	Pani amora(A)	Bark	Tr	Brownish black	20
26.	*Grewia optiva* Drumm.	Tiliaceae	Bhimal	Seeds	Tr	Black, brown	41
27.	*Haematoxylon campechianum* Linn.	Fabaceae	Logwood, Patang	Heartwood	Tr	Black	71, 72
28.	*Heliotropium indicum* Linn.	Boraginaceae	Hatta-juri	Leaves	H	Black	7
29.	*Heliotropium strigosum* Willd.	Boraginaceae	Chitiphul	Leaves	H	Black	1
30.	*Juglans regia* Linn.	Juglandaceae	Akhrot	Bark, roots, leaves, fruits	Tr	Black camel, brown, red	86, 2, 20, 4
31.	*Lagerstroemia parviflora* Roxb.	Lythraceae	Dhaura	Bark	Sh	Black	2, 6
32.	*Melastoma malabathricum* Linn.	Melastomataceae	Phutuka	Fruits, leaves, roots	Sh	Black, purple, pink	1, 6, 20, 98, 4

33.	*Nerium indicum* Mill.	Apocyanaceae	Indian Oleander, Kaner	Roots, bark	Sh	Black	20
34.	*Peganum harmala* Linn.	Zygophyllaceae	Harmal	Seeds	H	Brown to black, yellow, red	7
35.	*Peltophorum pterocarpum* Backer ex K. Heyne	Caesalpiniaceae	Copper pod	Wood, leaves, bark	Tr	Black, yellowish, brown	1
36.	*Perilla frutescens* (Linn.) Britton	Lamiaceae	Bhanjira	Leaves	H	Black, purple	109, 1, 4
37.	*Phyllanthus fraternus* Webster	Euphorbiaceae	Jungle amli	Leaves and stem	H	Black	96
38.	*Pinus roxburghii* Sarg.	Pinaceae	Chir Pine, Chir	Bark, needles	Tr	Black, brown	6, 111
39.	*Pinus wallichiana* A.B. Jackson	Pinaceae	Kal, Blue Pine	Bark	Tr	Black	2, 6
40.	*Polygonum hydropiper* Linn.	Polygonaceae	Water Peeper	Whole plant	H	Black, blue, yellow	20
41.	*Pyrus pashia* Buch. Ham.ex D.Don	Rosaceae	Mehal	Fruits	Tr	Black	20
42.	*Rauvolfia tetraphylla* Linn.	Apocynaceae	Barachandrika	Fruits	Sh	Black	96
43.	*Securinega virosa* (Roxb.ex Willd.) Pax &Hoffm.	Euphorbiaceae	Patala	Bark	Sh	Black	7, 4
44.	*Semecarpus anacardium* Linn.f.	Anacardiaceae	Bhilawa The Marking Nut Tree	Bark, fruits	Tr	Black	131, 2
45.	*Shorea robusta* Gaertn.f.	Dipterocarpaceae	Sal	Bark	Tr	Black, red	6, 132, 133
46.	*Spondias pinnata* (Linn.f.) Kurz	Anacardiaceae	Wild Mango	Bark	Tr	Black	7
47.	*Strychnos nux-vomica* Linn.	Loganiaceae	Snake Wood, Nux-Vomica	Seeds	Tr	Brown-black	7
48.	*Wedelia chinensis* Merrill	Asteraceae	Bhanra	Roots, leaves	H	Black, blue black	149a

Table 4.15 Blue colour imparting dye bearing plant species

S.No.	Name of plant species	Family	Popular name	Dye bearing parts	Pattern	Vital colour	References
1.	*Aronia melanocarpa* (Michx.) Elliott	Rosaceae	Black choke Berry	Fruits	Sh	Blue	4, 25
2.	*Chrozophora tinctoria* (Linn) Rafin.	Euphorbiaceae	Dyer's Litmus	Petals, fruits, sap	H	Blue, red	9, 4
3.	*Clitoria mariana* Linn.	Fabaceae	Sweta, Butterfly-Pea.	Flowers	Climber, Undershrub	Blue, violet	20
4.	*Clitoria ternatea* Linn.	Fabaceae	Aparajit, Conchflower	Flowers, seeds	Cl	Blue	1, 4, 48
5.	*Combretum pilosum* Roxb.ex G.Don	Combretaceae	Bhorel Loth	Leaves	Sh	Blue, black	7
6.	*Commelina benghalensis* Linn.	Commelinaceae	Benghal Day Flower	Juice of flowers	Serpentine leaf miner	Blue	9
7.	*Emblica officinalis* Gaertn.syn *Phyllanthus emblica* Linn.	Euphorbiaceae	Amla	Seedss, barks	Tr	Blue-black	2, 6, 58
8.	*Engalhardtia spicata* Lesch.ex. Blume	Juglandaceae	Corcorshing, Mahwa	Bark	Tr	Dark blue, Dark brown	20, 2
9.	*Entada pursaetha* DC.	Mimosaceae	Ghilla	Young pods	Cl	Dark blue	20
10.	*Genipa americana* Linn	Rubiaceae	Jagua	Fruits	Tr	Blue	66
11.	*Gentiana pneumonanthe* Linn	Gentiananceae	Gentiana	Flower	H	Blue	68, 4
12.	*Girardinia heterophylla* Decne	Urticaceae	Himalayan Nettle	Fibres	H	Blue	69
13.	*Gymnema tingens* Spreng.	Asclepiadaceae	Gurmar	Leaves	Cl	Blue	5
14.	*Indigofera arrecta* Hochst.	Fabaceae	Bengal Indigo	Leaves	Sh	Blue	5
15.	*Indigofera hirusta* Linn.	Fabaceae	Hairy Indigo	Leaves	H	Blue	5, 1
16.	*Indigofera tinctoria* Linn.	Fabaceae	Nil, Indian Indigo	Leaves, whole plant		Blue	6, 80, 4
17.	*Isatis tinctoria* Linn.	Cruciferae	Dyer's Woad	Leaves	H	Blue	83, 84
18.	*Jatropha curcas* Linn.	Euphorbiaceae	Safedarand	Bark, leaves	Tr	Blue	1, 85, 4

19.	*Ligustrum vulgare* Linn.	Oleaceae	European Privet	Seeds	Sh	Blue, yellow	1, 9
20.	*Malphigia glabra* Linn.	Malpighiaceae	Barbados Cherry	Flowers, fruits, bark	Sh	Blue, coffee tint, yellow	1, 5
21.	*Marsdenia tinctoria* R.Br.	Asclepiadaceae	Manri, Kali lara	Leaves	Cl	Blue	96
22.	*Moringa oleifera* Lam.	Moringaceae	Drumstick Tree	Wood	Tr	Blue	6, 103
23.	*Peristrophe bicalyculata* Nees	Acanthaceae	Atrilal	Leaves	H	Blue	109a
24.	*Polygonum hydropiper* Linn.	Polygonaceae	Water Peeper	Whole plant	H	Blue, black, yellow	20
25.	*Prinsepia utilis* Royle	Rosaceae	Bekkra	Fruits	Sh	Blue	2
26.	*Smilax glabra* Roxb.	Liliaceae	Barichob-Chini	Berries	Cl	Blue	5
27.	*Strobilanthes cusia* (Nees) Imlay	Acanthaceae	Assam Indigo, The Rum	Twigs	Sh	Blue	5
28.	*Tephrosia candida* DC.	Fabaceae	Lashtia	Leaves	Sh	Dark blue	20
29.	*Terminalia bellirica* Roxb.	Combretaceae	Bahera	Fruits	Tr	Blue	2, 6, 144
30.	*Wedelia chinensis* Merrill	Asteraceae	Bhanra	Roots, leaves	H	Blue-black, black	149a
31.	*Wrightia tinctoria* R.Br.	Apocyanaceae	Pala Indigo-Plant, Indrajau	Leaves	Tr	Blue	5, 4

Table 4.16 Brown colour imparting dye bearing Plant species

S.No.	Name of plant species	Family	Popular name	Dye bearing parts	Pattern	Vital colour	References
1.	*Acacia catechu* Willd.	Mimosaceae	Khair, Cutch Tree	Heartwood	Tr	Brown, red	3, 2, 4
2.	*Acacia dealbata* Link	Mimosaceae	Silver Wattle	Bark	Tr	Brown, black	1
3.	*Acacia farnesiana* Willd.	Mimosaceae	Vilayati Kikar	Bark, seeds, pods	Tr	Brown, black	6
4.	*Acacia nilotica* Delile subsp. *indica* (Benth.) Brenan	Mimosaceae	Babul	Fruits rinds, gum, leaves, and bark	Tr	Dark brown, black	2, 6
5.	*Acacia pennata* Willd.	Mimosaceae	Aila	Bark	Sh	Brown	5
6.	*Acacia sinuata* Willd. (Lour.) Merrill	Mimosaceae	Ritha	Bark	Sh	Brown	6, 7
7.	*Achyranthes aspera* Linn.	Amaranthaceae	Latjira, Prickly-Chaff Flower	Whole plant	H	Brown, black	2
8.	*Achyranthes bidentata* Blume	Amaranthaceae	Apamarga	Whole plant	H	Brown, black	2
9.	*Aesculus indica* Hook.	Hippocastanaceae	Pangar, Himalayan Chestnut	Bark	Tr	Brown	2
10.	*Albizia amara* Boiv.	Fabaceae	Krishnasirisha	Leaves, pods	Tr	Brown, grey, olive	8
11.	*Albizia lebbeck* Benth.	Fabaceae	Siris, Kokko	Bark	Tr	Brown	5
12.	*Albizia odoratissima* (L.f.) Benth.	Fabaceae, Mimosaceae	Black Siris	Bark	Tr	Brown	6
13.	*Aleurities moluccana* (Linn.)Willd.	Euphorbiaceae	Jungli Akrot	Roots	Tr	Brown	6
14.	*Alnus nitida* Endl.	Betulaceae	West Himalayan Alder, Seril	Bark	Tr	Brown-grey	4, 7
15.	*Alpinia galanga* (Linn.) Willd.	Zingiberaceae	Barakulanjar	Roots	H	Brown	9
16.	*Artemisia japonica* Thunb.	Asteraceae	Kunja	Leaves	H	Brown	2

17.	*Artemisia nilagirica* (C.B Clarke) Pamp.	Asteraceae	Indian Wormwood	Leaves	H	Brown	2
18.	*Artocarpus communis* J.R. & G.Forst.	Moraceae	The Bread Fruits	Leaves	Tr	Brown, pink	8
19.	*Averrhoa carambola* Linn.	Averrhoaceae	Kamrak, Star Fruits	Whole plant	Tr	Brown-yellow	2
20.	*Cassia fistula* Linn.	Fabaceae	Amaltas,Golden-Shower	Fruits, bark	Tr	Brown	2, 6
21.	*Cassia occidentalis* Linn.	Fabaceae	Kasonda, Coffee-Senna	Pods, seeds	H	Brown, yellow	5
22.	*Cassia tora* Linn.	Fabaceae	Panwar, Foetid Cassia	Seeds	H	Brown, red, blue	6, 41
23.	*Cassytha filiformis* Linn.	Lauraceae	Love-Vine	Stem	H	Brown	1
24.	*Casuarina equisetifolia* Linn.	Casuarinaceae	Jangli Saru, Casuarina	Bark	Tr	Brown	42, 6
25.	*Ceriops tagal* (Perr.) C.B.Robins	Rhizophoraceae	Goran, Guttia	Bark	Mangrove	Brown	9, 42
26.	*Cinnamomum camphora* (Linn.) Presl	Lauraceae	Camphor Tree, Kapur	Bark	Tr	Brown	46
27.	*Cinnamomum tamala* (F.Hamilt.) Nees & Eberm.	Lauraceae	Tejpat, Indian Cassia Lignea	Leaves	Tr	Brown	2
28.	*Crotalaria striata* DC.	Fabaceae	Can Janka	Seeds	Sh	Brown	5
29.	*Datura metel* Linn.	Solanaceae	Thorn Apple	Leaves	H	Brown, grey	8, 55a
30.	*Diospyros malabarica* Desr. syn. *D. peregrina* (Gartn.) Gurke	Ebenaceae	Gaub Persimmon, Gab	Unripe fruits	Tr	Brown	2, 6
31.	*Diospyros melanoxylon* Roxb.	Ebenaceae	Bidi Patta	Fruits	Tr	Brown, black	7
32.	*Dipterocarpus turbinatus* Gaertn.f.	Dipterocarpaceae	Common Gurjan Tree	Twigs, bark	Tr	Brown, yellow	1, 7
33.	*Erythrina suberosa* Roxb.	Fabaceae	Madar	Flowers, bark	Tr	Brown, dark	2

Contd...

S.No.	Name of plant species	Family	Popular name	Dye bearing parts	Pattern	Vital colour	References
34.	*Eucalyptus hybrid* (mainly *E.umbellata* Domin syn.*E.tereticornis* Sm.)	Myrtaceae	Safeda	Bark, leaves	Tr	Brown, yellow, black	41
35.	*Everniastrum cirrhatum* (Fr.) Hale ex Sipman	Parmeliaceae	Jhau, Charata	Whole plant	Lichen	Brown, red	2
36.	*Excoecaria agallocha* Linn.	Euphorbiaceae	Gengwa	Bark	Tr	Brown, black	5
37.	*Ficus benghalensis* Linn.	Moraceae	The Banynan	Leaves, bark, fruits	Tr	Brown, orange	62a
38.	*Ficus religiosa* Linn.	Moraceae	Pipal	Bark, leaves	Tr	Brown, red,	6, 63
39.	*Galium rotundifolium* Linn.	Rubiaceae	Caglio a foglie rotonde	Roots	Herb	Brown, yellow	1
40.	*Grewia optiva* Drumm.	Tiliaceae	Bhimal	Seeds	Tr	Brown, black	41
41.	*Impatiens balsamina* Linn	Balsaminaceae	Gul Mehndi	Flowers, leaves	H	Brown, red, orange	79, 9, 4
42.	*Juglans regia* Linn.	Juglandaceae	Akhrot	Bark, fruits, leaves	Tr	Brown camel, black, red	2, 20, 86, 4
43.	*Kandelia candel* (Linn.) Druce	Rhizophoraceae	Goria	Bark	Sh, Tree	Brown, red	6
44.	*Kigelia pinnata* DC.	Bignoniaceae	Common Sausage Tree	Woods, roots	Tr	Greyish brown	87
45.	*Lannea coromandelica* (Houtt.) Merrill	Anacardiaceae	Jhingan	Bark, resin	Tr	Brown, yellow	2
46.	*Lannea microcarpa* Engl. & K.Krause	Anacardiaceae	Microcarpa	Fruits, bark	Tr	Yellow, red-brown	88
47.	*Mimusops elengi* Linn.	Sapotaceae	Bulletwood	Bark	Tr	Brown	6, 100, 4
48.	*Murraya koenigii* (Linn.) Spreng.	Rutaceae	Kurry Patta	Leaves, fruits	Sh, Tr	Brown, grey	8, 103a
49.	*Nymphaea alba* Linn.	Nymphaeaceae	European White Nilofar, Water -Lily	Rhizomes	Aquatic H	Brown	1, 9, 7

50.	*Osbeckia stellata* Buch.-Ham ex Kew Gawler	Melastomataceae	Osbeckia	Fruits	Sh	Brown	2
51.	*Peganum harmala* Linn.	Zygophyllaceae	Harmal	Seeds	H	Brown to black, yellow, red	7
52.	*Peltophorum pterocarpum* Backer ex K. Heyne	Caesalpiniaceae	Copper Pod	Wood, leaves, bark	Tr	Brown, yellowish, black	1
53.	*Phaseolus mungo* Linn.	Fabaceae	Urd	Seeds covers	H	Brown, yellow	8
54.	*Pinus palustris* Mill.	Pinaceae	LongLeaves Pine	Flowers, needles	Tr	Brown, olive	8
55.	*Pinus roxburghii* Sarg.	Pinaceae	Chir Pine, Chir	Bark, needles	Tr	Brown, black	6, 111
56.	*Pistacia khinjuk* Stocks	Anacardiaceae	Bombay Mastic	Stem galls	Tr	Brown	2
57.	*Pongamia pinnata* Pierre	Fabaceae	Karanj	Leaves, bark, seeds covers	Tr	Brown, olive	8, 114a
58.	*Populus deltoides* Marsh.	Salicaceae	Poplar	Bark	Tr.	Brown, yellow	115
59.	*Prunus persica* Batsch.	Rosaceae	Peach, Aru	Leaves, roots bark	Tr	Brown	9
60.	*Psidium guajava* Linn.	Myrtaceae	Amrud	Bark	Tr	Black, brown	2
61.	*Pterocarpus marsupium* Roxb.	Fabaceae	Bijasal, Indian Kino Tree	Bark	Tr	Brown, red	6
62.	*Rubus fruticosus* Linn. var. *discolor*.	Rosaceae	Alish, Akhi	Fruits	Sh	Brown, red	9, 7
63.	*Sophora mollis* Grah. ex Baker	Fabaceae	The Himalayan Laburnum	Roots, flowers	Sh	Brown, yellow	2
64.	*Soymida febrifuga* A. Juss.	Meliaceae	Indian Redwood, Rohun	Bark	Tr	Brown	7
65.	*Spathodea campanulata* Beauv.	Bignoniaceae	African Tulip Tree	Leaves	Tr	Olive brown	134a

Contd...

S.No.	Name of plant species	Family	Popular name	Dye bearing parts	Pattern	Vital colour	References
66.	*Strychnos nux-vomica* Linn.	Loganiaceae	Snake Wood, Nux-Vomica	Seeds	Tr	Brown-black	7
67.	*Tectona grandis* Linn.f.	Verbenaceae	Sagwan, Teak	Leaves, bark, roots	Tr	Yellowish brown, reddish	2, 6, 143
68.	*Terminalia arjuna* (Roxb.) Wight &Arn.	Combretaceae	Arjuna	Wood, bark	Tr	Brown	143b
69.	*Terminalia catappa* Linn.	Combretaceae	Indian Almond-Tree, Deshi-Badam	Bark, leaves	Tr	Pinnkish red, light brownish, yellow	145, 146, 4
70.	*Thespesia populnea* Soland. ex Correa	Malvaceae	Parsipu, Portia Tree	Bark, wood, leaves, flowers	Tr	Brown, orange, yellow, green	7
71.	*Toona hexandra* (Wall. ex Roxb.) M. Roem. ssp. *hexandra* var.*gamblei* (DC.) Almedia & Almedia	Meliaceae	Toon	Flowers, seeds	Tr	Brown, yellow	2
72.	*Toona serrata* (Royle) M.Roem.	Meliaceae	Pahari-Toon	Flowers, seeds	Tr	Brown, yellow	2
73.	*Urtica dioica* Linn.	Urticaceae	Bichhu Booti	Roots	H	Brown black	2
74.	*Usnia verticillata* (Hoffm.) Schaer.	Usneaceae	Chulau	Whole plant	L	Brown, red	2
75.	*Zingiber officinale* Rosc.	Zingiberaceae	Adrak, Ginger	Rhizomes	H	Brown	151

Table 4.17 Green colour imparting dye bearing Plant species

S.No.	Name of plant species	Family	Popular name	Dye bearing parts	Pattern	Vital colour	References
1.	*Adhatoda zeylanica* Medic.	Acanthaceae	Vasaka, Arusa	Leaves	Sh	Yellow-green	2
2.	*Bacopa monnieri* (Linn.) Pannell	Scrophulariaceae	Brahmi	Whole plant	H	Green	8
3.	*Bauhinia racemosa* Lam.	Fabaceae	Kachnal	Bark	Tr	Light green	2
4.	*Brugmansia suaveolens* (Humb &Bonpl.ex.Willd.) Bercht. & J.Presl.	Solanaceae	Angel-Star	Leaves	Sh	Green	2
5.	*Eclipta alba* (Linn.) Hassk.	Asteraceae	Bhangra	Whole plant	H	Green, grey	57
6.	*Eupatorium glandulosum* Michex.	Asteraceae	Stickly Snake Root	Whole plant	Sh	Green	2, 60
7.	*Hedyotis capitellata* Wall ex G.Don	Rubiaceae	Tou hua er cao	Leaves	H or Sh	Green	7
8.	*Hypericum perforatum* Linn.	Hypericaceae	St. Jhon's Wort	Wood	H	Green	76a
9.	*Lantana camara* Linn.	Verbenaceae	Wild Sage	Leaves	Sh	Green	89
10.	*Lycopus europaeus* Linn.	Lamiaceae	Gipsywort	Fruits	H	Green	7
11.	*Rumex hastatus* D. Don	Polygonaceae	Amlora	Roots, bark	H	Yellow-green	2
12.	*Rumex nepalensis* Spreng.	Polygonaceae	Pahari Palang, Palak	Roots	H	Green, yellow	2
13.	*Sapium baccatum* Roxb.	Euphorbiaceae	Seleng, Mousedeer's Rubber-Tree	Leaves	Tr	Blackish green	20
14.	*Sapium indicum* Willd.	Euphorbiaceae	Hurna	Leaves	Tr	Greenish	96
15.	*Schleichera oleosa* (Lour.) Oken.	Sapindaceae	Kusum, The Lac Tree	Flowers	Tr	Greenish, yellow	7

Contd...

S.No.	Name of plant species	Family	Popular name	Dye bearing parts	Pattern	Vital colour	References
16.	*Spinacia oleracea* Linn.	Amaranthaceae	Palak, Garden Spinach	Leaves	H	Green	5
17.	*Tephrosia purpura* Pers. Baker	Fabaceae	Wild Indigo, Dhamasia	Leaves	H	Green	8, 143a
18.	*Thespesia populnea* Soland. and. ex Correa	Malvaceae	Parsipu, Portia Tree	Bark, wood, leaves, flowers	Tr	Orange, yellow, brown, green	7

Table 4.18 Grey colour imparting dye bearing plant species

S.No.	*Name of plant species*	*Family*	*Popular name*	*Dye bearing parts*	*Pattern*	*Vital colour*	*References*
1.	*Acacia chundra* Willd.	Mimosaceae	Red Cutch	Wood	Tr	Grey, black	5
2.	*Albizia amara* Boiv.	Fabaceae	Krishnasirisha	Leaves, pods	Tr	Grey, brown, olive	8
3.	*Alnus nitida* Endl.	Betulaceae	West Himalayan Alder, Seril	Bark	Tr	Brown-grey	4, 7
4.	*Delphinium grandiflorum* Linn.	Ranunculaceae	Sieberian Larkspur	Flowers	H	Grey, blue	56a
5.	*Ixora coccinea* Linn.	Rubiaceae	Jungle Flame, Ixora	Flowers	Sh, Tree	Grey, brown	84a
6.	*Quercus infectoria* Olivier	Fagaceae	Majuphal, Gall Oak	Galls	Tr	Grey, yellow, dark	9, 119

Table 4.19 Orange colour imparting dye bearing Plant species

S.No.	*Name of plant species*	*Family*	*Popular name*	*Dye bearing parts*	*Pattern*	*Vital colour*	*References*
1.	*Allium cepa* Linn.	Liliaceae	Piyaz	Skins	Bulbous biennial	Orange	15, 16
2.	*Calendula officinalis* Linn.	Asteraceae	Calendula	Inflorescences	H	Orange	37, 4
3.	*Convallaria majalis* Linn.	Lilaceae	May Lily	Leaves	H	Orange	1
4.	*Cotinus coggygria* Scop.	Anacardiaceae	Young Fustic	Wood, leaves	Sh	Orange, yellowish	52
5.	*Cudrania cochinchinensis* (Lour.) Kudo et Masam.	Moraceae	Pulikant	Leaves, seeds	Sh	Orange, yellow	20
6.	*Ficus benghalensis* Linn.	Moraceae	The Banynan	Leaves, bark, fruits	Tr	Orange, brown	62a
7.	*Grewia subinaequalis* DC.	Tiliaceae	Phalsa	Fruits	Tree	Orange-yellow	2
8.	*Impatiens balsamina* Linn.	Balsaminaceae	Gul Mehndi	Flowers, leaves	H	Orange, red, brown,	79, 9, 4
9.	*Lawsonia inermis* Linn.	Lythraceae	Mehndi, Henna	Leaves	Sh	Yellow-orange, red	6, 90, 91, 4, 2
10.	*Lotus corniculatus* Linn.	Fabaceae	Birdsfoot Trefoil	Flowers, leaves	H	Orange, yellow	92, 4, 96
11.	*Mallotus philippensis* Muell.Arg.	Euphorbiaceae	Kamala Tree, Sindur	Fruits	Tr	Orange	6, 95, 4
12.	*Nyctanthes arbortristis* Linn.	Oleaceae	Night Jasmine, Harsinghar	Flowers	Sh	Orange or yellow	2, 6, 105
13.	*Peristrophe bivalvis* Merr.	Acanthaceae	Bet-rang	Twigs	H	Yellow orange to deep red orange	5

14.	*Phlogacanthus thyrsiflorus* Nees	Acanthaceae	Banheka	Flowers	Sh	Orange-yellow	2
15.	*Rauvolfia serpentina* (L.) Benth.ex Kurz.	Apocynaceae	Sarpagandha, Rauvolfia	Roots	Sh	Orange pinkish	96
16.	*Rimelia reticulata* (Taylor) Hale & Fletcher	Parmeliaceae	Charata	Whole plant	Lichenous	Orange-yellow	2
17.	*Rubia cordifolia* Linn.	Rubiaceae	Indian Madder Manjit	Roots	Cl	Orange, scarlet-red,	2, 6, 126, 127, 4
18.	*Sorghum caudatum* (Hack.) Stapf.	Poaceae	Caudatum	Leaves sheaths	Stout Grass	Orange, yellow	134
19.	*Thespesia populnea* Soland. ex Correa	Malvaceae	Parsipu,Portia Tree	Bark, wood, leaves, flowers	Tr	Orange, yellow, brown, green	7

Table 4.20 Pink colour imparting dye bearing plant species

S.No.	Name of plant species	Family	Popular name	Dye bearing parts	Pattern	Vital colour	References
1.	*Indigofera heterantha* Brandis	Fabaceae	Sakina	Flowers	Sh	Pink-purple	2
2.	*Ipomoea asarifolia* R.et Schult.	Convolvulaceae	Sweet potato	Grains	H	Pink	4, 81
3.	*Prunus cerasus* Linn.	Rosaceae	Sour cherry, Gilas	Fruits	Tr	Pink	116
4.	*Pterocarpus dalbergiodes* Roxb.	Fabaceae, Papilionaceae	Padauk	Wood, bark	Tr	Pink, brown	8
5.	*Rhododendron lepidotum* Wall. ex G.Don	Ericaceae	Talisfur, Thyampa	Leaves, flowers	Sh	Pink-red	2
6.	*Salix caprea* Linn.	Salicaceae	Goat Willow, The Sallow	Bark	Tr	Pink	129, 4
7.	*Terminalia catappa* Linn.	Combretaceae	Indian Almond-Tree, Deshi-badam	Bark, leaves	Tr	Pinkish red, light brownish, yellow	145, 146, 4
8.	*Toona ciliata* M. Roemer	Meliaceae	Tun, Red Cedar	Flowers, sawdust	Tr	Pink, red	6, 4
9.	*Woodfrodia fruitcosa* Kurz.	Lythraceae	Dawi, Fire-Flame Bush	Flowers, leaves	Sh	Pink-red	2, 6, 150, 4
10.	*Ziziphus jujuba* Mill.	Rhamnaceae	Pitni-ber	Fruits	Tr	Reddish pink	9, 152
11.	*Zizyphus mauritiana* Lam.	Rhamnaceae	Indian Jujuba, Ber	Leaves	Tr	Pink-red	6

Table 4.21 Purple colour imparting dye bearing plant species

S.No.	Name of plant species	Family	Popular name	Dye bearing parts	Pattern	Vital colour	References
1.	*Abies spectabilis* (D.Don) *Spach*	Pinaceae	Indian Silver-Fir, Fir	Cones	Tr	Purple or violet	1
2.	*Alkanna tinctoria* (L.) Tausch	Boraginaceae	Dyer's Bugloss	Roots	H	Purple	4, 14
3.	*Arnebia benthamii* (Wall.ex G.Don) John.	Boraginaceae	Kashmiri kahzaban	Roots	H	Purple, red	2, 1
4.	*Arnebia nobilis* Reichb.f.	Boranginaceae	Ratanjot	Roots	H	Purple	5
5.	*Cynometra ramiflora* Linn.	Fabaceae	Cynometra	Roots	Tr	Purple	6
6.	*Desmodium multiflorum* DC.	Fabaceae	Bhatmase Phool	Flowers	Sh	Purple	20
7.	*Galium aparine* Linn.	Rubiaceae	Catch Weed	Roots	H	Purple	1, 7
8.	*Hibiscus rosa-sinensis* Linn.	Malvaceae	Jabaphool	Flowers	Sh	Purple	74, 75
9.	*Indigofera atropurpurea* Buch- Ham.	Fabaceae	Kala-Sakina	Leaves	Sh	Purple	2
10.	*Indigofera heterantha* Brandis	Fabaceae	Sakina	Flowers	Sh	Purple-pink	2
11.	*Ipomoea batatas* (Linn.) Lam.	Convolvulaceae	Sweet Potato	Grains	H	Purple	82
12.	*Perilla frutescens* (Linn.) Britton	Lamiaceae	Bhanjira	Fruits	H	Purple, black	109, 1, 4
13.	*Syzygium cuminii* (Linn.) Skeel	Myrtaceae	Jaman	Bark, fruits	Tr	Purple, black	2, 6, 20, 137

Table 4.22 Red colour imparting dye bearing plant species

S.No.	Name of plant species	Family	Popular name	Dye bearing parts	Pattern	Vital colour	References
1.	*Acacia catechu* Willd.	Mimosaceae	Khair, Cutch Tree	Heartwood	Tr	Red, brown	3, 2, 4
2.	*Acacia leucophloea* Willd.	Mimosaceae	Safed Kikar	Bark, leaves	Tr	Red	6
3.	*Adenanthera pavonina* Linn.	Fabaceae	Rakta Kambal, Red Wood	Woods	Tr	Red	5, 9
4.	*Alcea rosea* Linn.	Malvaceae	Hollyhock	Flowers	H	Red	13, 1, 9
5.	*Alnus nepalensis* D.Don	Betulaceae	Uitis, Indian Alder	Bark	Tr	Red	2, 6
6.	*Amaranthus tricolor* Linn.	Amaranthaceae	Garden Amaranth	Inflorescences, leaves	H	Red-violet	18, 4
7.	*Arnebia benthamii* (Wall.ex G.Don) John.	Boraginaceae	Kashmiri Kahzaban	Roots	H	Red, purple	2, 1
8.	*Arnebia densiflora* (Nordm.) Ledeb.	Boraginaceae	Ratanjot	Roots	H	Red	4, 22
9.	*Arnebia euchroma* (Royle) Johnston	Boraginaceae	Ratanjot	Roots	H	Red	23, 24, 4
10.	*Benthamidia capitata* (Wall.) Hara	Cornaceae	Bamora	Fruits	Tr	Red	2
11.	*Butea superba* Roxb.	Fabaceae	Palas Lata	Roots	H	Red	6
12.	*Caesalpinia pulcherrima* Sw.	Fabaceae	Peacock Flower	Flowers	Sh	Red	5
13.	*Caesalpinia sappan* Linn.	Fabaceae	Sappan Wood	Wood	Tr	Red	1, 6, 35
14.	*Capsicum annuum* Linn syn. *C. frutescens* C.B. Clarke	Solanaceae	Lalmirich, Chilli	Fruits	H	Red	38, 7, 14
15.	*Carthamus tinctorius* Linn.	Asteraceae	The Safflower, False Saffron	Flowers	H	Red	1, 6, 39, 4
16.	*Cassia tora* Linn.	Fabaceae	Panwar, Foetid Cassia	Seeds	H	Red, blue, brown	6, 41
17.	*Chrozophora tinctoria* (Linn) Rafin.	Euphorbiaceae	Dyer's Litmus	Petals, fruits, sap	H	Red, blue	9, 4

18.	*Chukrasia tabularis* A.Juss.var. *tabularis*	Meliaceae	Indian Mahogany Tree	Flower, leaves	Tr	Red, yellow	9, 7
19.	*Cladonia verticillata* (Hoffm.) Schaer	Cladoniaceae	Cup Lichen	Whole plant	Lichen	Red, yellow	2
20.	*Cordia dichotoma* Forster. f.	Boraginaceae	Lasura, Indian Cherry	Roots, leaves, fruits	Tr	Red, yellow, green	1, 5, 6, 7
21.	*Coriaria nepalensis* Wall.	Coriariaceae	Makola	Wood	Tr or Sh	Red	6, 4
22.	*Coscinium fenestratum* Clolebr.	Menispermaceae	Tree-Turmeric, Jharihuldi	Seeds, bark, wood	Sh	Red	1
23.	*Daphne papyracea* Wall.ex. Steud.emend.	Thymelaeaceae	Satpura	Bark, fruits	Sh	Red	2, 20, 55
24.	*Elaeodendron glaucum* (Rottb.) pers.	Celasteraceae	Chouri	Bark	Tr	Red	1
25.	*Embellia ribes* Burm.f.	Myrsinaceae	Barberang, False pepper	Fruits	H	Red	59
26.	*Erythrina variegata* Linn.	Fabaceae	Indian Coral (Tree)	Flowers	Tr	Red	5
27.	*Euphorbia tirucalli* Linn	Euphorbiaceae	Konpal	Wood	Sh	Red	1
28.	*Euterpe edulis* Mart.	Arecaceae	Jucara Palm	Fruits	Tr	Red	61
29.	*Euterpe oleracea* Mart.	Arecaceae	Acai Palm	Fruits	Tr	Red	61
30.	*Everniastrum cirrhatum* (Fr.) Hale ex Sipman	Parmeliaceae	Jhau, Charata	Whole plant	Lichen	Red, brown	2
31.	*Ficus religiosa* Linn.	Moraceae	Pipal	Bark, leaves	Tr	Red, brown	6, 63
32.	*Galium verum* Linn.	Rubiaceae	Cheese Rennet	Roots	H	Yellow, red	1, 4
33.	*Garuga pinnata* Roxb.	Burseraceae	Chogar	Leaves	Tr	Red	6, 65
34.	*Geranium nepalense* Sweet	Geraniaceae	Bhanda	Roots	H	Red	2
35.	*Geranium wallichianum* D. Don	Geraniaceae	Laljhari	Roots	H	Red-brown	2
36.	*Impatiens balsamina* Linn	Balsaminaceae	Gul Mehndi	Flowers, leaves	H	Red, brown Orange	79, 9, 4
37.	*Kirganelia reticulata* (Poir.) Baill.	Euphorbiaceae	Panjuli	Bark, roots	Sh	Red	1

Contd...

S.No.	Name of plant species	Family	Popular name	Dye bearing parts	Pattern	Vital colour	References
38.	*Lannea microcarpa* Engl. & K.Krause	Anacardiaceae	Microcarpa	Fruits, bark	Tr	Yellow, red-brown	88
39.	*Lithospermum arvense* Linn.	Boraginaceae	Corn Gromwell	Roots	H	Red	7
40.	*Lychnis chalcedonica* Linn.	Caryophyllaceae	Red Robin	Flowers	H	Red	7
41.	*Mahonia aquifolium* (Pursh) Nutt.	Berberidaceae	Holly Leaved Barberry	Fruits juice	Sh	Red	94
42.	*Mammea longifolia* Planch. & Triana	Clusiaceae	Nagkesar	Flowers buds	Tr	Red	5
43.	*Manilkara hexandra* (Roxb.) Dubard syn.*Mimusops hexandra* Roxb.	Sapotaceae	Khirni	Bark	Tr	Red, brown	7
44.	*Miliusa roxburghiana* Hook. f. & Thoms.	Annonaceae	Chhag-loti	Bark, ripe fruits	Tree	Red	20
45.	*Mirabilis jalapa* Linn.	Nyctaginaceae	Gulabans	Flowers	H	Red-pink	2
46.	*Morinda citrifolia* Linn.	Rubiaceae	Al, Ach	Roots bark	Tr	Red, yellow	6, 102
47.	*Myrtillocactus geometrizans* (Mart.) Console	Cactaceae	Blue Candle	Fruits	Shrubby Cactus	Red	104, 4
48.	*Nardostachys jatamansi* DC.	Valerianaceae	Jatamansi, Spikenard	Inflorescences	H	Red	2, 4
49.	*Oldenlandia umbellata* Linn. syn. *Hedyotis umbellata* (Linn.) Lam.	Rubiaceae	Indian Madder, Chay-Root	Roots	Woody Annual	Red	7, 106, 107
50.	*Papaver rhoeas* Linn.	Papaveraceae	Corn poppy, Lalpost	Petals	H	Red	1
51.	*Peganum harmala* Linn.	Zygophyllaceae	Harmal	Seeds	H	Red, brown to black, yellow,	7
52.	*Peristrophe bivalvis* Merr.	Acanthaceae	Bet-rang	Twigs	H	Yellow orange to deep red orange	5

53.	*Rhododendron arboreum* Sm.	Ericaceae	Burans, Rose-Tree	Flowers	Tr	Red	2, 125
54.	*Rhododendron lepidotum* Wall. ex G.Don	Ericaceae	Talisfur, Thyampa	Leaves, flowers	Sh	Pink-red	2
55.	*Rubia tinctorum* Linn.	Rubiaceae	European Madder Alizari	Roots	H	Red	128, 4
56.	*Rubus fruticosus* Linn. var. *discolor*.	Rosaceae	Alish,Akhi	Fruits	Sh	Red, brown	9, 7
57.	*Rumex dentatus* Linn.	Polygonaceae	Ambavah, Jungli palak	Roots	Deep rooted Annual	Red	1
58.	*Rumex acetosa* Linn.	Polygonaceae	Garden Soorel	Roots	H	Red	5
59.	*Sanguinaria canadensis* Linn.	Papaveraceae	Blood Root	Roots	H	Red	8
60.	*Taxus baccata* Linn.	Taxaceae	Thuner, Common Yew	Bark	Tr	Red	2
61.	*Toona ciliata* M. Roemer	Meliaceae	Tun, Red Cedar	Flowers, sawdust	Tr	Red, pink	6, 4
62.	*Usnia verticillata* (Hoffm.) Schaer.	Usneaceae	Chulau	Whole plant	L	Red, brown	2
63.	*Ventilago denticulata* Willd.	Rhamnaceae	Pitti	Bark, roots	Sh	Red	6, 149
64.	*Ventilago madraspatna* Gaertn.	Rhamnaceae	Raktavalli	Bark, roots	Cl	Red	149
65.	*Woodfrodia fruitcosa* Kurz.	Lythraceae	Dawi, Fire-Flame Bush	Flowers, leaves	Sh	Pink-red	2, 6, 150, 4

Table 4.23 Yellow colour imparting dye bearing plant species

S.No.	Name of plant species	Family	Popular name	Dye bearing parts	Pattern	Vital colour	References
1.	*Acalypha indica* Linn.	Euphorbiaceae	Indian Acalypha	Leaves	H	Yellow	8
2.	*Adhatoda zeylanica* Medic.	Acanthaceae	Vasaka, Arusa	Leaves	Sh	Yellow-green	2
3.	*Adina cordifolia* (Roxb.) Hook.f.ex Brandis	Rubiaceae	Haldu	Wood	Tr	Yellow	7
4.	*Aegle marmelos* (Linn.) Correa ex Roxb.	Rutaceae	Bel	Rinds of unripe fruits	Tr	Yellow	2, 6, 10, 4
5.	*Agrimonia eupatoria* Linn.	Rosaceae	Agrimony	Leaves, roots	H	Yellow	2
6.	*Ardisia solanacea* Roxb. syn. *Ardisia humilis* C.B. Clarke	Myrsinaceae	Bisi	Fruits	Sh	Yellow	5
7.	*Artocarpus heterophyllus* Lam .	Moraceae	Kathal, Jackfruit	Wood, roots, fruits	Tr	Yellow	2, 6, 26, 4
8.	*Artocarpus lakoocha* Roxb.	Moraceae	Monkey Jack	Fruits, wood, roots	Tr	Yellow	2
9.	*Averrhoa carambola* Linn.	Averrhoaceae	Kamrak, Star Fruit	Whole plant	Tr	Yellow-brown	2
10.	*Baptisia australis* (L.) R.Br.	Fabaceae	Blue Wild Indigo	Flowers	H	Yellowish	28, 4
11.	*Barleria prionitis* Linn.	Acanthaceae	Kala Bansa	Flowers	Sh	Yellow	7, 1
12.	*Bauhinia tomentosa* Linn.	Fabaceae	Bell Bauhina	Leaves	Sh	Yellow	1, 5
13.	*Berberis aristata* DC.	Berberidaceae	Kilmor	Bark, roots	Sh	Yellow	2
14.	*Berberis asiatica* Roxb. ex. DC.	Berberidaceae	Kilmor	Bark, roots	Sh	Yellow	2
15.	*Berberis chitria* Lindl.	Berberidaceae	Chetar	Bark, roots	Sh	Yellow	2
16.	*Berberis nepalensis* (DC.) Spreng.	Berberidaceae	Nepal Barberry	Roots	Sh	Yellow	6
17.	*Berberis vulgaris* Linn.	Berberidaceae	Barberry	Roots	Sh	Yellowish, yellow	2, 29, 4
18.	*Bidens bipinnata* Linn.	Asteraceae	Lumb	Leaves	H	Yellow	31, 32, 4

19.	*Bougainvillea glabra* Choisy	Nyctaginaceae	Paper Flowers	Flowers with ivory white bracts	Vine Sh	Yellow	9
20.	*Brassica nigra* (Linn.) Kuch	Brassicaceae	Black Mustard	Seeds	H	Yellow	8
21.	*Butea monosperma* (Lam.) Taub.	Fabaceae	Palas, Flame of the Forest	Flowers	Tr	Yellow-orange	2, 6, 34, 4
22.	*Caesulia axillaris* Roxb.	Asteraceae	Bangra	Whole plant	H	Yellow	4, 36
23.	*Careya arborea* Roxb.	Barringtoniaceae	Kumbi	Bark	Tr	Yellow	2, 6
24.	*Carthamus oxyancantha* Bieb.	Asteraceae	Wild Safflowers	Flowers	H	Yellow	5
25.	*Cassia auriculata* Linn.	Fabaceae	Tanner's Cassia	Flowers	Sh	Yellow	1, 6, 40, 4
26.	*Cassia obtusifolia* Linn.	Fabaceae	Banarh, Cakonda	Pods, seeds	H	Yellow	5
27.	*Cassia occidentalis* Linn.	Fabaceae	Kasonda, Coffee-Senna	Pods, seeds	H	Yellow, brown	5
28.	*Celtis australis* Linn.	Ulmaceae	European Nettle Tree	Bark	Tr	Yellow	2
29.	*Choiranthus cheiri* Linn.	Brassicaceae	Wall Flowers	Leaves, petals	H	Yellow	4, 43
30.	*Chlorophora tinctoria* (L.) Benth. & Hook.f.	Moraceae	Old Fustic	Wood	Tr	Yellow	44, 4
31.	*Chloroxylon swietenia* DC.	Flindersiaceae	Indian SatinWood Tree	Bark	Tree	Yellow	6, 45, 4
32.	*Chukrasia tabularis* A.Juss. var.*tabularis*	Meliaceae	Indian Mahogany Tree	Flowers, leaves	Tr	Red, yellow	9, 7
33.	*Citrus sinensis* (Linn.) Osbeck	Rutaceae	Malta	Peals	Tr	Yellow	47
34.	*Cladonia verticillata* (Hoffm.) Schaer	Cladoniaceae	Cholara	Whole plant	H	Yellow, red	2
35.	*Crocus sativus* Linn.	Iridaceae	Kesar, Saffron	Flowers	H	Yellow	6, 49, 50, 4
36.	*Cordia dichotoma* Forster. f.	Boraginaceae	Lasura, Indian Cherry	Roots, leaves, fruits	Tr	Yellow, red, green	1, 5, 6, 7

Contd...

S.No.	Name of plant species	Family	Popular name	Dye bearing parts	Pattern	Vital colour	References
37.	*Coreopsis tinctoria* Nutt.	Asteraceae	Calliopsis	Flowers	H	Yellowish	4, 51
38.	*Corylus avellana* Linn.	Betulaceae	Hazel	Flowers	Tr	Yellow	8
39.	*Cotinus coggygria* Scop.	Anacardiaceae	Young Fustic	Wood, leaves	Sh	Yellowish, orange	52
40.	*Cudrania cochinchinensis* (Lour.) Kudo et Masam.	Moraceae	Pulikant	Leaves, seeds	Sh	Yellow, orange	20
41.	*Cuminum cyminum* Linn.	Apiaceae	Cumin Seeds	Seeds	H	Bright yellow	52a
42.	*Curcuma amanda* Roxb.	Zingiberaceae	Amada, Am Haldi	Rhizomes	H	Yellow	5
43.	*Curcuma aromatica* Salisb.	Zingiberaceae	Jangli Haldi, Wild Turmeric	Rhizomes	H	Yellow	2
44.	*Curcuma longa* Linn. syn *C. domestica* Val.	Zingiberaceae	Haldi, Turmeric	Rhizomes	H	Yellow	2, 6, 53, 54, 4
45.	*Curcuma zedoaria* (Christm.) Rosc.	Zingiberaceae	Zedoary, Kachura	Rhizomes	H	Yellow	9, 7
46.	*Cyamopsis tetragonoloba* (Linn.) Tanb.	Fabaceae	Guar	Pods and seeds	H	Yellow	5
47.	*Datisca cannabina* Linn.	Datiscaceae	Akalbir, Bujr-Bhang	Roots	H	Yellowish-red	2, 6
48.	*Delphinium zalil* Aitch &Hemsl.	Ranunculaceae	Guljali, Larkspur	Flowers	H	Yellow	4, 5, 6
49.	*Dipterocarpus turbinatus* Gaertn.f.	Dipterocarpaceae	Common Gurjan Tree	Twig, bark	Tr	Yellow, brown	1, 7
50.	*Eucalyptus hybrid* (mainly *E.umbellata*Domin syn.*E.tereticornis* Sm.)	Myrtaceae	Safeda	Bark, leaves	Tr	Yellow, brown, black	41
51.	*Euonymus tigens* Wall.	Celastraceae	Dog Wood, Prick Wood	Bark	Tr	Yellow	2
52.	*Euphorbia hirta* Linn.	Euphorbiaceae	Dudhi	Whole plant	Tr	Yellow	8, 60a
53.	*Fagopyrum esculentum* Moench	Polygonaceae	Kotu,Common Buckwheat	Grains	H	Yellow	62
54.	*Ficus altissima* Blume	Moraceae	Gadgubar	Stem and roots	Tr	Yellow	20

55.	*Flemingia macrophylla* (Willd.)Merr.	Fabaceae	Varas	Leaves and bark	Sh	Black, yellow	20, 6, 1
56.	*Flemingia semialata* Roxb.	Fabaceae	Winged-Stalk Flemingia, Ban Chola	Pods	Sh	Yellow	7
57.	*Galium rotundifolium* Linn.	Rubiaceae	Caglio a foglie rotonde	Roots	Herb	Yellow, brown	1
58.	*Galium verum* Linn.	Rubiaceae	Cheese Rennet	Roots	H	Yellow, red	1, 4
59.	*Garcinia morella* Desr.	Clusiaceae	Indian Gamboge,Tamals	Bark (exudate)	Tr	Yellow	5
60.	*Garcinia xanthochymus* Hook. f.	Clusiaceae	Dephal, Vika	Unripe fruits	Tr	Yellow	6, 64, 4
61.	*Gardenia jasminoides* Ellis=*G.augusta* (Linn.) Merrill	Rubiaceae	Gandhraj	Fruits	Tr/Sh	Yellow	64a, 4
62.	*Genista tinctoria* Linn	Fabaceae	Dyer's Brown	Flowers	Sh	Yellow	67, 4
63.	*Gossypium herbaceum* Linn.	Malvaceae	Uppam Cotton	Flowers	Sh	Yellow	6, 70, 4
64.	*Grevillea robusta* A. Cunn.	Proteaceae	Silver-Oak	Flowers	Tree	Yellow	2
65.	*Grewia subinaequalis* DC.	Tiliaceae	Phalsa	Fruits	Tree	Yellow-orange	2
66.	*Hedera helix* C.B. Clarke, non Linn.=*H. nepalensis* Koch	Araliaceae	Common Ivy, Lablab	Bark	Cl	Yellow	73
67.	*Hedychium spicatum* Buch.-Ham.	Zingiberaceae	Bagaldu	Rhizomes	H	Yellow	2
68.	*Helianthus annuus* Linn.	Asteraceae	Sunflowers	Flowers	H	Yellow	73a
69.	*Hibiscus sabdariffa* Linn.	Malvaceae	Lalambari	Flowers	Sh	Yellow	76, 4
70.	*Hippophae salicifolia* D.Don	Elaeagnaceae	Chuk	Fruits	Sh	Yellow	2
71.	*Hypericum oblongifolium* Choisy	Hypericaceae	Basant	Flowers	H	Yellow	2

Contd...

S.No.	*Name of plant species*	*Family*	*Popular name*	*Dye bearing parts*	*Pattern*	*Vital colour*	*References*
72.	*Hyptis suaveolens* Poit.	Lamiaceae	Mint Weed,Wilayati	Whole plant	H	Yellow	4, 77
73.	*Illicium griffithii* Hook f.&Thomson	Illiciaceae	Lissi	Seeds, bark, leaves	Shrub, Tree	Yellow (traditional dye)	20, 78
74.	*Jasminum humile* Linn.	Oleaceae	Peeli Chameli	Roots	Sh	Yellow	6
75.	*Lannea coromandelica* (Houtt.) Merrill	Anacardiaceae	Jhingan	Bark, resin	Tr	Yellow, brown	2
76.	*Lannea microcarpa* Engl. & K.Krause	Anacardiaceae	Microcarpa	Fruits, bark	Tr	Yellow, red-brown	88
77.	*Lawsonia inermis* Linn.	Lythraceae	Mehndi, Henna	Leaves	Sh	Yellow-orange, red	6, 90, 91, 4, 2
78.	*Leucas aspera* spreng.	Lamiaceae	Chota Halkusa	Whole plant	H	Yellow	8, 91a
79.	*Ligustrum vulgare* Linn.	Oleaceae	European Privet	Fruits	Sh	Yellow, blue	1, 9
80.	*Lonicera periclymenum* Linn.	Caprifoliaceae	Woodbine, Honey Suckle	Flowers	Sh	Yellow	7
81.	*Lotus corniculatus* Linn.	Fabaceae	Birdsfoot Trefoil	Flowers, leaves	H	Yellow, orange	92, 4, 96
82.	*Maclura pomifera* Schneid.	Moraceae	Osage-orange	Wood, bark, roots	Tr	Yellow	93, 4
83.	*Madhuca longifolia* (Koenig) Macb.	Sapotaceae	South Indian Mahua	Bark	Tr	Yellow-brown	2, 6
84.	*Mahonia borealis* Takeda	Berberidaceae	Pualeng	Bark, roots, sap	Sh	Yellow	2
85.	*Malphigia glabra* Linn.	Malpighiaceae	Barbados Cherry	Flowers, fruits, bark	Sh	Yellow, blue, coffee tint	1, 5
86.	*Mangifera indica* Linn.	Anacardiaceae	Aam, The Mango	Bark	Tr	Yellow	6, 96, 97, 4
87.	*Memecylon umbellatum* Burm.f.	Melastomataceae	Iron Wood Tree	Leaves	Sh, Tree	Yellow	6, 7, 99
88.	*Mentha piperita* Linn. emend-Huds.	Lamiaceae	Peppermint	Whole plant	H	Yellow, green	8, 99a

89.	*Mesua ferrea* Linn.	Clusiaceae	Nagkesar, Mesua	Wood, flowers, buds	Tr	Yellow	7
90.	*Michelia champaca* Linn.	Magnoliaceae	Champak	Flowers	Tr	Yellow	7
91.	*Michelia montana* Blume	Magnoliaceae	Pan Sopa	Bark, flowers	Tr	Yellow	20
92.	*Morinda angustifolia* Roxb.	Rubiaceae	Asho	Roots	Sh, Tree	Yellow	7, 101, 4
93.	*Morinda citrifolia* Linn.	Rubiaceae	Al, Ach	Roots bark	Tr	Yellow, red	6, 102, 4
94.	*Morinda umbellata* Linn. Hook.f.	Rubiaceae	Vomit Vine	Roots, bark	Sh	Yellow	6
95.	*Myrica esculenta* Buch.-Ham.	Myricaceae	Kaiphal, Box Mrytle	Bark	Tr	Yellow	2, 6, 7, 4
96.	*Nelumbo nucifera* Gaertn.	Nymphaeaceae	Sacred Lotus, Kamal	Leaves	Aquatic H	Sandal yellow	8
97.	*Peganum harmala* Linn.	Zygophyllaceae	Harmal	Seeds	H	Yellow, brown to black, red	7
98.	*Peltophorum pterocarpum* Backer ex K. Heyne	Caesalpiniaceae	Copper pod	Wood, leaves, bark	Tr	Yellowish, brown, black	1
99.	*Pfaffia iresinoides* (Kunth.) Spreng.	Amaranthaceae	Brazil Gingseng	Roots	H	Yellow	110
100.	*Phaseolus mungo* Linn.	Fabaceae	Urd	Seeds covers	H	Yellow, brown,	8
101.	*Phlogacanthus thyrsiflorus* Nees	Acanthaceae	Banheka	Flowers	Sh	Orange-yellow	2
102.	*Phytolacca dioica* Linn.	Phytolaccaceae	Ombu	Leaves	Sh	Yellow	7
103.	*Pistacia integerrima* Stewart. ex. Brandis	Anacardiaceae	Kakara	Galls of leaves and petioles	Tr	Yellow	1, 2, 112
104.	*Pithecellobium dulce* Benth.	Fabaceae	Jangle Jalebi, Quamachil	Bark	Tr	Yellow	5
105.	*Plecospermum spinosum* Trec.	Moraceae	Gumbenfong	Bark, wood	Sh or Tr	Yellow	113

Contd...

S.No.	Name of plant species	Family	Popular name	Dye bearing parts	Pattern	Vital colour	References
106.	*Polygonum hydropiper* Linn.	Polygonaceae	Water Peeper	Whole plant	H	Yellow, blue, black	20
107.	*Polygonum tortuosum* D.Don.	Polygonaceae	Niala	Flowers, bark	Sh	Yellow	114
108.	*Populus deltoides* Marsh.	Salicaceae	Poplar	Bark	Tr.	Yellow, brown	115
109.	*Prunus cerasoides* D.Don	Rosaceae	Himalayan Wild Cherry, Paddam	Fruits	Tr	Yellowish	2
110.	*Quercus infectoria* Olivier	Fagaceae	Majuphal, Gall Oak	Galls	Tr	Yellow, dark grey	9, 119
111.	*Quercus robur* Linn.	Fagaceae	English Oak	Bark	Tr	Yellow	8, 119a
112.	*Reseda luteola* Linn.	Resedaceae	Weld, Dyer's Rocket	Whole plant	H	Deep yellow	120, 121, 4
113.	*Rheum emodi* Wall.ex Meissn.	Polygonaceae	Dolu, Himalayan Rhubarb	Rhizomes	H	Reddish yellow	122, 123
114.	*Rheum moorcroftianum* Royle	Polygonaceae	Luan guo da huang	Roots	H	Yellow	2
115.	*Rheum webbianum* Royle	Polygonaceae	Archa	Roots	H	Yellow	2
116.	*Rhus coriaria* Linn.	Anacardiaceae	Sicilian Sumac	Leaves, twigs	Sh	Yellow	7
117.	*Rhus cotinus* Linn.	Anacardiaceae	Royal Purple	Leaves, twigs	Sh	Yellow	7
118.	*Rhus paniculata* Wall. ex. Hook.f.	Anacardiaceae	Benwe, Khaung-bin	Bark	Tr	Yellow	96
119.	*Rimelia reticulata* (Taylor) Hale & Fletcher	Parmeliaceae	Charata	Whole plant	Lichenous	Orange-yellow	2
120.	*Rumex hastatus* D. Don	Polygonaceae	Amlora	Roots, bark	H	Yellow-green	2
121.	*Rumex nepalensis* Spreng.	Polygonaceae	Pahari Palang, Palak	Roots	H	Yellow, green	2
122.	*Saraca asoca* (Roxb.) De Wilde	Fabaceae	Ashoka	Leaves and bark	Tr	Yellow, brown	129a
123.	*Sarothamnus scoparius* (Linn.) Wimm. ex Koch.	Fabaceae	Broom	Flowers	Sh	Yellow	130, 4

124.	*Schleichera oleosa* (Lour.) Oken.	Sapindaceae	Kusum, The Lac Tree	Flowers	Tr	Greenish, yellow	7
125.	*Serratula tinctoria* Linn.	Asteraceae	Saw Wort	Whole plant juice except roots	H	Yellow	120, 4
126.	*Sesbania bispinosa* W.F. White	Fabaceae	Dhaincha	Pods, seeds	H	Yellow	5
127.	*Solanum indicum* Linn.	Solanaceae	Barhanta, Poison-Berry	Half ripe fruits	Sh	Yellow, black	20
128.	*Solidago canadensis* Linn.	Asteraceae	Goldenrod	Whole plant	Sh	Golden yellow	19
129.	*Sophora japonica* Linn.	Fabaceae	Chinese Scholar Tree	Flowers	Tr	Yellow	5
130.	*Sophora mollis* Grah. ex Baker	Fabaceae	The Himalayan Laburnum	Roots, flowers	Sh	Yellow, brown	2
131.	*Symplocos paniculata* Miq .	Symplocaceae	Sweet Leaves, Ludh	Bark, leaves	Sh	Yellow	2, 20, 135
132.	*Symplocos ramosissima* Wall.ex G.Don	Symplocaceae	Lodh	Leaves, bark	Sh, Tr	Yellow	2
133.	*Symplocos laurina* Wall.ex Rehd.&E.H.Wils	Symplocaceae	Bholia, Chungan	Leaves	Tr	Yellow	5
134.	*Symplocos sumunita* Buch-Ham.ex D.Don	Symplocaceae	Lodh, Pathani Lodh	Leaves, bark	Tr	Yellow	5
135.	*Tagetes erecta* Linn.	Asteraceae	Genda, Azetac or African Marigold	Flowers	H	Yellow	2, 138, 139, 140, 4
136.	*Tagetes minuta* Linn.	Asteraceae	Jangli Genda, Stinking-Roger	Flowers, whole plant	H	Yellow	141
137.	*Tagetes patula* Linn.	Asteraceae	French Marigold, Genda	Flowers	H	Yellow	142, 4
138.	*Tamarindus indica* Linn.	Caesalpiniaceae	Imli, Amli	Leaves	Tr	Reddish-yellow	2

Contd...

S.No.	Name of plant species	Family	Popular name	Dye bearing parts	Pattern	Vital colour	References
139.	*Tectona grandis* Linn.f.	Verbenaceae	Sagwan, Teak	Leaves, bark, roots	Tr	Yellowish, brown, reddish,	2, 6, 136
140.	*Terminalia catappa* Linn.	Combretaceae	Indian Almond-Tree, Deshi-badam	Bark, leaves	Tr	yellow, pinkish red, light brownish	145, 146, 4
141.	*Terminalia chebula* Retz.	Combretaceae	Harra, Myrobalan	Fruits, barks	Tr	Yellow, brown	2, 6, 147, 148, 4
142.	*Thespesia populnea* Soland. and. ex Correa	Malvaceae	Parsipu, Portia Tree	Bark, wood, leaves, flowers	Tr	Yellow, green, orange, brown	7
143.	*Toddalia asiatica* (L.) Lam.	Rutaceae	Forest Peeper, Jangli-Kalimirch	Roots bark	Sh	Yellow	1
144.	*Toona hexandra* (Wall. ex Roxb.) M. Roem. ssp. *hexandra* var.*gamblei* (DC.) Almedia & Almedia	Meliaceae	Toon	Flowers, seeds	Tr	Yellow, brown	2
145.	*Toona serrata* (Royle) M.Roem.	Meliaceae	Pahari-Toon	Flowers, seeds	Tr	Yellow, brown	2
146.	*Vitex altissima* Linn.f.	Verbenaceae	Milla	Wood, bark	Tr	Yellow	7
147.	*Wrightia tomentosa* Roem. & Schult.	Apocynaceae	Dudhi	Yellow milky juice	Tr	Yellow	7, 4

References

1. Siva, R. (2007). Status of natural dyes and dye-yielding plants in India. *Current Science*, 92(7), 916-925.
2. Gaur, R.D. (2008).Traditional dye yielding plants of Uttarakhand. *Natural Product Radiance*, 7(2), 154-165.
3. Khan, M.I., Khan, S.A., Yusuf, M., Shahid, M., Mohammad, F., Khan, M.A. (2010). Eco-friendly shades on wool using mixed mordants with *Acacia catechu. Colourage*, 57(8), 81-86.
4. Kumar, J.K., Sinha, A.K. (2004). Resurgence of natural colourants: A holistic view. *Natural Product Letters*, 18(1), 59-84.
5. Farooqui, M.I.H. (1993). Indian vegetables dyes prospects and possibilities, In Proceedings of the Inter-Regional Workshop on Natural dyes, National Handloom Development Corporation in Collaboration with UNDP, November 1-4, New Delhi, pp. 1-8.
6. Tewari, D.N. (1994). *Tropical Forest Produce.* International Book Distributors, Rajpur Road, Dehra Dun (India), pp. 317-358.
7. Saxena, B.N. (1989). In *Application Methodology for Dyeing/Printing with Natural Dyes.* Proceedings of the National Seminar on Natural Dyes, Jaipur, October 20-21. A paper entitled "Dye-yielding Flowering Plants of India: A Partial Catalogue and Suggestion for a Taxonomic Review," pp. 12-17.
8. Bharathan., Daniel, M. (2006).In Daniel, M., Bhattacharya, S.D., Arya A., Raole, V. M. (Editors). *Natural Dyes Scope and Challenges.* Scientific Publishers, Jodhpur (India), pp. 245-249, 1-34.
9. Gokhale, S.B., Tatiya A.U., Bakliwal, S.R., Fursule, R.A. (2004). Natural dye yielding plants in India. *Natural Product Radiance*, 3(4), 228-234.
10. Hussain, A., Jahan, S. (2011). Extraction of natural dye from medicinal plant and dyeing of silk. *China Textile Science*, (2), 30-34.
11. Dayal, R., Dobhal, P.C., Kumar, R., Onial, P., Rawat, R.D. (2007). Natural dye from *Ageratum conyzoides. Colourage*, Vol. LIV, No. 12, 42-45.
12. Vinod, K.N., Puttaswamy, Gowda, K.N.N., Sudhakar, R. (2010). Kinetic and adsorption studies on Indian siris (*Albizia lebbeck)* natural dye on silk. *Indian Journal of Fibreand Textile Research*, 35(2), 159-163.
13. Guinot, P., Roge, A., Gargadennec, A., Garcia, M., Dupont, D., Lecoeur, E., Candelier, L., Andary, C. (2006). Dyeing plants screening: an approach to combine past heritage and present development. *Coloration Technology,* 122(2), 93-101
14. Tiwari, V., Vankar, P.S. (2001). Unconventional natural dyeing using microwave and sonicator with alkanet root bark. *Colourage,* 48(5), 25-26,28.
15. Vankar, P.S., Shankar, R., Wijayapala, S. (2009).Dyeing of cotton, wool and silk with extract of *Allium cepa. Pigment and Resin Technology*, 38(4), 242-247.

16. Gumrukcu, G., Ozgur, M.U., Gultekin, C. (2008). Extraction of anthocyanin pigments from red onion *(Allium cepa)* and dyeing woolen fabrics. *Asian Journal of Chemistry*, 20(4), 2891-2902.
17. Kikuzaki, H., Tesaki, S. (2002). New flavonol-phenylbutadiene adducts from the Leaves of *Alpinia flabellate. Journal of Natural Products*, 65(3), 389-391.
18. Biswas, M., Das, S. S., Dey, S. (2013). Establishment of a stable *Amaranthus tricolor* callus line for production of food colourant. *Food Science and Biotechnology*, 22 (Suppl. 1), 1-8.

19 Padma, V.S. (2000). Chemistry of natural dyes. *Resonance*, 5(10), 73-80.

19a. Bechtold, T., Kaulfuss, P., Mahmut, A., Geissler, S., Ganglberger, E. (2004). Natural dyes in Central Europe-raw material sources and dyeing quality. *Melliand Textilberichte*, 85(4), 268-272.

20. Mahanta, D., Tiwari, S.C. (2005). Natural dye-yielding plants and indigenous knowledge on dye preparation in Arunachal Pradesh, Northeast India. *Current Science*, 88(9), 1474-1480.
21. Sudhakar, R., Gowda, K.N.N., Padaki, N.V. (2006). Natural dyeing silk with the nut extract of *Areaca catechu. Colourage*, 53(7), 61-66.
22. Orhan, I., Senol, F. S., Koca, U., Ercetin, T., Toker, G.(2011). Evaluation of the antioxidant and acetylcholinesterase inhibitory activities of *Arnebia densiflora* L. *Turkish Journal of Biology*, 35(1), 111-115.
23. Haosheng, L.(2012). Process for preparation of plant dye from *Arnebia euchroma. Fanning Zhuanli*. Shenqing, CN 1025 16801 A20120627.
24. Papageorgiou, V.P., Assimopoulou, A.N., Couladouros, E.A., Hepworth, D., Nicolaou, K.C. (1999). The chemistry and biology of alkannin, shikonin, and related napthazarin natural products. *Angewandte Chemie, International Edition*, 38 (3), 271-300.
25. Perevertkina, I. V., Titova, N. N. (2014). Method for production of anthocyanin colouring agent from refuses of berries of dark varieties. *Russ*. RU 2515478 C1 2014051
26. Vankar, P.S., Shankar, R., Wijayapala, S. (2011). Dyeing of cotton, silk and wool with bark extract of *Artocarpus heterophyllus. Research Journal of Textile and Apparel*, 15(2), 52-60.
27. Mathur, J.P., Mehta, A., Karnawat, R., Bhandari, C.S. (2003). Use of neem bark as wool colourant Optimum conditions of wool dyeing. *Indian Journal of Fibre and Textile Research*, 28(1), 94-99.
28. Liberton, P., Markham, K.R., Swift, W.T., Oung, B., Mabry, T.J. (1967). Flavonoids of *Baptisia australis. Phytochemistry*, 6(12), 1675.
29. Paul, S., Grover, E., Sharma, A. (2003). Application of natural dye *(Berberis vulgaris)* on wool and its colour fastness. *Man-Made Textiles in India*, 46(8), 311-316.
30. Vygodin, V.A., Babodei, A.P., Ivanov, V.V., Frolov, V.L., Chizhik Yu, L. (1995). Colouring agent from beet root. *Molochnaya Promyshlennost*, 1, 16.

31. Cao, Y., Qu, H., Yao, Y., Liu, Zhi-hui., Fang, Zhu-yuan (2013). Chemical constituents from *Bidens bipinnata. Zhongcaoyao,* 44(24), 3435-3439.
32. Li, S., Kuang, Hai-Xue., Okada, Y., Okuyama, T. (2005). New flavanone and chalcone glucosides from *Bidens bipinnata L. Journal of Asian Natural Products Research,* 7(1), 67-70.
33. Sinha, K., Chowdhury, S., Saha, P.D., Datta, S. (2013). Modeling of microwave-assisted extraction of natural dye from seeds of *Bixa orellana* using response surface methodology (RSM) and artificial neural network (ANN). *Industrial Crops and Products,* 41, 165-171.
34. Sinha, K., Saha, P.D., Datta, S. (2012). Extraction of natural dye from petals of flame of forest (*Butea monosperma)* flower: Process optimization using response surface methodology (RSM). *Dyes and Pigments,* 94(2), 212-216.
35. Gupta, D. In Gulrajani, M. L. and Gupta, D. (editors.). *Natural Dyes and Their Application to Textiles.* IIT, New Delhi, India.
36. Geda, A. (2000). A review: chemistry of essential oils *Caesulia axilaris. Acta Ciencia Indica,* 26(4), 127-131.
37. Mohamed, R. M. (2009). Dyeing of nylon 6 fibers with *Calendula officinalis* & *Casuarina cunninghamiana* extracts. *Colourage,* 56(1), 49-56.
38. Kulkarni, S.S., Bodake, U.M., Pathade, G.R. (2011). Extraction of natural dye from Chili (*Capsicum annum*) for textile colouration. *Universal Journal of Environmental Research and Technology,* 1, 58-63.
39. Basalma, D., Kayabasi, N., Gurbuz, B. (2008). Determination of dyeing properties in petals of different safflower *(Carthamus tinctorius)* varieties. *Asian Journal of Chemistry,* 20(8), 6358-6364.
40. Reddy, K. R. S., Srimannarayana, G., Subba Rao, N. V. (1972). Proanthocyanidin dimer from *Cassia auriculata* flowers. *Indian Journal of Chemistry,* 10(9), 956-7
41. Dayal, R. and Dobhal, P.C. (2001). *Natural dye from Indian plants. Colourage,* Vol. XLVIII (8), 33-38.
42. Anon. (1981).The Wealth of India: *A Dictionary of Indian Raw Materials and Industrial Products. Raw Materials,* Vol. II, pp. 101–103, 124-125. Council of Scientific and Industrial Research, Publications and Information Directorate, New Delhi, India.
43. Tatsuzawa, F., Saito, N., Shinoda, K., Shigihara, A., Honda, T. (2006). Acylated cyanidin 3-sambubioside-5-glucosides in three garden plants of the Cruciferae. *Phytochemistry,* 67(12), 1287-1295.
44. Gasman, T.A. (1962). *The Chemistry of Flavonoid Compounds,* p. 423. Pergmon Press, Oxford.
45. Prabakaran, R., Arivoli, S., Hema, A., Kamatchi, C. (2011) Isolation and characterization of flavonoids from *Chloroxylon swietenia. Journal of Chemical and Pharmaceutical Research,* 3(3), 805-813.
46. Zhang, Yu-Ping, Yuan, De-hong, Shen, Mei-Yun. (2012).Microwave extraction of natural dye from camphor leaves and its fabric. *Maofang Keji,* 40(4), 48-52.

47. Padron Pereira, C.A., Moreno Alvarez, M.J. (2000). Extraction of colouring with Orange peel (*Citrus sinensi* var. *valencia*) from conventional methods and utilization colour fortified beverages. *Revista unellez de Cienciay Technologia Production Agricola*, 17(1), 125-140.

48. Sinha, K., Das, P.S., Datta, S. (2012). Natural blue dye from *Clitoria ternatea*: extraction and analysis methods. *Research Journal of Textile and Apparel* 16(2), 34-38.

49. Alonso, G.L., Salinas, M.R., Saez, J.R. (1998). Crocin as coloring in the food industry. *Recent Research Development in Agricultural and Food Chemistry*, 2(1), 141-154.

50. Raja, A.S.M., Pareek, P.K., Shakyawar, D.B., Wani, S.A., Nehvi, F.A., Sofi, A.H. (2012). Extraction of natural dye from Saffron flower waste and its application on Pashmina fabric. *Advances in Applied Science Research*, 3(1), 156-161.

51. Rose, N. M., Jeet, S. S., Khanna, S., Gaba, G. (2004). Optimization of dyeing variables using double *coreopsis* flowers. *Man-Made Textiles in India*, 47(8), 300-303.

52. Mayer, F. (1943). *The Chemistry of Natural Colouring Matters*, Translated and revised by A.H. Cook, p. 184. Reinhold Publishing Corporation, New York.

52a. Tayade, P.B., Adivarekar, R.V. (2013). Dyeing of silk fabric with *Cuminum Cyminum* as a source of natural dye. *International Journal of Chem Tech Research*, 5(2), 699-706.

53. Verghese, J. (1993).Isolation of curcumin from *Curcuma longa* rhizome. *Flavour and Fragrance Journal*, 8(6), 315-319.

54. Kamel, M.M., El-Hossamy, M., Helmy, H.M., El Hawary, N.S. (2008). Some studies on dyeing properties of cotton fabrics with *Curcuma longa* (Turmeric) roots using ultrasonic method (Part II). *Man-Made Textiles in India*, 51(8), 267-273.

55. Vankar, P.S., Shankar, R., Dixit, S., Mohanta, D., Tewari, S.C.(2009).Chemical characterization of extract derived from *Daphne papyracea* and sonicator dyeing of cotton, silk and wool with the extract. *Pigment and Resin Technology*, 38(3), 181-187.

55a. Yang, Bing-you., Li, T., Guo, R., Wang, Qiu-hong., Kuang, Hai-xue. (2013). Chemical constituents from leaves of *Datura metel* (I). *Zhongcaoyao*, 44(20), 2803-2807.

56. Vankar, P.S., Shankar, R. (2009). Eco-friendly pretreatment of silk fabric for dyeing with *Delonix regia* extract. (2009). *Coloration Technology*, 125(3), 155-160.

56a. Honda, K., Tsutsui, K., Hosokawa, K. (1999). Analysis of the flower pigments of some *Delphinium* species and their interspecific hybrids produced via ovule culture. *Scientia Horticulturae (Amsterdam)*, 82(1,2), 125-134.

57. Vankar, P.S, Shankar, R, Srivastava, J. (2007). Ultrasonic dyeing of cotton fabric with aqueous extract of *Eclipta alba*. *Dyes and Pigments*, 72(1), 33-37.

58. Prabhu, K.H, Teli, M.D, Waghmare, N.G. (2011). Eco-friendly dyeing using natural mordant extracted from *Emblica officinalis* fruit on cotton and silk with antibacterial activity. *Fibres and Polymers*, 12(6), 753-759.

58a. Joshi, A. B., Satyanarayana, D., Chandrashekar, K. S., Subrahmanyam, E.V.S. (2004). Isolation and characterization of indole alkaloids from the flowers of *Ervatamia coronaria. Natural Product Sciences,* 10(3), 96-98.

59. Samatha, S., Vasudevan, T.N. (1996). *Embellia ribes* as a source of red colourant. *Journal of Scientific and Industrial Research,* 55, 888–889.

60. Kumar, R, Onial, P, Dobhal, P.C., Dayal, R. (2009). Natural dye from *Eupatorium glandulosum. Asian Dyer,* November issue, 52-56.

60a. Galvez, J., Zarzuelo, A., Crespo, M.E., Lorente, M.D., Ocete, M.A., Jimenez, J. (1993). Antidiarrhoeic activity of *Euphorbia hirta* extract and isolation of an active flavonoid constituent. *Planta medica,* 59(4), 333-6.

61. Iaderoza, M., Baldini, V.L.S, Draetta I dos, S., Bovi, M.L.A. (1992). Anthocyanins from fruits of acai (*Eterpa oleraceae,*) and Jucara (*E. edulis*). *Tropical Science,* 32(1), 41–46.

62. Dietrych-Szostak, D. and Oleszek, W. (1999). Effect of processing on the Flavonoid content in buckwheat (*Fagopyrum esculentum*). *Journal of Agriculture and Food Chemistry,* 47(10), 4384.

62a. Hussain, A., Jahan, S. (2011). Extraction of natural dye from medicinal plant and dyeing of silk.*China Textile Science,* (2), 30-34.

63. Saravanan, P., Chandramohan, G. (2011). Dyeing of silk with eco-friendly natural dye obtained from barks of *Ficus religiosa. Universal Journal of Environmental Research and Technology,* 1(3), 268-273.

64. Baggett, S., Protiva, P., Mazzola, E.P., Yang, H., Ressler, E. T., Basile, M.J., Weinstein, I. B., Kennelly, E.J. (2005). Bioactive benzophenones from *Garcinia xanthochymus* fruits. *Journal of Natural Products,* 68(3), 354-360.

64a. Yaun, H. (2008). Method for Dyeing *Gardenia jasminoides* on natural fiber fabrics. *Faming Zhuanli Shenqing,* CN 101307572 A 2008 1119.

65. Ansari, F. R., Ansari, W. H., Rahman, W. (1978).A biflavone from *Garuga pinnata. Indian Journal of Chemistry (Section B),* 16B(9), 846.

66. Tonyama, R., Takeda, Y., Inone, K., Kawamura, I., Yatsuzuka, M., Ikumoto, T., Shingu, T., Yokoi, T., Inonye, H. (1994). Studies on the blue pigment produced from genipap and methylamine. I. Structure of the brownish-red pigments, intermediates leading to the blue pigments. *Chemical and Pharmaceutical Bulletin,* 42(3), 668-673.

67. Quye, A. (1999). Really wild dyes. *Chemistry Review,* 9(2), 2-6.

68. Burret, F., Chulia, A.J., Debelmas, A.M., Hostettmann, K. (1978). Presence of isoscoparin, isoscoparin-7-O-glucoside and saponarin in *Gentiana pneumonanthe. Planta Medica,* 34(2), 176-9.

69. Paul, S., Madan, S., Jaiswal, G. (2007). Effect of bleaching and degumming of *Girardinia heterophylla. Asian Textile Journal,* 16(1), 73-76.

70. Rao, P.S., Seshadri, T.R. (1943). Pigments of cotton flowers. IX. Occurrence of populnetin in Indian cotton flowers. *Proceedings - Indian Academy of Sciences,* Section A 18A, 204-5.

71. Waheed, S., Alam, A. (2004). Studies of some natural dyes. *Journal of the Chemical Society of Pakistan*, 26(3), 255-263.
72. Sakata, K. (2008). A consideration of the dyeing of silk with hematoxylin in long wood extracts. *Seni Gakkaishi*, 64(9), 229-235.
73. Tronchet, J.(1964). Evidence for a biochemical dorsiventrality in young branches of *Hedera helix*. *Compt. Rend.* 258(8), 2390-2.

73a. Mell, C. D. (1930). Interesting sources of natural dyestuffs. *Textile Colourist*, 52, 49-53,122-5,190-3.

74. Nakamura, Y., Hidaka, M., Masaki, H., Seto, H., Uozumi, T. (1990). Major anthocyanins of the flowers of hibiscus (*Hibiscus rosa-sisnensis*). *Agricultural and Biological Chemistry*, 54(12), 3345-3346.
75. Vankar, P.S., Shukla, D. (2011). Natural dyeing with anthocyanins from *Hibiscus rosa-sinensis* flowers. *Journal of Applied Polymer Science*, 122(5), 3361-3368.
76. Bora, M. M., Dutta, N. N., Rao, P. G. (2009). Extraction of Roselle dye from Roselle (*Hibiscus sabdariffa*) fruits by adsorption onto activated carbon: equilibrium and kinetic studies. *Journal of Scientific & Industrial Research 68(6), 555-559.*

76a. Dadashev, M. N., Kukharenko, A. A., Fedorenko, G. I. (2004). Study of the extraction of dyestuff from renewable plant raw material. *Oboronnyi Kompleks--Nauchno-Tekhnicheskomu Progressu Rossii*, (4), 71-75.

77. Singh, A.K. (1999). *Hyptis suaveolens*—A source of natural dye. *Journal of Economic and Taxonomic Botany*, 23(3), 719-722.
78. Vankar, P.S., Dixit, S. (2011). Natural dyeing of cotton, wool and silk with stem and leaves of *Illicium griffithii*. *Research Journal of Textile and Apparel*, 15(2), 77-83.
79. Tiwari, V., Ghorpade, B., Vankar, P.S. (2000). Ultrasonic dyeing with *Impatiens balsamina* (balsam) using eco-friendly mordants on cotton. *Colourage* 47(3), 21-22, 26.
80. Ujjin, S., Setthayanond, J. (2014). Optimized vatting condition study for natural indigo dyeing on silk yarn. *Applied Mechanics and Materials 535 (Energy Engineering and Environment Engineering)*, 749-752, 5.
81. Pale, E., Nacro, M., Vanhaelen, M., Vanhaelen, F.R., Ottinger, R. (1998). Acylated anthocyanins from the flowers of *Ipomoea asarifolia*. *Phytochemistry*, 48(8), 1433-1437.
82. Peng, L., Mao, H., Xiu, Y., Zhang, K., Wang, C. (2012). Dyeing and anti-ultraviolet properties of anthocyanins extracted from purple sweet potatoes for silk. *Advanced Materials Research (Durnten-Zurich, Switzerland) 441(Eco-Dyeing, Finishing and Green Chemistry)*, 371-375.
83. Kokubun, T., Edmonds, J., John, P. (1998).Indoxyl derivatives in woad in relation to medieval indigo production. *Phytochemistry*, 49(1), 79-87.
84. Potsch, W. R. (2002). Dye production from woad - the stench as a feature. *Melliand Textilberichte*, 83(3), E35-E36,170-171.

84a. Torres, D.V., Cabrera, S.P., Saavedra, C. M.A., Torres, C. G.M.(2009). Phytochemical screening of alcoholic, ether and aqueous extracts from leaves, stems and flowers of *Ixora coccinea*. *QuimicaViva*, 8(3), 185-191.

85. Gahlot, M., Fatima, N., Papnai, N. (2008). *Jatropha* flowers Natural colourant for dyeing of cotton, wool and silk. *Colourage*, 55(2), 96-100.
86. Gungor, M., Cesme, M., Golcu, A. (2012). Isolation of colour component from *Juglans regia*: chromatographic determination, spectrophotometric investigation and dyeing properties. *Research Journal of Textile and Apparel*, 16(1), 1-9.
87. Anon. (1985).The Wealth of India: *A Dictionary of Indian Raw Materials and Industrial Products. Raw Materials*, Vol. (V), pp. 318–319. Council of Scientific and Industrial Research, Publications and Information Directorate, New Delhi, India
88. Sereme, A., Millogo-Rasolodimby, J., Guinko, S., Nacro, M. (2008).Tannin concentration of tannin-producing plants of Burkina Faso. *Journal de la Societe Ouest-Africaine de Chimie*, 13(25), 55-61.
89. Dayal, R., Dobhal, P.C., Kumar, R., Onial, P., Rawat, R.D. (2006).Natural dye form *Lantana camara* Leaves. *Colourage*, Vol. LIII, No. 12, 53-56.
90. Zohourian, T.H., Quitain, A.T., Sasaki, M., Goto, M. (2012).Extraction of Bioactive compounds from leaves of *Lawsonia inermis* by green pressurized fluids. *Separation Science and Technology*, 47(7), 1006-1013.
91. Mondal, M.I.H., Abu Sayeed, M., Newaz, M.O., Salam, M.A.(2009). Isolation of natural dye from the leaves of *Lawsonia inermis* and its dyeing. *Colourage*, 56(7), 52-56.
91a. Karthikeyan, S., Sivakumar, A., Anbalagan, M., Nalini, E., Gothandam, K. M. (2013). Finger printing of alkaloids, steroids and flavonoids using HPTLC of *Leucas aspera* whole plant methanolic extract. *Journal of Pharmaceutical Sciences and Research*, 5(3), 67-71.
92. Buxton, D. R., Russell, J. R. (1988). Lignin constituents and cell-wall digestibility of grass and legume stems. *Crop Science*, 28(3), 553-8.
93. Furry, M.S. (1945).Some natural dyes give long life to cotton fabric. *Rayon Textile Monthly*, 26, 603-5.
94. Sorokopudov, V. N., Chlebnikov, V. A., Deineka, V. I. (2005). Anthocyanins of some Berberidaceae plants. *Khimiya Rastitel'nogo Syr'ya*, (4), 57-60.
95. Khan, S.A., Khan, M.I.,Yuauf, M.,Shahid, M., Mohammad, F., Khan, M.A.(2011). Natural dye shades on woolen yarn dyed with Kamala (*Mallotus philippensis*) using eco-friendly metal mordants and their combination. *Colourage*, 58(11), 38-44.
96. Anon., (2009).The Wealth of India: *A Dictionary of Indian Raw Materials and Industrial Products. Raw Materials*, Vol. VI, pp. 176, 265, 306 Vol. IX, p.23,229 Vol. VIII, p.119, 377,35,390. Council of Scientific and Industrial Research, Publications and Information Directorate, New Delhi, India.
97. Bhattacharya, S.D., Mathew, S. (2002). Wool dyeing by natural colourants. *Colourage*, 49(9), 19-24.
98. Vankar, P.S., Tewari, V., Singh, L.W., Potsangbam, L. (2009).Sonicator dyeing of cotton fabric and chemical characterization of the colorant from *Melastoma malabathricum. Pigment and Regin Technology*, 38(1), 38-42.

99. Killedar, S. G., More, H. N. (2011). Seasonal variation of tannin content in different parts of *Memecylon umbellatum*. *Research Journal of Pharmacy and Technology*, 4(1), 78-81.

99a. Calis, A., Celik, G.Y., Katircioglu, H.(2009). Antimicrobial effect of natural dyes on some pathogenic bacteria. *African Journal of Biotechnology*, 8(2), 291-293.

100. Bhuyan, R., Saikia, C. N. (2004). Isolation of colour components from native dye-bearing plants in northeastern India. *Bioresource Technology*, 96(3), 363-372.

101. Bhuyan, R.and Saikia, C.N. (2003). Extraction of natural colourants from roots of *Morinda angustifolia*. *India Journal of Chemical Technology*, 10(2), 131-136.

102. Ngo, J. S. K., Ong, W. F., Ahmad, F. B., Bujang, K. B. (2013). A study of soluble-powdered natural dyes. *Research Journal of Textile and Apparel*, 17(1), 104-112.

103. Beltran-Heredia, J., Sanchez-Martin, J., Delgado-Regalado, A. (2009). Removal of dyes by *Moringa oleifera* seed extract. Study through response surface methodology. *Journal of Chemical Technology and Biotechnology*, 84(11), 1653-1659.

103a. Mallavarapu, G.R., Rao, L., Ramesh, S. (2000). Volatile constituents of the leaf and fruit oils of *Murraya koenigii* . *Journal of Essential Oil Research*, 12(6), 766-768.

104. Herrera-Hernandez, M.G., Guevara-Lara, F., Reynoso-Camacho, R., Guzman-Maldonado, S.H. (2011). Effects of maturity stage and storage on cactus berry (*Myrtillocactus geometrizans*) phenolics, vitamin C, betalains and their antioxidant properties. *Food Chemistry*, 129(4), 1744-1750.

105. Mukherjee, A., Mitra, S., Konar, P., Mukherjee, A.(2008). Modification of natural yellow dyes for improvement of washing and light fastness. *Man-Made Textiles in India*, 51(12), 429-433.

106. Ramamoorthy, S., Mudgal, G., Rajesh, D., Nawaz Khan, F., Vijaya Kumar, V., Rajasekaran, C. (2009). Characterization of novel pH indicator of natural dye *Oldenlandia umbellata*. *Natural Product Research*, 23(13), 1210-1217.

107. Siva, R., Mayes, S., Behera, S.K., Raja sekaran, C. (2012). Anthraquinones dye production using root cultures of *Oldenlandia umbellata*.*Industrial Crops and Products*, 37(1), 415-419.

108. Dayal, R., Dobhal, P. C., Kumar, R., Onial, P., Rawat, R.D. (2008). Natural dye from *Parthenium hysterophorus*. *Colourage*, Vol. LV, No. 8, 75-78.

109. Jiao, L., Liu, Shu-hua.(2011). Study on mordant-free dyeing of flax fabrics with pigment from purple Perilla. *Yinran Zhuji* 28(7), 47-50.

109a. Wu, Z. (2011). Preparation process of pure natural dye. *Faming Zhuanli Shenqing*, CN 102040858 A 20110504,

110. Shiobara, Y., Inone, S., Nishiguchi, Y., Kato, K., Takemoto, T., Nisimoto, N., De Oliveira, F., Akishue, G., Akishue, M.K., Hashimoto, G. (1992). Iresinoside, a yellow pigment from *Pfaffia iresinoides*. *Phytochemistry*, 31(3) 953-956.

111. Dayal, R., Dobhal, P.C., Parcha, V. (2001).Natural dye from *Pinus roxburghii bark* and needles. *Colourage*, Vol., XLVIII (3), 27-32

111a. Rajopadhye, A.A., Upadhye, A.S., Mujumdar, A.M. (2011). HPTLC method for analysis of *piperine* in fruits of Piper species. *Journal of Planar Chromatography--Modern TLC, 24(1), 57-59.*

112. Badoni, R., Semwal, D.K., Sati, S.C., Rawat, U.(2009).Optimization of dyeing procedure on wool yarn with *Pistacia integerrima* dye using synthetic and natural mordants. *Colourage*, 56(9), 67-70.

113. Anon. (1982). *The Wealth of India: A Dictionary of Indian Raw Materials and Industrial Products. Raw Materials*, Vol. VIII, p. 158. Council of Scientific and Industrial Research, Publications and Information Directorate, New Delhi, India.

114. Anon. (2009). *The Wealth of India: A Dictionary of Indian Raw Materials and Industrial Products. Raw Materials*, Vol. VIII, p. 203. Council of Scientific and Industrial Research, Publications and Information Directorate, New Delhi, India.

114a. Marzouk Mohamed, S.A., Ibrahim Magda, T., El-Gindi Omima, R., Abou Bakr Marwa, S. (2008). Isoflavonoid glycosides and rotenoids from *Pongamia pinnata* leaves. *Zeitschrift fur Naturforschung. C, Journal of Biosciences*, 63(1-2), 1-7.

115. Dayal, R. and Dobhal P.C. (2000). Natural dye from *Populus deltoides* Bark. *The Indian Textile Journal*, CX(7), 60-63.

116. Chandra, A., Nair, M.G., Lezzoni, A.F. (1993). Isolation and stabilization of anthocyanins from tart cherries (*Prunus cerasus*).*Journal of Agriculture and Food Chemistry*, 41(7), 1062-1065.

117. Singh, S.S., Srivastava, A., Saxena, R., Pandey, S.C., Sharma, V. (2005). Red sandar *(Pterocarpus santalinus)* Chemistry, biological activities and uses a review. *Journal of Medicinal and Aromatic Plant Sciences*, 27(2), 303-308.

118. Kulkarni, S.S., Gokhale, A.V., Bodake, U.M., Pathade, G.R. (2011). Cotton dyeing with natural dye extracted from pomegranate (*Punica granatum*) peal. *Universal Journal of Environmental Research and Technology*, 1(2), 135-139.

119. Sahid, M., Ahmad, A., Yusuf, M., Khan, M.I., Khan, S.A., Manzoor, N., Mohammad, F.(2012).Dyeing, fastness and antimicrobial properties of woolen yarns with gallnut *(Quercus infectoria)* extract. *Dyes and Pigments*, 95(1), 53-61.

119a. Gudzenko, A. V. (2013). Development of methods to standardization of *Quercus robur* in multicomponent plant composition for maintenance ellagic acid using HPLC method. *Voprosy Biologicheskoi, Meditsinskoii Farmatsevticheskoi Khimii*, (10), 9-12.

120. Kaiser, R. (1993). Quantitative analyses of flavonoids in yellow dye plant species Weld (*Reseda luteola*) and Sawwort (*Serratula tinctoria*). *Angewandte Botanik*, 67(3-4), 128-131.

121. Deveoglu, O., Sahinbaskan, B.Y., Torgan, E., Karadag, R. (2011). Dyeing properties and analysis by RP-HPLC-DAD of silk fibres dyed with weld (*Reseda luteola*) and Walloon Oak. *Asian Journal of Chemistry*, 23(12), 5441-5446.

122. Maulik, S.R., Pradhan, S.C. (2005). Dyeing of wool and silk with jujube bark, Himalayan Rhubarb and Hinjal bark. *Man-Made Textiles in India*, 48(10). 396-400.

123. Srivastava, M., Udawat, P., Gupta, P., Mishra, S.K., Gupta, K.C. (2011). Dyeing and silk with natural dye extracted from *Rheum emodi* with different mordants. *Man-made Textiles in India*, 8, 273-278.

124. Rohini, R. M., Das, Amit K. (2011). Determination of lupeol, β-sitosterol and quercetin from ethyl acetate extract of *Rhizophora mucronata* bark by HPTLC technique. *Asian Journal of Pharmaceutical and Clinical Research*, 4(1), 103-105.

125. Purohit, P., Sati, S.C., Purohit, M.C.(2008). Optimization of procedure for dyeing of wool and cotton fibres with *Rhododendron arboretum*, a source of eco-friendly natural dyes. *Asian Journal of Chemical and Environmental Research*, (1), 13-18.

126. Itokawa, H., Qiao, Y., Takeya, K. (1989).Anthraquinones and naphtha hydroquinones from *Rubia cordifolia*. *Phytochemistry*, 28(12), 3465-3468.

127. Deshkar, N., Tilloo, S., Pande, V. (2008).A comprehensive review of *Rubia cordifolia*. *Pharmacognosy Reviews*, 2(3), 124-134.

128. Goodarzian, H., Ekrami, E. (2010).Extraction of dye from madder plant (*Rubia tinctorum*) and dyeing of wool. *World Applied Sciences Journal*, 9(4), 434-436.

129. Guinot, P., Roge, A., Gargadennec, A., Garcia, M., Dupont, D., Lecoeur, E., Candelier, L., Andary, C. (2006). Dyeing plants screening: an approach to combine past heritage and present development. *Colouration Technology*, 122(2), 93-101.

129a. Baliarsingh, S., Panda, A.K., Jena, J., Das, T., Das, N.B. (2012). Exploring sustainable technique on natural dye extraction from native plants for textile: identification of colourants, colourimetric analysis of dyed yarns and their antimicrobial evaluation. *Journal of Cleaner Production*, 37, 257-264.

130. Ma, W. (2014). Dye formula and production method for peach puff wheat straw. *Faming Zhuanli Shenqing* CN 103773067 A 20140507.

131. Gowda, K. N. N.; Padaki, N.V., Sangappa, S., Mishra, S. N., Sudhakar, R., Munshi, R. (2010). Natural black dye from the nuts of *Semecarpus anacardium*. *Asian Dyer*, 7(3), 50-52.

132. Jain, A., Bhartiya, H.P., Vishwakarma, A.N. (1982). A Chalcone glycoside from the heartwood of *Shorea robusta*. *Phytochemistry*, 21(4), 957-958.

133. Dayal, R., Dobhal, P. C. (2001).Natural dyes from *Shorea robusta* bark. Convention Proceedings Natural Dyes, Edited by Deepti Gupta and M.L. Gulrajani, Indian Institute of Technology Delhi. December 17-18, 2001, p. 47-52.

134. Kouda-Bonafos, M., Czyzewska, E, Nacro, M., Oehlschlager, A. C. (1994). Isolation of apigeninidin from leaf sheaths of *Sorghum caudatum*. *Journal of Chemical Ecology*, 20(8), 2123-5.

134a. Kumaresan, M., Palanisamy, P. N., Kumar, P. E. (2013). Dyeing of silk fabric with eco-friendly natural dyes obtained from *Spathodea campanulata* and *Cordia sebestena*: a comparison. *Asian Journal of Chemistry*, 25(15), 8513-8516.

135. Badoni, R., Semwal, D.K. (2012). Dyeing performance of *Symplocos paniculata* and *Celtis australis*. *Asian Dyer*, 8(6), 51-58.

136. Singh, S.V., Purohit, M. C. (2014). Evaluation of colour fastness properties of natural dye extracted from *Symplocos racemosa* (Lodh) on wool fibres using combination of natural and synthetic mordants. *Indian Journal of Fibre & Textile Research* 39(1), 97-101.

137. Sari, P., Wijaya, C.H., Sajuthi, D., Supratman, U. (2012). Colour properties, stability, and free radical scavenging activity of Jambolan (*Syzygium cuminii*) fruit anthocyanin in a beverage model system Natural and copigmented anthocyanins. *Food Chemistry*, 132(4), 1908-1914.

138. Dedhia, E.M. (2006). Eco-friendly process of dyeing textiles material with the extract of *Tagetes erecta* flower. *Indian Pat. Appl.* In 2005 MU 00338 A 20060224.

139. Pan, N.C., Chattopadhyay, S.N., Day, A. (2004). Dyeing of jute fabric with natural dye extracted from marigold flower. *Asian Textile Journal*, 13(7), 80-82.

140. Sowbhagya, H.B., Sampathu, S.R., Krishnamurthy, N. (2004). Natural colourant from *Marigold* chemistry and technology. *Food Reviews International*, 20(1), 33-50.

141. Kumar, R., Dayal, R., Onial, P. (2013). Utilization of *Tagets minuta* aerial parts as a source of natural dyes for textile colouration. *Waste and Biomass Valorization* , September 27, pp. 1-9

142. Jadhao, N.U., Rathod, S.P. (2013). The extraction process and antioxidant properties of patuletin dye from wasted temple French marigold flower. *Asian Journal of Plant Science and Research*, 3(2), 127-132.

143. Mahale, G., Sunanda, R.K., Sakhi, V. (2006). Dyeing wool with *Tectona grandis* and colour fastness. *Asian Textile Journal*, 15(7), 56-60.

143a. Rao, N. V. Subba. (1956). Chemical components of the pods of *Tephrosia purpurea var. maxima.Current Science*, 25, 396.

143b. Kamel, M.M., Abdelghaffar, F., El-Zawahry, M.M.(2011). Eco-friendly dyeing of wool with a mixture of natural dyes. *Journal of Natural fibres*, 8(4), 289-307.

144. El-Zawahry, M.M. and Kamel, M.M. (1999). Dyeing of wool cloth with *Terminalia bellerica* fruit extract. *Indian Journal of Fibre and Textile Research*, 24(2), 126-130.

145. Chitnis, K. S. (2013). Extraction, characterization and application as natural dyes of extracts from *Terminalia catappa* leaf and seed pericarp. *Research Journal of Chemistry and Environment*, 17(9), 94-99.

146. Prusti, A.K., Purohit, A., Das, N.B., Nayak, A. (2009). *Terminalia catappa* as a suitable natural dye for silk and cotton yarns. *Colourage*, 56(11), 84-85.

147. Patel, B. H. (2009). Printing of jute fabric with natural hydrolysable tannins dye from *Terminalia chebula*. *Asian Dyer*, 6(4), 27-30.

148. Zhu, L., Deng, Y., Dong, X., Fu, S. (2010). Research on dyeing of cotton fabric with natural dye extracted from *Terminalia chebula*. *Tianran Chanwu Yanjiu Yu Kaifa*, 22(1), 149-153.

149. Kesava Rao, B., Hanumaiah, T., Rao, C.P., Rao, G.S.R., Rao, K.V.J., Ronald, H.T. (1983). Anthraquinones in *ventilago* species. *Phytochemistry*, 22(11), 2583-2585.

149a. Kamalam, M., Jegadeesan, M. (1999). Estimation of bioactive principle, wedelolactone in *Wedelia chinensis. Indian Drugs*, 36(7), 484-486.

150. Mahangade, R.R., Varadarajan, P.V., Verma, J.K. (2012). A new process of dyeing of cotton with natural dyes. *Indian Pat. Appl.*, IN 2009MU02861 A 20120224.

151. Popoola, A.N., Ipinmoroti, K.O., Adetuyi, A.O., Ogunmoroti, T.O. (1994). Extraction, separation and colouring potential studies of the dye in ginger rhizome (*Zingiber officinale*). *Pakistan Journal of Scientific and Industrial Research*, 37(5), 217-220.

152. Zhan, P. (2009). Preparation of jujube read dyes from *Ziziphus jujube* skin. *Farming Shenqing Gongkai Shuomingshu*, CN 101555358 A 20091014.

PART B: Two Versatile Indian Natural Dyes

Lac dye and indigo or indigotin are the two famous natural dyes, which were and are being obtained from the resinous secretion of tiny lac insects *Laccifer lacca*, and *Indigofera tinctoria* respectively. These were widely used in the past for colouring fabrics and other materials. A brief description of both the dyes is given in this chapter. Indigotin is now prepared synthetically on a large scale.

LAC DYE

Introduction

Lac is one of the most valuable gifts of nature to the mankind. It was known as *laksha* during the Vedic period. A brief description of lac, lac insect and its medicinal uses has been provided in the *Atharvaveda*. The origin of the word "lac" can be traced back to the Indian word "lakh", meaning "one hundred thousand", which indicates that a large number of insects are required to produce 1 kg of lac. In India, uses of lac can be traced to the ancient times of *Mahabharata*, if not earlier. It is mentioned in the *Mahabharata* that the Kauravas asked the renowned architect Purochak to build a palace that could be easily set on fire in order to eliminate the Pandavas. The architect obeyed the order and built a beautiful palace of lac (Yatugriha or Lakshagriha), which got completely burnt as planned, but the Pandavas escaped before it was set ablaze (Anon, 1972).

Lac dye ranks as the most ancient animal dye of the world. The Chinese used lac dye, about 4000 years ago, for imparting a very attractive and lustrous red colour to silk. Leather dyeing was another Chinese craft in which lac dye was widely used. The Vinaya text of Buddhists describes the extraction method of lac dye and its application in dyeing. The lac dye was valued in India since ancient times, but was recognized as an export item for Europe from the seventeenth century onward, due to the high price of cochineal dye; lac dye was its cheaper substitute. In the middle ages, lac was used in several countries of Europe and Asia for impressing seal on documents. The *Ain-i-Akbari* of the sixteenth century records the use of pigmented lac varnishes for painting screens. The lac dye industry flourished till 1871-1872, and thereafter, its trade declined due to the commercial availability of cheap synthetic dyes (Prasad, 1998).

Lac is the resinous secretion of a tiny insect *Laccifer lacca* (syn. *Kerria lacca*), commonly known as the lac insect. The resin is always formed on plants as the plant sap forms the sole source of food for the lac insect. The minute larvae of the insect attach themselves in their myriads to the succulent twigs and young branches of the host plant and draw their nourishment from the sap, which they suck through their long proboscis inserted into the bark of tender shoots. The exudation, which the bodies of the closely settled larvae coalesce, appears as hard resinous encrustations over the twigs. The lac in this form is known as sticklac. On an average, about 3,00,000 lac insects form 1 kg of the resin. The trees on which lac insects thrive are known as lac host trees. Lac insect can be reared on about 160 tree species, but Palas (*Butea monosperma*), Ber (*Zizyphus mauritiana*), and Kusum (*Schleichera oleosa*) are the prominent host trees. India and Thailand are the two major lac-producing and exporting countries of the world. China produces lac mainly for its domestic consumption. Bangladesh, Myanmar, Vietnam, and Sri Lanka are minor producers of lac. In India, lac is grown mainly in Bihar, Orissa, Madhya Pradesh, West Bengal, and Gujarat. Lac has various uses as it is non-toxic and non-odoriferous.

The Lac Insect

Description: The lac insect (*Laccifer lacca*; syn. *Kerria lacca*) belongs to the family Lacciferidae, which constitutes a specialized and isolated group in the super family Coccoidea of the order Hemiptera. The family contains more than 60 species, some of which produce commercial lac (Anon, 1972). Of the several genera known, the genus *Laccifer* contains the lac-secreting insects, the most important being *Laccifer lacca* (Grover, 1973; Anon, 1972), which produces a bulk of commercial lac. Its two strains, namely Rangeeni and Kusumi, are common in India. Each strain completes the lifecycle twice a year, but in each case, the duration of the life cycle and the seasons of maturity differ considerably. A number of researchers have made substantial contributions in the field (Grover, 1931; Grover, 1937).

The insect begins its life as a crimson-coloured, oval-shaped larva that is 0.6 mm long (excluding the antennae and caudal setae) and 0.25 mm wide across the thorax. Each female insect has the capacity to produce 300 to 1000 larvae. A small patch of encrustation contains a few hundred females. Thus, a large number of the larvae emerge, and the phenomenon of the large-scale emergence is generally called "swarming of larvae". After a short stay inside the mother's lac cell, the young larvae leave the cell in search of proper spots for settlement. The underside of succulent and twigs of a host tree is the proper place for their settlement. After settling, they insert their long, slender proboscises through the thin bark of the twigs and proceed to suck the plant sap. After sometime, they start secreting a thin, shiny layer of lac distributed under the cuticle all over the body; expect the mouth parts, the two breathing pores, and anus. Once settled, the female never moves during its life time, while the male moves out only during the last phase of its life, which is spent in procreation. One square centimeter can accommodate 150 to 180 larvae. Birds and winds are helpful in spreading them from one tree to another (Anon, 1972)

Lac Hosts

The quantity and quality of the lac depends mainly on the host trees used for the rearing of lac insect. The trees should be fairly fast-growing, adapted to pollarding, and have the strength to withstand heavy infestation of lac insect. The lac insect host trees may be grouped into three categories, namely, the common or major hosts, the occasional hosts, and the rare hosts, depending on their performance on lac cultivation. The species that are excellent hosts throughout the year at the place of their occurrence are covered under the category of the common hosts (Mukhopadhyay and Muthana, 1962; Roonwal et al., 1958). The second category of hosts are good in certain restricted regions of India only. Also, there are hosts that are a major host only for certain specific purposes and in certain specified seasons. The plant species that are used as common hosts are given in Table 4.1(b).

Table 4.1(b) Plant species used as common hosts (Tewari, 1994)

S.No.	Name of the species	Family	Popular name	Pattern	States in which used
1.	*Acacia nilotica* Delile subsp. *indica* (Benth.) Brenan syn. *Acacia arabica* Wild.var. *indica* Benth.	Mimosaceae	Babul, Kikar	Tree	Punjab, Uttar Pradesh, Gujarat, Maharashtra, Tamil Nadu
2.	*Acacia catechu* Wild.	Mimosaceae	Khair, Cutch Tree	Tree	Jammu & Kashmir, Punjab, Gujarat, Maharashtra
3.	*Butea monosperma* (Lam.) Taub.	Fabaceae	Dhak, Palas	Tree	Bihar, Madhya Pradesh, West Bengal, Uttar Pradesh, Gujarat, Maharashtra, Andhra Pradesh, Tamil Nadu.
4.	*Cajanus cajan* (Linn.) Millsp.	Fabaceae	Arhar; Tur, Red gram	(Herb or undershurb)	Assam
5.	*Ficus benghalensis* Linn.	Moraceae	Bar (Banyan tree)	Tree	West Bengal, Punjab, Rajasthan.
6.	*Ficus cunia* Buch.-Ham.	Moraceae	Jaharphali, Khain	Tree	West Bengal, Rajasthan
7.	*Ficus lacor* Buch.-Ham	Moraceae	Kahimal, Keol	Tree	West Bengal, Rajasthan
8.	*Ficus racemosa* Linn.	Moraceae	Gular, Umar	Tree	West Bengal, Rajasthan
9.	*Ficus religiosa* Linn.	Moraceae	Pipal, Pipli	Tree	Punjab, West Bengal, Rajasthan

Contd...

10.	*Leea crispa* Linn.	Vitaceae	Ban chalita	Under Shrub	Assam
11.	*Leea robusta* Roxb.	Vitaceae	Galeni	Shrub	Assam
12.	*Schleichera oleosa* (Lour.) Oken.	Sapindaceae	Kusum, The Lac Tree	Tree	Bihar, Madhya Pradesh, West Bengal
13.	*Zizyphus mauritiana* Lam.	Rhamnaceae	Ber, Indian Jujuba	Tree	Bihar, Madhya Pradesh, Punjab, Rajasthan, Uttar Pradesh, Gujarat, Maharashtra, Tamil Nadu.
14.	*Zizyphus xylopyra* Willd.	Rhamnaceae	Kat-ber, Ghont	Shrub or Tree	Madhya Pradesh, Uttar Pradesh.

The three hosts, namely, *Butea monosperma, Schleichera oleosa,* and *Ziziphus mauritiana,* are widely used for lac cultivation throughout the year in the areas where they occur.

Collection of Sticklac and Isolation of Seedlac

The lac collection is carried out in two forms, namely, Ari and Phunki. The former is cut from the host plant, while the latter is generally collected from the brood lac after it has been used for infection. The lac obtained after the removal of sticks is known as "sticklac". It is dried in the shade and winnowed to remove the foreign matter, like bit of wood, sand, etc. (Anon, 1972).

Lac Dye From Sticklac

The sticklac is first cleaned by hand picking and then ground in a corn crusher or stone chakki to yield granular material able to pass through a 5 to 10 mesh sieve. Further, winnowing and sieving is essential to make it free from remaining foreign material. The ground material is then washed either in mechanically operated large drums or in hemispherical vats, about 0.75 m in diameter, operated manually. The stone or cement vats with a rough inner surface having the capacity to hold about 20 kg of crushed sticklac are used for the isolation of dye. The lac is covered with water and allowed to remain in that state for about half an hour. Thereafter, a man stands in the vat and rubs the lac with his feet against the rough side of the vat. During the process of rubbing, the quantity of water is optimized so as to give the resulting mixture the consistency of loose mortar. This rubbing operation crushes the lac cells and washes out the dye, which comes out of the bodies of the dead females. The process of washing is repeated till the complete removal of lac dye. The residual product in the vat is then taken out and dried on a cement floor. Dried material is winnowed and sieved to obtain different grades of "Seedlac"(Anon, 1972). The process has been modified as per the requirement.

The convenient methods for the isolation of lac dye have been developed by Indian Lac Research Institute, Ranchi. The common steps used for the isolation of the lac dye are given below (Prasad et al., 1998).

Common Method for the Isolation of Lac Dye:

(i) The sticklac water washings from the factory were collected and filtered through the cloth bags.
(ii) The filtrate obtained was acidified and filtered again.
(iii) The residue obtained after filtration was treated with hot water and filtered again. Both the filtrates were mixed together.
(iv) The calcium salt precipitated from the filtrate was removed by filtration.
(v) The calcium salt was treated with the sodium carbonate solution and filtered.
(vi) The sodium salt solution was passed through cation exchange resin to obtain the lac dye solution.
(vii) The dye solution was concentrated in vacuum and crystallized to obtain the lac dye.

The crimson-coloured lac dye was costly in the nineteenth century, and at one time, it was so much in demand that the importance of shellac became secondary. This dye formed a waste product in the twentieth century because of the availability of cheap synthetic dyes. In the present century, it is again gaining importance due to its non-toxic nature and natural occurrence. India and China are the major producers of lac dyes.

The lac dye is acidic in nature and is generally present as sodium/potassium salts, which are completely soluble in water. The pure dye was obtained in 0.6%to 0.7% on the weight of sticklac and is sparingly soluble in water. The physicochemical properties of the lac dye are given in Table 4.2(b) (Prasad et al., 1998).

Table 4.2(b) Physicochemical properties of lac dye

S.No	Properties	Value
1.	Melting point	>230°C with decomposition
2.	Acid value	234-240
3.	Ash content (%Max)	1%
4.	Toxicity	Non-toxic
5.	UV spectrum (a) Water (b) Methanol	 λ_{max} 292, 490 nm λ_{max} 230, 292, 340, 500, 530 nm
6.	Solubility	(a) Pure dye sparingly soluble in cold water but soluble in boiling water. (b) Soluble in acetic acid, acetone, amyl alcohol, formic acid and methyl alcohol. (c) Insoluble in benzene, chloroform and diethyl ether (d) Dissolves slowly in ethyl alcohol
7.	Colour of solution	(a) Aqueous acidic solution is orange-red (b) Alkaline solution is reddish-violet

The chemical analysis of lac dye indicated it to be a mixture of five closely related anthraquinone derivatives. Their structures were elucidated and named as laccaic acids A, B, C, E, and laccaic acid D (Mayer and Cook, 1943; Prasad et al., 1998; Thomson, 1987, 1971 Ch. 5B). India has the potential of producing about 200 tonnes of lac dye annually (Prasad et al., 1998).

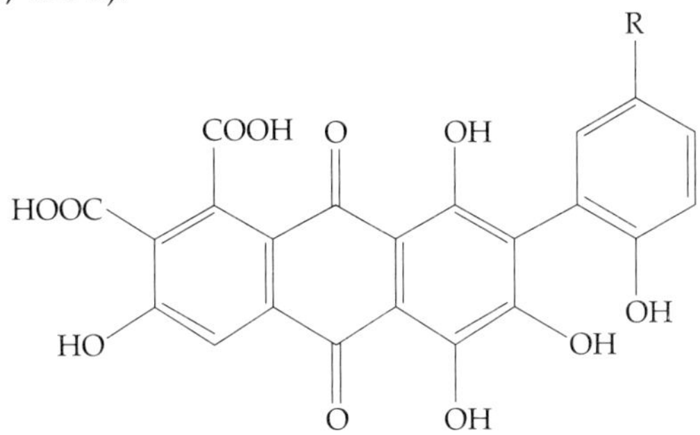

		R
Laccaic acid A	:	$CH_2CH_2NHCOCH_3$
Laccaic acid B	:	CH_2CH_2OH
Laccaic acid C	:	$CH_2CH(NH_2)COOH$
Laccaic acid E	:	$CH_2CH_2NH_2$

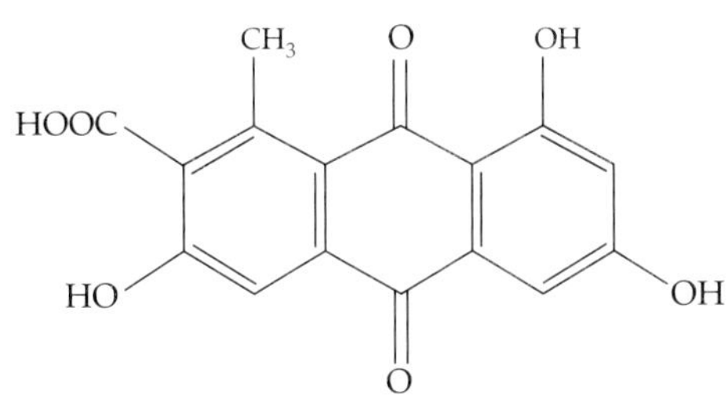

Laccaic acid D

Dyeing with Lac Dye

The dye solution acidified with hydrochloric acid imparts orange-red shade to silk and wool, with good light-fastness property. A number of varying shades can be obtained using different mordants (Prasad et al., 1998).

The dyeing trials using lac dye were also carried out on wool and silk, respectively (Khan et al., 2004; Suri et al., 2000; Khan et al., 2012). Alum, ferrous sulphate, potassium dichromate, and stannous chloride were used as mordants. A number of shades with good colour-fastness properties were obtained. Lac dye is non-toxic in nature and thus has the potential of its use in colouring the food items, besides the colouring of silk and wool.

INDIGO

Natural dyes were used by humans since time immemorial, and indigo is believed to be one of the first dyes used by them. The dye has a religious significance also as the colour imparted by this dye resembles with the colour of Lord Krishna's skin. The use of this dye was found about 4000 years ago. The traces of indigo were detected in the Indian textiles of 1600 B.C., when it was used for wrapping Egyptian mummies (Padmini, 1999). In the ancient times, all the things that were imported into Europe from India were known as *Indicum* in Latin and *Indikos* in Greek. As the blue dye exported from India was very famous, with the passage of time, this word got associated with blue only, which eventually became indigo. The dyeing process to obtain the blue colour on fabric was seen by Marco Polo in the thirteenth century, at Colium in the State of Travancore, India. The indigo produced in the sixteenth century was very costly as its price was comparable to pearls. It imparted light blue to deep blue colour on different fabrics. Indigo is the strongest and fastest blue dye that nature has produced. It was used in European countries to impart black colour on different yarns and fabrics till the beginning of the seventeenth century.

Sources of Indigo

Fourteen different plant species were used for the isolation of the blue dye, of which 50% plant species belongs to the genus *Indigofera* while rest are covered under different genus. About 54 *Indigofera* species have been recorded in India, of which *Indigofera arrecta, Indigofera tinctoria, Indigofera suffruticosa,* and *Indigofera sumatrana* were used for the isolation of indigo (Anon, 2010). Two other plant species, namely, *Strobilanthes flaccidifolius* and *Wrightia tinctoria,* were also cultivated for the blue dye.

The Indian *Indigofera tinctoria* yielded superior blue dye to that obtain from *Istatis tinctoria* (woad) grown in Europe, as a result of which, Indian indigo started replacing the one obtained from woad. The anti-Indian campaign was started by woad cultivators, manufactures of indigo, and merchants. Indigo was given the name as "devil's food" and branded to be poisonous. This strong resistance to indigo led to its prohibition in France in 1598, in Germany in 1607, and in England in the beginning of the seventeenth century. The use of indigo during the regime of Henry IV in France became capital offence and any person found using indigo was liable to death sentence. However, the belief was discarded and East India Company again started exporting indigo to Britain from India in 1664.

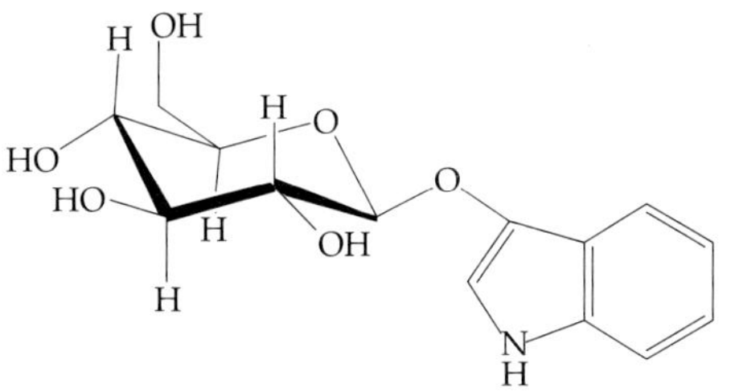

Indican

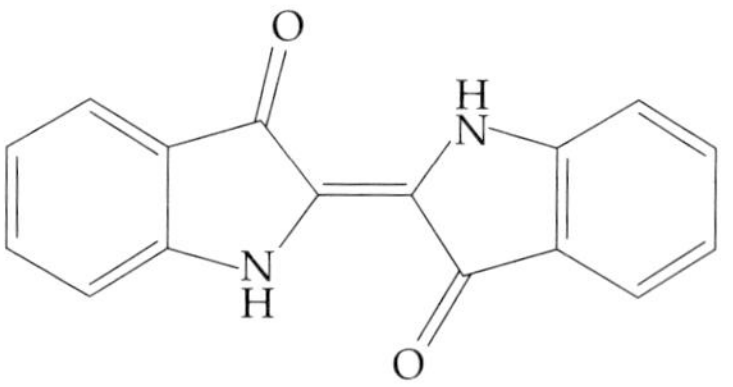

Indigotin

Uses and Benefits of Indigo

1. Indigo is sole vat dye among natural dyes and is mainly obtained from *Indigofera* species.
2. It was extracted from the plants by steeping them, draining, and beating the water of the steeped plants, followed by draining, boiling, and compressing the residue to form cakes.
3. The cake possesses a long shelf-life, is easy to transport, and ensures regular supply throughout the year.
4. Dye vats are prepared for dyeing using indigo cakes. Traditionally, these vats were underground vats prepared using natural reducing agents and alkalis, which varied from place to place. Blue shades depend on the method of fermentation, ingredients used, and the strength of the dye. These traditional vats are continuous vats, and older the vat, the better it dyes.
5. Fabrics dyed in natural indigo are believed to be non-suffocating when worn, as this dye allows skin to "breathe", unlike chemical dyes. Therefore, they are cool when worn in summer and warm in winter.
6. The indigo dye-yielding plants (*Indigofera* species) are ecologically useful and can be cultivated easily.
7. The growing plants and the plants material left after extraction of the dye are useful to soil as they enrich the soil with nitrogen.
8. The indigo dye imparts pleasing blue shades on different fabrics and has excellent light fastness. The indigo plants in India are used for imparting blue and black colours of its own and brown, green, and purple in combination with other dyes.
9. Indigo is also used in preparing pigments of paints, lacquers, rubber, and painting inks.
10. The traditional indigo manufacture process is eco-friendly as it does not involve any pollutants.
11. In the recent times, chemical vats, which use caustic soda and hydro sulphate, are required to be used with utmost care as the excess of caustic soda could be harmful to dyers and fabric to be dyed.

Indigo Synthesis and Civil Disobedience Movement in India

The structure of indigo was elucidated by Baeyer and his coworkers in 1880. Its synthesis by several methods was accomplished by Baeyer in 1882, mainly using o-nitro cinnamic acid as the starting compound. The BASF company purchased the patent rights of these processes by paying \$1, 20,000, as royalty to Baeyer but these processes could not be commercialized due to higher production cost in comparison to natural indigo. The cost-effective process for the synthesis of indigo was later developed by Heumann by using anthranilic acid as the starting material. The same could be possible after getting a very good yield of phthalic anhydride from naphthalene, which Heumann used for the preparation of anthranilic acid. Saper was heating naphthalene with sulphuric acid and while monitoring the temperature, the thermometer broke by accident and mercury fell into the reaction mixture, thereby completing the reaction at low temperature, with an excellent yield of phthalic anhydride. Therefore, the seeds of Civil Disobedience Movement were sown in the BASF laboratory at Germany, because the cost-effective synthesis of indigo could be accomplished, forcing Indian farmers to stop the cultivation of *Indigofera tinctoria* in Bengal and the Champaran region of Bihar, as it turned into a loss-making business. A number of other highly convenient methods were developed later by using phenyl glycine. It appears that the BASF spent about one million pounds at that time for developing synthetic indigo. This indicates the huge profit that the BASF was hoping to earn by selling synthetic indigo in the market (Nagendrappa, 2011 and 2014).

The financial condition of farmers was very bad even before the availability of synthetic indigo as the zamindars and indigo factory owners used cruel methods to control them. They provided very little money for purchasing the seedlings, with the condition of compulsory indigo plantation in all their land. The system existed was such that all indigo cultivation would have to be given to the factory owner at his will when the crop was ready. The price was too meager even for buying rice against it. As the farmers were not allowed to grow any other food-yielding crop, they had nothing edible that could be cultivated, nor did they have enough money to buy the required quantity of it. As a consequence, farmers in Bihar were facing starvation deaths and were kept under control of the indigo owners by whips and other cruel means.

Gandhiji was against this system and wanted to give farmers the right to grow the crop of their choice. Excess indigo cultivation in Bihar brought indigo prices down, while scarcity of paddy cultivation considerably increased the price of rice. This resulted in an imbalance and starvation of poor people. The synthesis of indigo added to their misery manifold. The tenants had to give up indigo farming. However, this did not eliminate their obligation due to the existence of the *tinkathia* system. The tenants were pressurized to compensate for the losses to the planters (landlords), subjecting them to further hardships. The local British government had no interest in their grievances. The situation became worse by 1916, and in 1917, Gandhiji intervened on behalf of *raiyats* by going to Champaran to study the situation and conduct an enquiry. The British government objected to such an enquiry but Gandhiji did not fear his arrest. Later, they allowed him to conduct the enquiry. This was the first Civil Disobedience Movement by Gandhiji (Nagendrappa, 2011).

References

1. Anon. (2010). *The Wealth of India, Raw Materials,* Vol. V, 182. Council of Scientific and Industrial Research, New Delhi, India.
2. Anon. (1972). *Indian Forest Utilization*, Vol. II. Manager of Publications, Delhi, p. 801.
3. Grover, P.M. (1931). *A Practical Manual of Lac Cultivation*. Indian Lac Research Institute, Ranchi, India. p. 88.
4. Grover P.M. (1937). *Lac Cultivation in India*. Indian Lac Research Institute, Ranchi, India, p. 8, 147.

5. Grover, P.M. (1973). *Lac Cultivation in India*, India Lac Research Institute, Ranchi, India.
6. Khan, M.A; Khan, M., Srivastava, P.K. and Mohammad, F. (2004). Extraction of lac dye and its application on wool. *Colourage*, 27-30.
7. Khan, M.I., Shahid, M., Khan, S.A., Yusuf, M., Mohammad, F. and Khan, M.A.(2012). Studies on application of lac natural dye on wool using eco-friendly metal mordants. *Colourage*, 59(3), 42-51.
8. Mayer, F. and Cook. A.H. (1943). *The Chemistry of Natural Colouring Matters.* Reinhold Publishing Corporation, U.S.A, p. 147.
9. Mukhopadhyay, B. and Muthana, M.S. (1962) *A Monograph on Lac.* Indian Lac Research Institute, Ranchi, India. p. 378.
10. Nagendrappa, Gopalpur, (2011). Indigo in India Independence: Role of a broken thermometer. *Resonance*, 16(12), 1168-1174.
11. Nagendrappa, Gopalpur, (2014). Johann Friedrich Wilhelm Adolf von Baeyer, A pioneer of synthetic organic chemistry. *Resonance*, 19(6), 489-522.
12. Padmini, T.B.(1999). Use of natural indigo in Asia., Some myths and their clarification about Indigo dye (Separate lecture). *Convention on Natural Dyes*, Indian Institute of Technology, New Delhi, India. pp. 63-66.
13. Prasad, K.M., Prasad, N. and Aggarwal, S.C. (1998). Lac dye, in: *Commemoration of Golden Jubilee of India's Independence.* Indian Lac Research Institute, Ranchi, India.
14. Roonwal, M.L., Raizada, M.B., Chattergi, R.N. and Singh, B. (1958). *Descriptive Account of the Host Plants of the Lac Insect, Laccifer Lacca (Kerr), and the Allied Plants in The Indian Region.* Indian Lac Cess Committee, Ranchi, India. p. 139.
15. Suri, M., Sethi, B., Bedi, D.K. and Anand, S. (2000). Application of selected eco and lac dyes on silk. *Colourage*, 13-18.
16. Tewari, D.N. (1994). *Tropical Forest Produce.* International Book Distributors, Dehradun, India. p. 477.

PART C: Herbal Colours

INTRODUCTION

Life without colours would have been monotonous and unexciting. This colourful world is amazing. The human eye is empowered with the ability to recognize electromagnetic waves within a certain specific range of electromagnetic spectrum. It covers the wavelengths ranging from about 4×10^{-7} m or 400 nm (violet) to about 7×10^{-7} m or 700 nm (red). It can be expressed as the frequency in the range varying from about 7.5×10^{14} Hz (violet) to about 4.3×10^{14} Hz (red). This range is known as the visible part of the spectrum, as light is nothing but an electromagnetic wave of the wavelength belonging to this region of the spectrum. Electromagnetic radiation can be defined as the energy which is propagated by means of electric and magnetic fields while oscillating in the directions right angle to the direction of travel of energy. It is mainly through light and sense of vision that we know and recognize the world around us. The sunlight, also known as white light, includes the different wavelengths of visible light which gives a continuous spectrum on observing, as in a rainbow. The other components of the sunlight besides visible light, are the shorter-wavelength and longer-wavelength radiation known as ultraviolet and infrared radiation respectively. The human eye cannot detect either of the two. These may be detected and recorded photographically or by specific detectors developed for this specific purpose. The electromagnetic radiation, thus consists of different wavelengths of varying speed which becomes 2.9979×10^8m/s in vacuum. Mathematically, the wavelength (λ) and frequency of electromagnetic radiation (ν), after rounding up c can be expressed as $\lambda\nu = c = 3.00 \times 10^8$m/s (Whitten et al., 1996).

It is well known that when a narrow beam of white light (sunlight) falls on a glass prism, the emergent light consists of seven main colours, namely, violet, indigo, blue, green, yellow, orange, and red, which is abbreviated as VIBGYOR. Violet light at the short wavelength end (400 nm) bends the maximum, while red light long wavelength end (700 nm) bends the minimum, implying that red light travels faster than violet light in a glass prism. Dispersion of light is caused due to variation in the refractive index of the medium (glass) for different wavelengths.

Why an object has a colour?

When the sunlight (white light) falls on a coloured object, it may either be totally reflected or totally absorbed. The object appears white in the former case and black in the latter. In addition, light is partially absorbed and partially reflected. In this case, the object has the colour of reflected light. If only a single band is absorbed, the object has the complementary colour of the absorbed band. In case an object absorbs the whole visible light except one band, it will have the colour of the reflected band. Thus, the colour of a substance may be blue due to solo absorbance of the yellow portion of the spectrum., or due to absorbance of whole visible spectrum except blue. The other complementry colours will be observed in accordance with absorbance of other portion of the spectrum (band). The human eye detects the reflected light in the visible region, and therefore, each wavelength in the region appears as a different colour. The absence of all wavelengths in the visible region is called blackness. The colourless substances also have absorption spectra, but for them, absorption takes place in the infrared or ultraviolet region and not in the visible region of the spectrum (Finar, 1996).

HERBAL HOLI COLOURS

It is the general belief of people that bright colours represent energy, life, and joy, and life should be full of colours. All colours emanate from white, and when they are mixed together again, they become black. Each colour is meant to be seen and enjoyed separately. Pink colour is the colour of universal love. The emotions of a person burn him like a fire. But the presence of the fountain of colours adds charm to the life. In ignorance, emotions are a trouble, in knowledge, the same emotions add colour. The colours may be associated with different effects as given below (Nature Energies, 2014):

1. Blue with communication, decisiveness, personal expression.
2. Green with harmony, love, peace.
3. Indigo with imagination, serenity, stillness, understanding.
4. Magenta with balancing of emotions
5. Orange with happiness, resourcefulness, sociability, success.
6. Red with courage, power, feeling of security and safety, self-confidence, vitality.
7. Violet with beauty, creativity, inspiration.
8. Yellow with cheerfulness, inspiration, mental clarity, stimulating curiosity and interest.

The colours may possess the following therapeutic functions (Nature Energies, 2014):

1. Blue: Beneficial in reducing the blood pressure, possesses anti-inflammatory and muscle relaxing effects and other benefits.
2. Green: Beneficial in regulating the pituitary gland, fights depression, useful in calming the nervous system and other benefits.
3. Red: Beneficial in energizing, stimulating and other benefits.
4. Yellow: Beneficial in digestion, purifying blood, strongly stimulating happiness and other benefits.
5. White: Beneficial for regeneration, gives energy and stabilizes the chrono-biological rhythm and other benefits.

According to Robert Krulwich, pink is a combination of red and violet colours, which appear on opposites sides of the spectrum in the rainbow. As colour is a construct of people's eyes and brains, when one looks at a pink object, one is not actually seeing pink wavelengths of length. It appears pink as certain wavelengths of light are reflected, while others are absorbed, quenched by the object. Therefore, pink (made-up colour) is a reflective colour and not a transmissive colour, as one can see it, because the brain translates the light rebounding off the objects. Pink paint can be made by mixing red and white (PTI, 2012).

Skin is a very protective part of our body as it saves it from foreign attacks. It is dangerous to use chemical-laced colours for celebrating Holi. The colours contain harmful chemicals, and excessive penetration or absorption of these chemicals into the skin not only harms it, but also leads to serious lifestyle issue. In case, if any synthetic colour falls into the eyes, it can cause redness, irritation, and may damage the cornea of the eyes, leading to complete blindness. The oxidized metals, industrial dyes, engine oils, and the inorganic compounds like copper sulphate, chromium iodide, aluminium bromide, lead oxide, and mercury sulphide are used for making bright colours by using maize starch and arrowroot as the base material. The colours obtained by using different chemicals ensure lethal health problems, affecting your body and mind. The end users of the colours suffer, depending on the quantity and toxicity of the chemicals used. Some of the health hazards are given in Table 4.1(c) (Varshney, 2002).

Table 4.1(c) Health hazards of Holi colours prepared using some inorganic chemicals

S.No.	*Chemical used*	*Colour of Gulal*	*Harmful effects*
1.	Aluminium bromide (Al_2Br_6)	Silver	(i) May cause burn on skin on contact (ii) Shortness of breath on inhaling (iii) Carcenogenic
2.	Chromium iodide (CrI_2)	Purple	(i) May cause skin irritation. (ii) Mucous membrane and respiratory tracts affected on inhaling. (iii) Long-term effect can cause dermatitis. (iv) Bronchial asthma and allergies.
3.	Copper sulphate ($CuSO_4$)	Green	(i) Skin contact may result in itching or eczema (TOXENT) (ii) Eyes contact may cause conjuctivities, inflammation of the eyelid lining, ulceration and clouding of cornea (Clayton & Clayton, 1981).
4.	Lead oxide (PbO)	Black	(i) Affects renal failure (ii) Application on body makes brain blunt. (iii) It causes irritation to skin, eyes, and respiratory tract. (iv) It also affects gum tissue, central nervous system, blood, and reproductive system. (v) It can bioaccumulate in plants and in mammals (Anon, 2009).
5.	Mercury sulphide (HgS)	Red	(i) May cause skin cancer (ii) Mental retardation (iii) Paralysis and impaired vision

It is evident that artificial colours are not only harmful to human health but also for the biodiversity, particularly for plants, water, and soil. The use of herbal colours and gulal prepared by using the dyes obtained from flowers, leaves, fruits, wood, bark, and roots of different plants will overcome all the above problems. Some of the plant species that can be used for the preparation of herbal Holi colours and gulals are given in Table 4.2(c).

Table 4.2(c) Some dye yielding plant species for making herbal colours

S.No.	*Name of the species*	*Family*	*Common name*	*Pattern*	*Parts used*	*Colour obtained*
1.	*Alium cepa* Linn.	Liliaceae	Onion	Bulbous biennial	Peels	Pink
2.	*Beta vulgaris* Linn.	Amaranthaceae	Beetroot	H	Fruits	Magenta
3.	*Bixa orelleana* Linn.	Bixaceae	Annatto	Tr	Seeds	Red

Contd...

4.	*Butea monosperma* (Lam.) Taub.	Fabaceae	Tesu or Flame of the forest	Tr	Flowers	Saffron colour
5.	*Curcuma longa* Linn. syn. *C. domestica* Val.	Zingiberaceae	Haldi	H	Rhizome	Yellow
6.	*Indigofera tinctoria* Linn.	Fabaceae	Nil	Sh	Leaves	Blue
7.	*Lawsonia inermis* Linn.	Lythraceae	Mehndi	Sh	Leaves	Green
8.	*Mallotus philippensis* (Lamk.)Muell.-Arg.	Euphorbiaceae	Kamala	Tr	Seeds	Red
9.	*Pterocarpus santalinus* Linn. f.	Fabaceae	Rakta-chandan	Tr	Wood	Red
10.	*Punica granatum* Linn.	Punicaceae	Anar	Tr	Rind of fruits	Brown/Red
11.	*Tagetes erecta* Linn.	Asteraceae	Genda	H	Flowers	Yellow

(H = Herb., Tr = Tree., Sh = Shrub)

Herbal Water Colour

Herbal water colours (TNN, 2002) can be prepared at home, as given below:

(i) Pink colour may be obtained by boiling the peels of onions in water.

(ii) Magenta colour may be obtained by warming the pieces of beetroot in water.

(iii) Genda flowers on soaking in water overnight and subsequent boiling yields a yellow colour solution.

(vi) Green colour may be obtained on warming mehndi powder with water in optimum material-to-liquor ratio.

(v) Turmeric powder may be mixed with water in optimum concentration for preparing yellow colour.

(vi) The *Butea monosperma* (Tesu) flowers, which are available around Holi period, on boiling with water may produce saffron like colour.

The material-to-liquor ratio depends on the percentage of dye in the material and has to adjusted suitably for making different colours.

Herbal Gulal

The use of gulal containing inorganic salts may cause harmful effects on humans. On the contrary, herbal gulal has the following advantages:

(i) Soft in nature

(ii) Biodegradable: No disposal problem

(iii) Non-toxic: No health hazard

(iv) Eco-friendly
(v) Aesthetically appealing
(vi) Commercial production will lead to employment generation

The maize starch or arrowroot may be used as the base material for the preparation of herbal gulal. A number of vegetable dyes isolated from different plant species (Table 4.2c) may be used for the preparation of herbal gulal of varying colours by using the optimum dye concentration and an appropriate amount of starch.

References

1. Anon (2009). Lead (ii) oxide. *International Occupational Safety and Health Information Centre*. Retrieved 2009-06-06
2. Clayton, G.D., Clayton, F.E., eds. (1981). Patty's industrial hygiene and toxicology. Third Edition Vol. 2, Part 6. *Toxicology*. NY: John Wiley and Sons. ISBN 0-471-01280-7.
3. Nature Energies, Colour Therapy Resources and Equipment (2014). http://www.natures-energies.com/natural-therapies/colour-liq-therapy – chromotherapy – resources – and – equipment (2014).
4. Finar, I.L. (1973) (Reprinted 1996). Organic Chamistry, Vol. 1, pp. 868–878, Longman Singapore Publishers (Pvt) Ltd., Singapore.
5. PTI. (2012). Hue and Cry: Pink just a pigment of our imagination? The *Times of India*, March, 2012, New Delhi.
6. TNN. (2002). The natural colours this Holi, they' are cheaper, better, The *Times of India*, March 2002, New Delhi.
7. Varshney, Vibha. (2002). Colours of pure radiance, *Down to Earth*, April 30, p. 22.
8. TOXNET. (1975-1986). *Natural Library of Medicines Toxicology Data Network*. Hazardous Substances Data Bank (HSDB), Public Health Service National Institute of Health, U.S. Department of Health and Human Services. Bethesda, M.D: NLM.
9. Whitten, K.W., Davis, R.E., Peck, M.L. (1996). General Chemistry, Saunders College Publishing, USA, 169.

5 Chemistry of Natural Dyes

Part A

INTRODUCTION

Plants are the greatest organic chemists on the earth. They possess unique quality of photosynthesis and are able to synthesize a number of complex organic compounds with meticulous details by using the greenhouse gas carbon dioxide, water, and the energy supplied by the sun. Forest plants provide two types of products, namely, wood (including fuelwood) and non-wood forest products (NWFPs), besides other tangible and intangible benefits. Wood consists mainly of cellulose, hemicelluloses, and lignin. The NWFPs are defined as the goods of biological origin other than wood, derived from forests, other wooded land, and trees outside forests. These include mainly medicinal plants, essential-oil-yielding species, tanning and natural dye materials, edible plants, and other products. The organic compounds occurring in plants may be classified as primary or secondary metabolites, with former being involved in basic cellular metabolism and the latter being of much limited taxonomic distribution and without apparent metabolic functions. Primary metabolites include the common sugars, protein amino acids, purines and pyrimidines of nucleic acids, chlorophyll, and so on. Secondary metabolites, which are biosynthetically derived from primary metabolites, make up all the remaining chemicals, which vary in their distribution from plant to plant (Ref. 199, 200; Harborne, 1999; Heath, 2002) and appear to have ecological roles, such as serving to ward off animals and plants predators or to attract insect pollinators. These metabolites consist of different types of chemicals, namely alkaloids, anthocyanins, coumarins, flavonoids, phenolics, and other classes of compounds, and their secretion depends on a number of factors such as light, moisture availability, and stress. It has been observed that high-intensity light or ultraviolet light leads to the formation of anthocyanins, flavones, isoflavonoids, psoralens, and sinapyl esters; wounding to coumesterol, coumarins, chlorogenic acid, ferulate esters, lignin, and seburin, wall-bound phenolic acids; pathogen attack to aurones, coumarins, 3-deoxyanthocyanidins, flavonols, furanocoumarins, prenylated isoflavonoids, pterocarpans, isoflavans, and stilbenes; low temperature to anthocyanins; low nitrogen, phosphate or iron content in the soil to flavonoids/isoflavonoids, anthocyanins, and phenolic acids, respectively

(Ref. 200a, Singh, 2007; Ref. 200b; Pichersky and Gershenzon, 2002; Ref. 200; Heath, 2002). The extensive research in plant chemistry using modern spectroscopic and separation techniques led to the elucidation of structures of at least 1,00,000 secondary constituents produced during the growth and development of more than 2,00,000 species of flowering plants (Ref. 201, 199; Metcalf and Metcalf, 1992; Harbone, 1999). The structures of more than 10,000 alkaloids (Ref. 202; Southon and Buckingham, 1989), 20,000 terpenoids (Ref. 203; Hill and Connolly, 1991), 4000 flavonoids (Ref. 204; Harborne, 1994), 8000 phenolics, including non-flavonoids (Ref. 199; Harborne, 1999), 500 quinones, and 650 polyacetylenes (Ref. 201; Metcalf and Metcalf, 1992) have already been elucidated. The number has been increasing exponentially by the use of latest spectroscopic and separation techniques in the structure elucidation of natural products.

CHEMICAL CONSTITUENTS OF 69 DYE-YIELDING PLANT SPECIES

Natural dyes are mainly obtained from different parts of plants, namely, barks, flowers, fruits, leaves, seeds, roots, and wood. The dyes obtained from these parts consist of organic compounds belonging to different classes, namely, cartenoids, indigoids, quinones, and others. These compounds occur in different plants as secondary metabolites. Natural dyes are also obtained from animal and mineral sources.

A list of dye-yielding plant species has been provided in Chapter 4 (Part A) of the book. It contains 392 plant species along with their families, plant parts used, and the possible colours imparted by their use. These plant species contain chemical compounds belonging to different classes of organic compounds. The chemical constituents isolated and characterized from the dye-bearing parts of 69 prominent dye-yielding plant species and 4 insect species are given in Table 5.1. About 570 compounds were isolated in the pure state by using the latest chromatographic techniques and their structures were elucidated by using sophisticated spectroscopic and separation techniques (IR, UV, NMR, MS, GC-MS, LC-MS, etc). The compounds belonging to different classes of organic compounds, mainly alkaloids, anthocyanins, carotenoids, flavonoids, indigoids, phenolic acids, quinones, tannins, and other classes, are responsible for imparting different colours to fabrics and other items. The colouring principles of some well-known natural dye sources used in the past were found to be curcumin from *haldi* or turmeric (*Curcuma longa*), crocin from saffron (*Crocus sativus*), lawsone from henna (*Lawsonia inermis*), indigotin from *Indigofera tinctoria*, alizarin from madder (*Rubia tinctoria*), munjistin and alizarin from manjith (*Rubia cordifolia*), laccaic acids A, B, C, E, and laccaic acid D from the lac dye of lac insect (*Laccifer lacca*), carminic acid from the cochineal insect (*Dactylopius coccus*) dye, kermesic acid from kermes (*Kermes ilicis*) dye, and 6,6′-dibromoindigotin from the shellfish (*Murex brandaris*) dye.

Table 5.1 Chemical Constituents of Dye-Yielding Parts of Different Plants

S. No.	*Name of the species*	*Compounds characterized*	*Parts*	*References*
1.	*Abrus precatorius*	Abranin.; 7,3′5′-Trimethoxy-4′-hydroxyflavone-3-O-β-D-galactosyl-(1→4)-α-L-xyloside.; Precatorin I, II, III [6-C-glycosides] and saponins.; Abrusin and its 2′-O-apioside.; Abrectorin; Desmethoxycentaureidin-7-O-rutinoside	Seeds	1-5
2.	*Acacia catechu*	Quercetin.; Catechin; Epicatechin; Epicatechin-3-O-gallate; Epigallocatechin-3-O-gallate.; (+)-Catechin.; (+)-Homo-isocatechin.; (+)-Catechin.; (–)-Epicatechol	Heartwood	6-11
3.	*Adina cordifolia*	Chrysin, Gossypetin; Kaempferol; 7,4′-Di-O-methyl-6-methylflavanone.; 3′-Hydroxy-5,7,4′-tri-O-methyl flavanone.; 7,4′-Dimethoxy-5-hydroxyflavanone; 5,7-Dimethoxy-4′-hydroxyflavanone	Heartwood	12-14
4.	*Albizia lebbeck*	Albizia saponnins A, B, and C [acacia acid lactone 3-O-β-D-xylopyranosyl-(1→2) α-L-arabinopyranosyl-(1→6)-β-D-glucopyranoside; 3-O-β-D-Glucopyranosyl-(1→2)-O-[-L-arabinopyranosyl-(→6)]-β-D-glucopyranoside and 3-O-β-D-xylopyranosyl-(1→2)-α-L-arabinopyranosyl-(1-→6)-O-[β-D-glucopyranosyl-(1→2)]-β-D-glucopyranoside	Bark	15
5.	*Alkanna tinctoria*	Alkannin; angelylalkannin.; Acetylalkannin; angelylalkannin; isovalerylalkannin; 2-methylbutanoylalkannin.; Deoxyalkannin; alkannin acetate.; Alkannin or shikonin isovalerate; alkannin angelate.; 3-Methyl-3-acetoxy-butanoic ester of alkannin (β-acetoxy- isovaleric ester of alkanin).; Alkannin esters of beta, beta-dimethylacrylic acid; beta-acetoxy-isovaleric acid, angelic acid; isovaleric acid	Roots	16 -21
6.	*Allium cepa*	Quercetin 4′-O-β-D-glucopyranoside.; Cepaic acid.; Kaempferol; quercetin	Red skin, yellow skin, light pink skin	22-24
7.	*Anacardium occidentale*	Occidentoside or (–) salipurposide (Tetrahydrohinokiflavone-C-glucoside); β-sitosterol	Cashewnut shells	25
8.	*Anogeissus latifolia*	Gallic acid; 4,18-dimethyl,22-methyl,15-hydroxy,11,12-naphtha phenanthrene	Leaves	26
9.	*Areca catechu*	(+)-Catechin;(–)-epicatechin;procyanidin A-1;procyanidin B-1; procyanidin B-2; procyanidin B-7; procyanidin trimers (three); procyanidin tetramers (two); procyanidin pentamers (two)	Fresh betel nuts	27

Contd...

S. No.	Name of the species	Compounds characterized	Parts	References
10.	*Arnebia nobilis*	Alkannin β, β-dimethylacrylate [5,8-dihydroxy-2-(1′-β, β-dimethylacryoxy-4′ methylpent-3′ -enyl)-1,4-naphthoquinone]; alkannin acetate [2-(1′-acetoxy-4′-methylpent-3′-enyl)-5,8- dihydroxy-1,4-naphthoquinone]; shikonin [5,8,-dihydroxy-4′-methylpent-3′-enyl)-1,4-naphthoquinone]	Roots	28
11.	*Artocarpus heterophyllus*	Artocarmins A-D; artocarmitins A-C.; Artocarpfuranol; dihydromorin; steppogenin; norartocarpetin; artocarpanone; artocarpesin; artocarpin; cyloartocartocarpin; cycloartocarpesin; artocarpetin; brosimone; cudraflavone B; carpachromene; isoartocarpesin; cyanomaclurin	Wood	29-30
	Artocarpus heterophyllus	Heterophylol.; Heteroflavanones A and B.; Artonins J, K and L.; Artonins C and D.; Artonins A and B; heterophyllin, cycloheterophyllin.; Artoflavonone; ursolic acid, betulnic acid; cycloartenone; β- sistosterol	Root bark, roots	31-36
12.	*Berberis aristata*	Aromoline; berbamine; berberine chloride; oxyberberine; oxyacanthine.; Karachine (Protoberberbine-acetone adduct)	Root bark	37-38
13.	*Beta vulgaris*	Betavulgarosides V-X.; Betavulgarosides I-V.; Betanin; isobetanin; β-D-fructofuranosyl-α-D-(6-O-(E)feruloylglucopyranoside.	Roots	39-41
14.	*Bischofia javanica*	Betulinic acid; betulonic acid; 3β-O-(Z)-coumaroylbetulinic acid; 3β-O-(E)-coumaroylbetulinic acid	Bark	42
15.	*Bixa orellana*	Bixin; norbixin	Seeds	43
16.	*Butea monosperma*	Lanceoletin, butein; monospermoside; sulfurein.; 5,7-Dihydroxy-3,6,4′-trimethoxyflavone-7-0-α-L-xylopyranosyl-(1→3)-0-α-L-arabinopyranosyl-(1→4)-β-D-galactopyranoside.; Dihydromonospermoside; butein; monospermoside; isoliquiritigenin; 7,3,′4′-trihydroxyflavone; (–)-butin; (–)-butrin; (+)-isomonospermoside; (–)-liquiritigenin; formononetin; afromosin; formonoetin-7-0-beta-D-glucopyranoside.; Butaspermin (3′-hydroxy2,3-dihydroflavone,7,4′-di-O-β-glucopyranoside.; Butrin; isobutrin; coreopsin; isocoreopsin; sulphurein; monospermoside; isomonospermoside	Petals, flowers	44-48
17.	*Caesalpinia sappan*	Caesalpiniaphenols A-D.; Caesalpiniaphenol E[(3S, 4 R)-3,7,2′,3′-tetrahydroxy-3,4-dihydro-9H-indeno[6,5-c] chromene; caesalpiniaphenol F [(3R, 4S)-3,7-dihydroxy-3-(3′methoxy-4′-hydroxyphenyl)-4-methoxychroman.; Caesappanin A and B	Heartwood	49-51

18.	*Carthamus tinctorius*	Hydroxysafflor yellow A;6-hydroxykaempferol-3,6,7-triglucoside.; 2,3-Dimethoxy-5-methylphenyl-1-0-β-D-glucopyranoside (Carthamosides B 1); 2,6-dimethoxy-4-methylphenyl-1-0-β-D-glucopyranoside (carthamosides B 2); ethyl-3-(4-O-β-D-glucopyranosyl-3-methoxyphenyl) propionate (carthamosides B-3); methyl-3-(4-O-β-D-glucopyranosyl-3-methoxyphenyl) propionate; ethylsyringin; methylsyringin.; Carthamosides B4-B8; 4′-(hydroxyphenacyl)-β-D-glucopyranoside, benzyl -O-α-L-rhamnopyransoyl-(1→6)-β-D-glucopyoanoside; 4-(methoxybenzyl)-0-β-D-glucopyranoside.; (2R)- 4′,5-Dihydroxyl-6, 7-di-O-β-D-glucopyransyl flavanone; methyl-3-(4-O-β-D-glucopyranosyl phenyl) propionate; 2(S)-4′,5-dihydro-6,7-di-O-β-D-glucopyranosyl flavanone; 6-hydroxykaempferol-3,6-di-O-β-D-glucopyranoside; 6, 4-O-β-D-glucosyl-trans-p-coumaric acid; 4-O-β-D-glucosyl-cis-p-coumaric acid.; Safflor yellow B (Quinochalcone C-glycoside).; Carthamin; safflor yellow A.	Flowers	52-57
19.	*Cassia tora*	Cassiaside; rubrofusarin-6-β-gentiobioside and two other naphthopyrone glycosides.; 1-[(β-D-Glucopyranosyl-(1→3)-O-β-D-glucopyranosyl-(1→6)-O-β-D- glucopyranosyl)oxy]-8-hydroxy-3-methyl-9,10-anthraquinone and two other anthraquinone glycosides.	Seeds	58-59
20.	*Crocus sativus*	Crocin (Crocetin digentiobiose).; Crocetin; Kaempferol -3-O-sophoroside; quercetin; methyl-quercetin glycosides.; Kaempferol	Dried trifid stigma, stigma, stamens, sepals, fresh flower petals	60-62
21.	*Curcuma longa*	Curcumin; demethoxycurcumin;bis-demethoxy curcumin.; 6,8,3′-Trimethoxy-5,4′-dihydroxy-5′-methyl-rhamnoglucosyl) isoflavone.; Thirteen curcuminioids namely 1,5-dihydroxy- 1,7-bis (4-hydroxyphenyl)-4,6-heptadiene-3-one; 1,5,-dihydroxy-1-(4-hydroxy-3-methoxyphenyl)-7-(4-hydroxythenyl)-4,6-heptadiene-3-one; 1,5-dihydroxy-1-(4-hydroxyphenyl)-7-(4-hydroxy-3-methoxyphenyl)-4,6-heptadiene-3-one; 3-hydroxy-1,7-bis-(4-hydroxy phenyl)-6-heptene-1,5-dione and other compounds.; Curcumin; demethoxycurcumin; bis-demethoxycurcumin.	Rhizomes	63-66

Contd...

S. No.	Name of the species	Compounds characterized	Parts	References
22.	*Delonix regia*	Peonidin-3-O-glucoside; petunidin-3-O-acetyl-glucoside; astaxanthin (ketocarotenoid); cyanidin-3-O-rutinoside.; Gallic acid; procatechuic acid, quercetin; 2-hydroxy 5-[(3,4,5-trihydroxyphenyl)carboxyl oxy]benzoic acid; rutin; quercetin 3-O-glucoside; quercetin 3-O-galactoside;quercetin trihexoside; quercetin 3-O-robinobioside; kaempferol rhamnosylhexoside; isorhamnetol rhamnosyl-hexoside	Floral petals, red flowers	67-68
23.	*Eclipta alba*	Wedelolactone.; Verazine[(20S) (25S)-22,26-imino-cholesta-5,22(N)-dien-3β-ol]and seven other steroidal alkaloids	Whole plant	69-70
24.	*Emblica officinalis*	Gallic acid.; Ellagic acid; 1-O-galloyl-β-D-glucose; 3,6-di-O-galloyl-D-glucose; chebulagic acid; quercetin; chebulinic acid; corilagin; 3-ethylgallic acid; isostrictiniin; 1,6-di-O-galloyl-β-D-glucose.; Emblicanin-A; emblicanin-B; punigluconin; pedunculagin.	Fruit, fresh pericarp	71-73
25.	*Eucalyptus* hybrid	Gallic acid.; Ursolic acid.; Gallic acid, 7β-D-glucosyloxy-5-hydroxy-2-methylchromone.; Ursolic acid, ursolic acid lactone, 2α-hydroxy ursolic acid.; Ursolic acid.	Leaves	74-78
26.	*Ficus bengalensis*	Methyl ethers of Delphinin-3-O-α-L-rhamnoside; leucocyanidin-3-O-β-D-galactosylcellobioside; pelargonidin-3-O-α-L-rhamnoside	Stem bark	79
27.	*Haematoxylon campechianum*	Hematoxylin; hematoxylol A; tetra-O-methylhematoxylot B(7-hydroxy-3,4,10, 11-tetramethoxy-7,8-dihydro-6H-dibenz[b,d] oxocin-7-methanol.; Hematoxylin	Heartwood	80-81
28.	*Indigofera tinctoria*	Indigotin; Indican	Leaves	82
29.	*Isatis tinctoria*	2,4(1 H, 3H)- quinazolinedione; 5-hydroxy-2-indolinone; indigo; indirubin; β-rosasterol.; Isatan A [(1H-indol-3-yl6′-O-(carboxyacetyl)-beta-D-ribohex-3′-ulopyranoside]; isatan B(1H-indol-3-yl beta-D-ribohex-3-ulopyranoside).; Indigotin	Leaves	83-85
30.	*Juglans regia*	Gallic acid; 5-hydroxy-2-methoxy-1,4-napthoquinone; salidroside and other compounds.; Juglone.; casuarinin; 1,2,3,4-tetra-O-galloyl-β-D-glucose; 1,2,6-tri-O-galloy-β-D-glucose;1,2,6-tri-O-galloyl-α-D-glucose; heterophylliin A; heterophylliin F; strictinin	Seeds coats, husks of walnuts	86-88

31.	*Lawsonia inermis*	Lawsone (2-hydroxy-1,4-napthoquinone).; Afzelin (3-O-α-L-rhamnopyranosyl 5,7,4- trihydroxyflavone).; Lawsone; apigenin; luteolin; cosmosiin; p-coumaric acid; 2-methoxy-3-methyl- 1,4-naphthoquinone and apiin.; Luteolin; luteolin-7-O-glucoside; acacetin-7-O-glucoside; laxanthone 1; laxanthone 11; lawsone; β- sitosterol-β-D-glucoside	Leaves	89-92
32.	*Maclura pomifera*	Osajaxanthone(6′6′-dimethyl-1,7-dihydroxypyrano-(2',3',3,2)xanthone	Root bark	93
33.	*Mallotus philippensis*	Rottlerin; isorottlerin; homorotterlin and other compounds.; Mallotophilippen A (1-[5,7-dihydroxy-2,2-dimethyl-6-(2,4,6-trihydroxy-3-isobutyryl-5-methylbenzyl)-2 H- chromen-8-yl]-2-methyl-butan-1-one; mallotophilippen β(1-[6-(3-acteyl-2,4,6-trihydroxy-5-methylbenzyl)-5,7, [dihydroxy-2,2-dimethyl-2H-chromen-8-yl]-2-methyl-butan-1-one	Fruits	94-95
34.	*Mangifera indica*	Mangiferin (1,3,6,7-tetrahydroxy-xanthone-2-C-glucoside).; Daphnoretin; mangiferin(main constituent); myricetin; myricetrin; quercetin; rutin; β-sitosterol.; 3,4-Dihydroxy benzoic acid; benzoic acid;(+)-catechin; (–)-epicatechin; gallic acid; methyl ester of gallic acid; mangiferin.; Mangdesisterol; mangfarnasoic acid; mangeudesmenone; amentoflavone.; Mangiferin (synthesis)	Bark	96-100
35.	*Memecylon umbellatum*	Ursolic acid.; Apigenin; quercetin; rutin	Leaves, roots	101-102
36.	*Morinda citrifolia*	2-Ethoxy-1-hydroxyanthraquinone; 2-(ethoxy-ethyl)-1-hydroxyanthraquinone; 1-hydroxy-2-methylanthraquinone; damnacanthal; nordamnacanthal; 2-formyl-1-hydroxyanthraquinone; morindone-6-mehylether.; Damnacanthal; damnacanhol; 1,3dihydroxy-2-methoxymethylanthraquinone; 3-hydroxy-1-methoxy-2-methoxymethylanthraquinone; lucidin; morindone; nordamnacanthal; rubiadin; rubiadin 1-methyl ether;sorandidiol.; Damnacanthol-11-O-beta-primeveroside;digiferruginol-1-methylether -11-O-beta- gentiobioside; digiferruginol-11-O-beta-primeveroside; 1-hydroxy-2-primeverosyl-oxymethyl-anthraquinone-3-olate; 1- hydroxy-5,6-dimethoxy-2-methyl-7-primeverosyloxy- anthraquinone.; Asperulosidic acid; deacetylasperulosidic acid; damnacanthol-3-O-beta-D-primeveroside mordindone-6-O-beta-D-primeveroside.; 1,2-Dihydroxyanthraquinone;1,3-dihydroxy-2-methylanthraquinone;2-hydroxy-3-(hydroxymethyl) anthraquinone; morintrifolins A and B [Benzophenones]; 1,3,6-trihydroxy-2-methylanthraquinone.; Morenone-1; morenone-2.; Damnacanthal; rubiadin- 1-methyl ether; alizarin; morindone	Roots	103-109

Contd...

S. No.	Name of the species	Compounds characterized	Parts	References
37.	*Morinda tinctoria*	Morindonin, Morindone	Root bark	110
38.	*Morinda umbellata*	5,7-dihydroxy-2-methylanthraquinone; morindone; morindin	Root bark	111
39.	*Oldenlandia umbellata*	1,2,3-Trimethoxyanthroquinone; 1,3-dimethoxy-2-hydroxyanthraquinone; 1,2-dimethoxyanthraquinone; 1-methoxy-2-hydroxyanthraquione; 1,2-dihydroxyanthraquinone	Roots	112
40.	*Parthenium hysterophorus*	3,5,7,4′-Tetrahydroxy-3′-methoxyflavone-3-O-β-L-galactopyranosyl-(1→3)-O-β-D-arabinopyranosyl-7-O-β-L-rhamnopyranoside; cirsilineol; lutexin.; Coronopilinl; isoparthenin; parthenin; stigmasterol	Stems, aerial parts	113-114
41.	*Pinus roxburghii*	Quercetin; myrcetin; kaempferol; isorhamr.etin	Bark and needles	115
42.	*Polygonum hydropiper*	Gallic acid; quercetin; quercitrin; rutin; ursolic acid and other compounds.; Isorhamnetin 3,7-disulfate; quercetin 3-sulphate; tamarixetin.; 3′-Methylquercetin; 7,4′-dimethylquercetin.; 6-Hydroxyluteolin; 3′,4′,5,6,7-pentahydroxyflavone and other flavonoids	Whole plant, Leaves	116-119
43.	*Populus deltoides*	Grandidentatin; grandidentoside; 6-methyldihydroquercetin; populoside; pyrocatechol; salicin; salicortin; salicyl alc; trichocarposide	Bark	120
44.	*Psidium guajava*	Ellagic acid.; Amritoside(4β-gentiobioside of ellagic acid); hexahydroxy-diphenic acid; leucocyanidin	Bark	121-122
45.	*Pterocarpus santalinus*	Pterolinus K (Phenanthrenedione); Pterolinus L (chalcone).; 6-Hydroxy-5-Me-3′,4′,5′-trimethoxy aurone-4-O-α-L-rhamnopyranoside.; 6,4′-dihydroxyaurone-4-O-rutinoside.; 4′,5-dihydroxy-7-O-methylisoflavone3′-O-β-D-glucoside.; Santalins A and B	Heartwood	123-127
46.	*Punica granatum*	Cyanidin-3-glucose; gallic acid; myricetin; pelargonidin-3-galactose; quercetin.; caffeic acid; ellagic acid; luteolin; punicic acid.; Punicanolic acid; tricetin; β-sitosterol laurate and myristate.; granatin B	Fruits, rind	128-130a
47.	*Quercus infectoria*	Gallic acid; m-digallic acid; p-digalloyl acid; methyl gallate, ethyl gallate; 1,2,3,6-tetra-O-galloyl-β-D-glucose; 1,2,3,4 6- penta-O-galloyl-β-D-glucose.; Ellagic acid-4-O-[β-D-glucopyranosyl]-10-O-[β-D-glucopyranosyl] (4→1)-β-D-rhamnopyranoside; 2-methyl-3-hydroxymethylene-4,5,6,7,8-pentahydroxynaphthalene.; Hexagalloylglucose(3-O-digalloyl-1,2,4,6-tetra-O-galloyl-β-D-glucose	Galls	131-133
48.	*Reseda luteola*	Apigenin; apigenin 7-O-glucoside.; Luteolin; luteolin 7-O-glucoside; luteolin 3′, 7-O-diglucoside; luteolin 4′-O-glucoside	Leaves, stem	134-135

49.	*Rheum emodi*	Chrysophanol; catechin; emodin; daucosterol; physcion; piceatannol; β-sitosterol; emodin-8-O-β-D-glucopyranoside; chrysophanol-8-O-β-D-glucopyranoside; piceatannol-4′-O-β-D-glucopyranoside; piceatannol-4′-O-β-D-(6″-O-galloyl)-glucopyranoside.; 6-Methylrhein; 6-methyl-aloe-emodin.; Revandchinone-1,2,3,4.; Rheinal; rehin -11-O-β-D-glucoside.; Rhein; physcion, aloe-emodin; chrysophanol	Rhizomes	136-140
50.	*Rhododendron arboreum*	Epicatechin;quercitrin; quercetin-3-O-galactoside; syringic acid	Leaves	141
51.	*Rubia cordifolia*	Rubiacordone A(6-acetoxy-1-hydroxy-2-methylanthraquinone-3-O-α-L-rhamnopyranoside); 1-acetoxy-6-hydroxy-2-methylanthraquinone-3-O- [α-L-rhamnopyransoyl-(1→2)-β-D-glucopyranoside.; 1-Hydroxy-2,7-dimethylanthraquinone; 2- hydroxy-6-methylanthraquinone; 2,6-dihydroxyanthraquinone; 3,5-di-(p-hydroxybenzyl) phenol; 8′-hydroxy-n-pentadecanyl decan-4-en-1-oate; n-octacosyanyl octa-1-oate; n- nonadecane.; 1-Hydroxy-3-ethyl-9, 10-anthraquinone; 1,8-dihydroxy-11,20-(15-butylnaphthaquinonyl) phenantherene.; Rubiasins A-C (New anthracene derivatives) and mollugin.; 3′-Carbonmethoxy-4′-hydroxy-naphtho[1,2′-2,3] pyran-6-one; 3′-carbomethoxy-4′-hydroxy-naptho[1,2′-2,3] furan; dihydromollugin; 3-carbomethoxy-2-(3′-hydroxy)-isopentyl-1,4-naphthohydroquinone -1-O-β-D-glucoside.; Alizarin; 1-hydroxy-2-methyl-9,10-anthraquinones; 1,3,6-trihydroxy-2-methyl-9-10-anthraquinone-3-O-(6′-O-acetyl)-α-L-rhamnosyl (1→2)-β-D-glucoside; 1,3,6-trihydroxy-2-methyl-9,10-anthraquinone-3-O-α-L-rhamnosyl(1→2)-β-D-glucoside; 1,3,6-trihydroxy-2-methyl-9,10-anthraquinone-3-O-(6′-acetyl)-β -D-glucoside; 2-carbmethoxy-3-prenyl-1,4 naphthohydroquinone di- β-D-glucoside; rubimallin; daucosterol.;1,3-Dihydroxy-2-methylanthraquinone; 1-methoxy-2-methoxymethyl-3-hydroxyanthraquinone; 4-hydroxy-2-carboxyanthraquinone; 1,4-dihydroxy-2-hydroxymethylanthraquinone; 1- hydroxy-2-hydroxymethylanthraquinone.; 1,4-Dihydroxy-2-carboethoxyanthraquinone; 1-hydroxy-2-carboxy-3-methoxyanthraquinone; 1-hydroxy-2-methyl-6 or-7-methoxyanthraquinone, oleanolic acid acetate; β-sitosterol; scopoletol.; Mollugin; 1-hydroxy-2-methyl-9,10-anthraquinone; alizarin; ruberythric acid; 1,3-dihydroxy-2-ethoxymethyl-9,10-anthraquinone; lucidin primeveroside; 2-methyl-1,3,6-trihydroxy-9,10-anthraquinone; 2-methyl-1,3,6- trihydroxy-9,10-anthraquinone 3-O-(6′-	Roots	142-153

Contd...

S. No.	Name of the species	Compounds characterized	Parts	References
		O-acetyl)-α-rhamnosyl-(1→2)-β-glucoside; 2-methyl-1,3,6-trihydroxy-9,10-anthraquinone 3-O-α-rhamnosyl-(1→2)- β-glucoside.; 1-Hydroxy-2-methoxyanthraquinone; 1,4-dihydroxy-2-methyl-5-methoxyanthraquinone; 1,3-dimethoxy-2-carboxyanthraquinone; rubiadin.; 1,4-Dihydroxy-2-methylanthraquinone; 1,5-dihydroxy-2-methylanthraquinone; 3-prenyl-5(or 8)-methoxy-1,4-naphthoquinone.; 1-Hydroxy-2-methylanthraquinone; nordamnacanthal; physcion; 1,4-dihydroxy-6-methylanthraquinone.; 1-Hydroxy-3-ethyl-9,10-anthraquinone [Cordifoliol]; 1,8-dihydroxy-11,20(15-butyl-napthaquinonyl) phenanthrene[Cordifodiol]		
52.	*Rubia tinctorum*	Alizarin; purpurin; lucidin	Roots	154
53.	*Semecarpus anacardium*	Anacardium(1-O-β-D-glucopyranosyl-(1→6)-β-D-glucopyranosyloxy-3-hydroxy-5-methylbenzene (Anacardoside)	Seeds	155
		Anacarduflavanone.; Semecarpetin.; Semecarpulflavanone.;Galluflavone.; Jeediflavanone.; Tetrahydroamentoflavone; tetrahydrobustaflavone	Nut shells	156-161
54.	*Shorea robusta* and *Shorea talura*	Hopeaphenol	Heartwood	162
55.	*Symplocos paniculata*	3′,4′,5′,6′,-tetramethoxy flavone 7-O-β-D-glucopyranosyl(1→3)-β-D-glucopyranoside and other compounds	Stem bark	163
56.	*Syzygium cuminii*	(7α,8α,2′α)-3,4,5-trimethoxy-7,3′,1′,9′, diepoxylignan(Cuminiresinol) and eight other compounds	Bark	164
57.	*Tagetes minuta*	Patuletin; quarcetagetin; quercetin; patuletin 7-glucoside; quarcetagetin 3-arabinogalactoside; quercetin 5-glucoside.; compounds of essential oil	Aerial parts	165-166
58.	*Tamarindus indica*	Apigenin; 7-methoxy luteolin; 4′-methoxy apigenin; cis-ferulic acid; trans-ferulic acid.; Vitexin; isovitexin [8-and 6-C-glucopyranoside of apigenin]; orientin and iso-orientin (homo-orientin) [8-and 6-C-glucopyranoside of luteolin]	Leaves	167-168
59.	*Tectona grandis*	Napthotectone; anthratectone.; Gallic acid, ellagic acid; rutin; quercetin.; Tectone (3,8-dihydroxy-2-Methyl anthraquinone.; Tectograndone.; Tectoleafquinone	Leaves	169-173
60.	*Tephrosia purpurea*	Isoglabratephrin; tephropurpulin A[prenylated flavonoids]; glabratephrin	Aerial parts	174

61.	*Terminalia chebula*	Tri-ethyl ester of chebulic acid, trans-cinnamic acid; arjunglucoside; arjunic acid; arjungenin; daucosterol.; Chebulinic acid; ethyl gallate; gallic acid; tannic acid.; 2,3-(S)-(Hexahydroxydiphenyldicarboxyl)- D-glucose; 3,6-di-O-galloyl-D- glucose; 6-O-galloyl-D-glucose; (–)-shikimide 4-O-gallate; (–) shikimic acid 3-O-gallate; shikimic acid 5-O-gallate; methyl gallate; 1,2,6-tri-O-galloyl-β-D-glucose.	Fruits	175-177
62.	*Thespesia populnea*	Kaempferol; populnin (kaempferol-7-glucoside); herbacetin; populneol;(+) gossypol	Flowers	178
63.	*Toona ciliata*	Siderin(4,7,8-trimethoxy-5-methylcoumarin); 4,8-dimethoxy-7-hydroxy-5-methylcoumarin and three steroids.; Caffeic, chlorogenic, ferulic, gallic, p-hydroxybenzoic, protocatechuic, syringic, vanillic acids.; Siderin	Bark, leaves, wood	179-181
64.	*Ventilago denticulata* syn. *Ventilago calyculata*	Emodin-1-O-α-l-rhamnoside; emodin-8-O-α-L-rhamnoside; emodin-8-O-β-D-glucoside.; ventilatones A and B; ventileins A and B.; Ventiloquinones I, J, and K.; 2-methoxystypandrone; 1-hydroxy-6-methoxy-3-methylxanthone-8-carboxylic acid (Calyxanthone).; 2,4,8-Trihydroxy-1-methoxy-3-methyl-anthraquinone; 2,4,8-trihydroxy-1,6-dimethoxy-3-methylanthraquinone.; Xanthorin 5-methyl ether; 2-hydroxyislandicin	Root bark	182-187
65.	*Ventilago maderaspatana*	Ventiloquinones A, B, C, D, E, F, G, and H.; Cordeauxione; isocordeauxione; maderone[2(2′-acetoxypropyl)-3-hydroxy-5,7,8-trimethoxy-1,4-naphthoquinone; 1,3-dihydro-6,9-dihydroxy-7,8-dimethoxy-1-methylnaphtho[2,3-C]furan-3-one (Ventilagol); 2-acetyl-6,7-dimethoxy-3-methyl-1,8-methylenedioxynaphthalene.; Ventilone A, B, C, and D.; Ventinone A(1,4,5-trihydroxy-7-methoxy-2,3-dimethyl-9,10-anthraquinone); Ventinone B(1,4-dihydroxy-5,7-dimethoxy-3-methyl-9,10-anthraquinone).; Ventinone- A and-B; chrysophanol; physcion; islandician; emodin; xanthorin; xanthorin-5-methyl ether	Root bark	188-189, 184, 185, 187
66.	*Wedelia chinensis*	Astragalin; 3,5,7,3′,4′-pentahydroxy-6-methoxy-flavone-7-O-α-l-rhamnopyranosyl-3′-O-β-D-xylopyranosyl(1→3)-O-β-D-galactopyranoside.	Leaves	190
67.	*Wrightia tinctoria & Wrightia tomentosa*	Anthranillate; indigotin; indirubin; isatin; trypanthrin; rutin	Leaves	191
68.	*Zingiber officinale*	(E)-3-hydroxy-1-(4′-hydroxy-3′,5′-dimethoxyphenyl)-tetradecan-6-en-5-one; eightphenylakanoids and thirteen other compounds	Rhizomes	192

Contd...

S. No.	Name of the species	Compounds characterized	Parts	References
69.	*Ziziphus jujuba*	Rutin; maslinic acid; 2α-hydroxyursolic acid, and other compounds.; Apigenin 7-methoxy-6-C-glycoside	Fruits, seeds	193-194
		Animal (Insect) Dyes		
70.	(a) *Bolinus brandaris* (*Murex brandaris*) (b) *Hexaplex trunculus* (*Murex trunculus*)	Indigotin; 6-monobromoindigotin; 6,6′-dibromoindigotin	Hypobranchial glandular secretion of the snails	195-196
71.	*Dactylopius coccus*	Carminic acid	Dried insect bodies of the female insects	197-198
72.	*Kermes vermilio, Kermes ilicis*	Kermesic acid; flavokermesic acid	Dried bodies of the female scale insects	197-198
73.	*Laccifer lacca*	Laccaic acids A, B, C, E, and laccaic acid D	Resinous secretion of the lac female insects	197-198

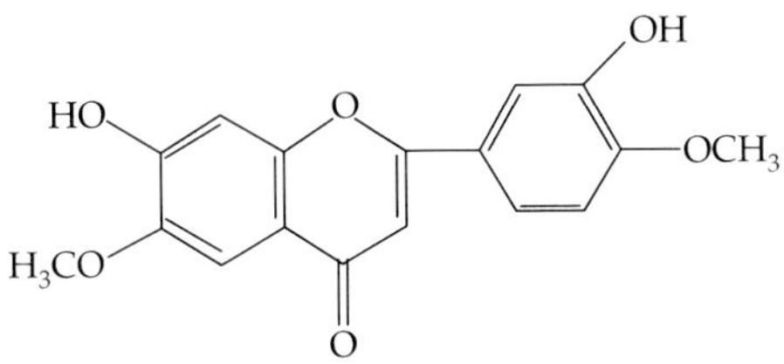

Abrectorin

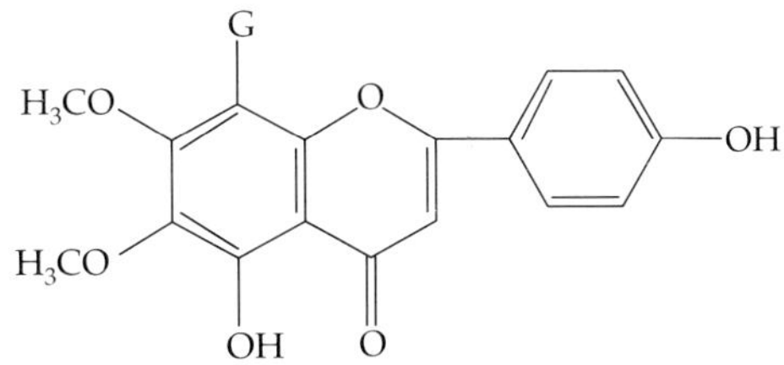

Abrusin, G = Glucose

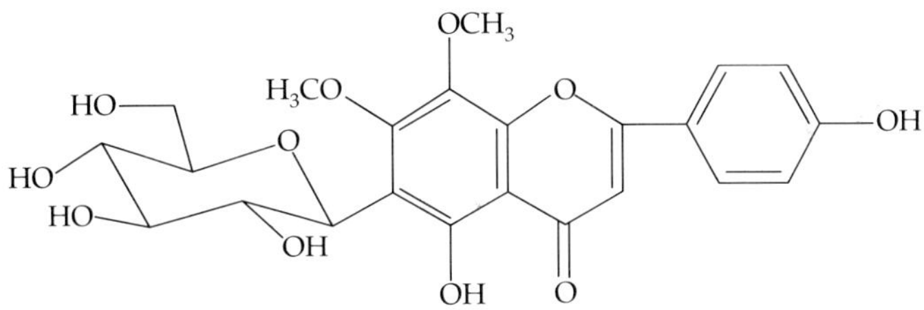

Percatorin I

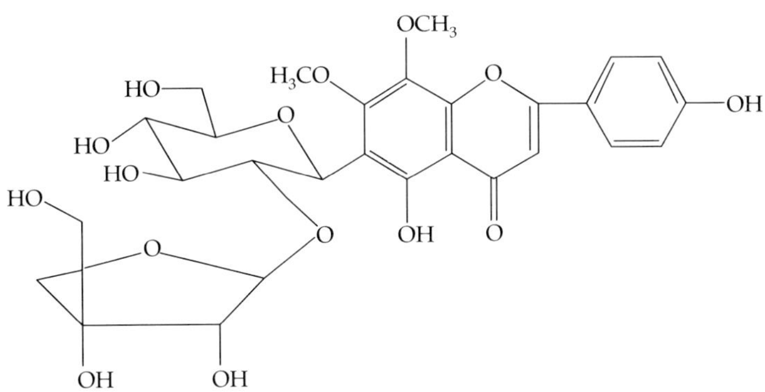

Percatorin II

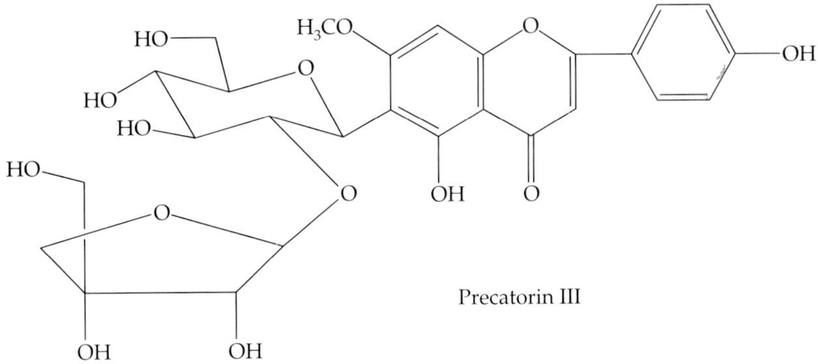

Precatorin III

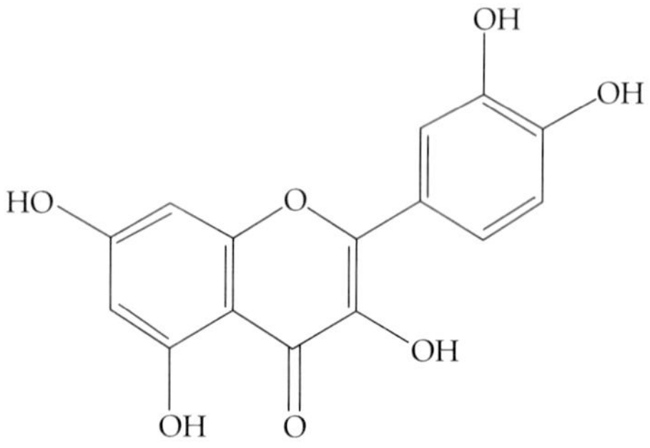

Quercetin

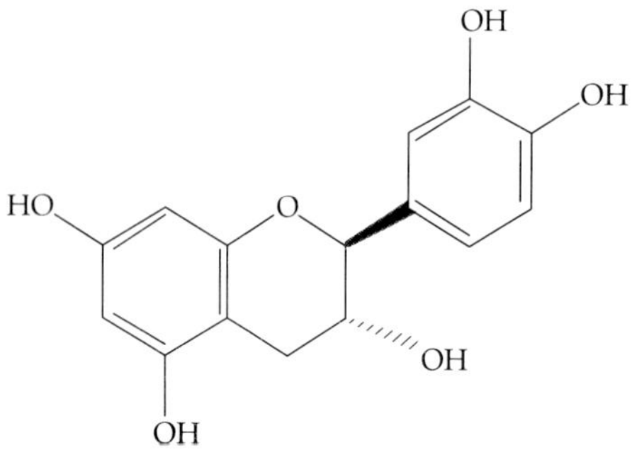

(–) Catechin

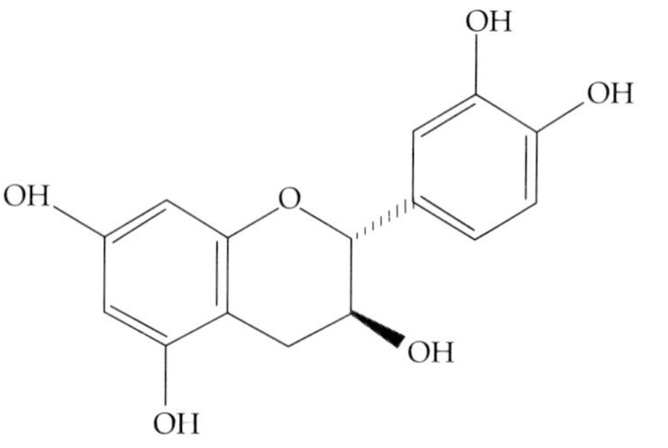

(+) Catechin

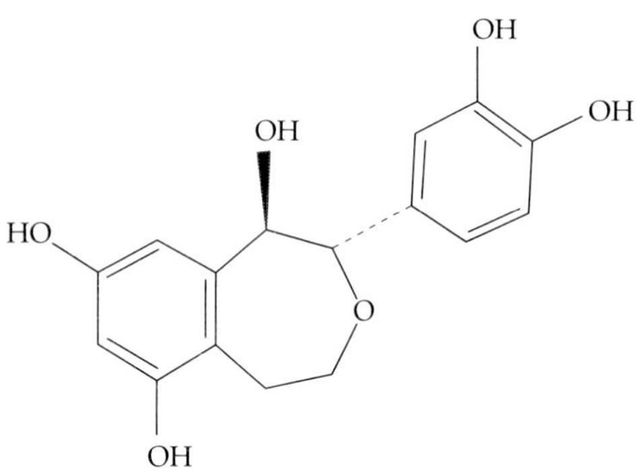

(+)-homo-isocatechin

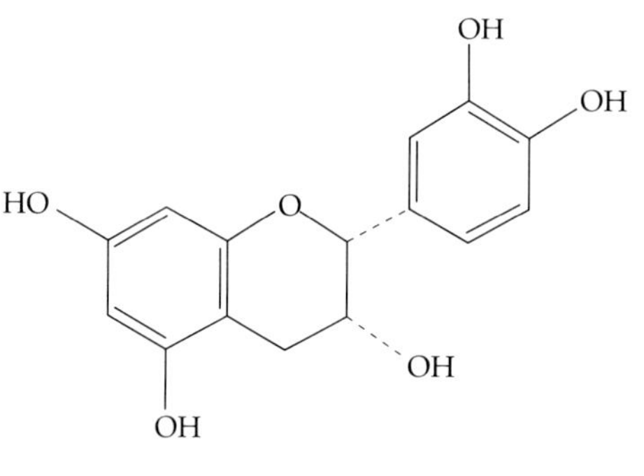

(–)-Epicatechin

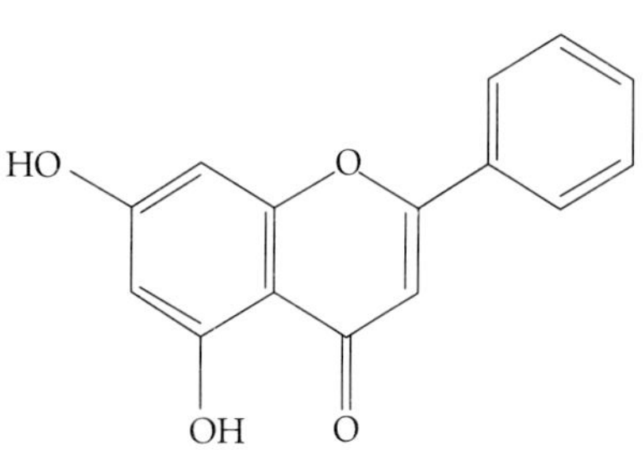

Chrysin

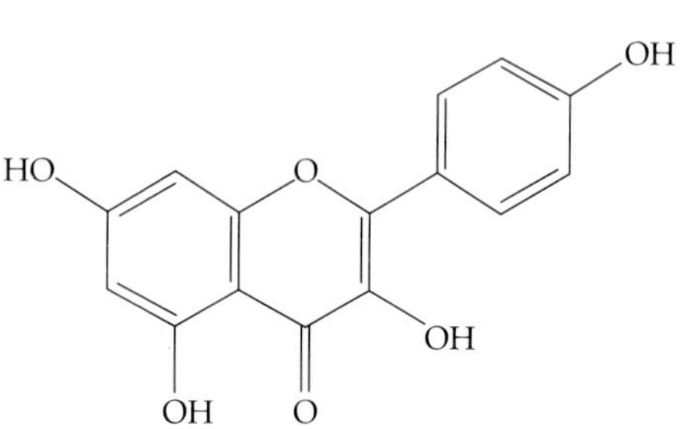

Kaempferol

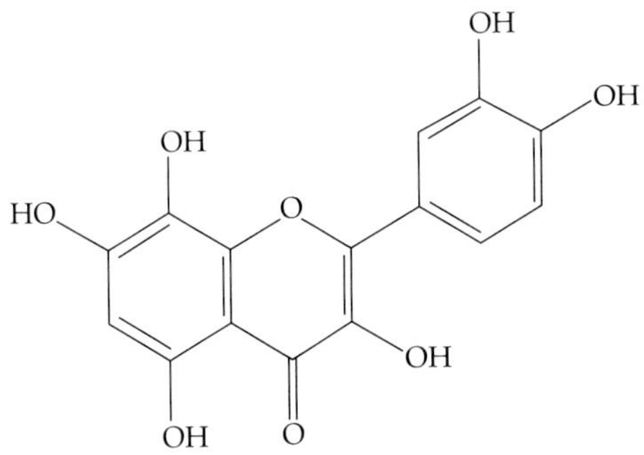

Gossyptein

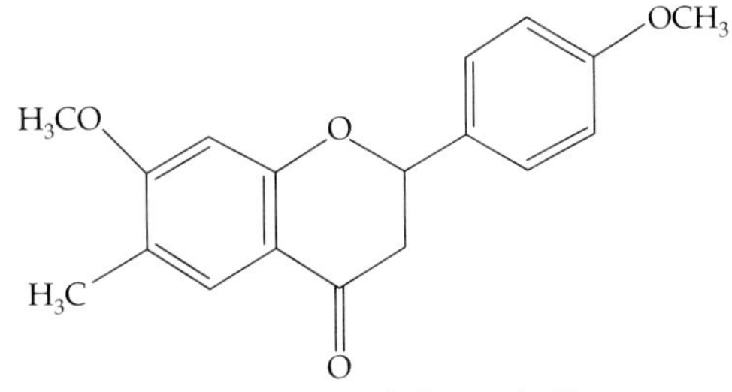

7,4′-di-O-methyl-6-methylflavanone

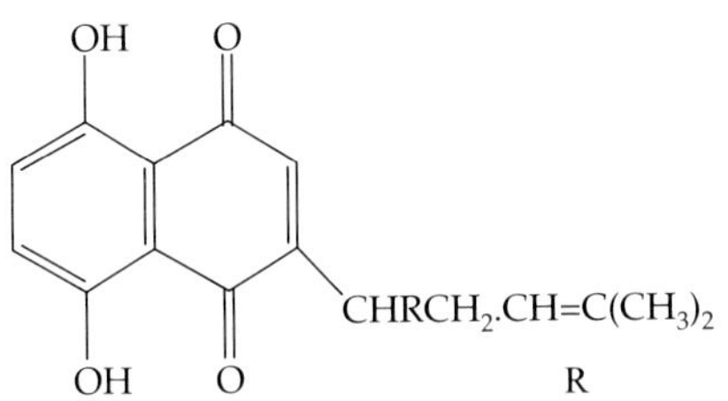

	R
Deoxyalkannin	H
Alkanninacetate	OAc
Alkannin	OH

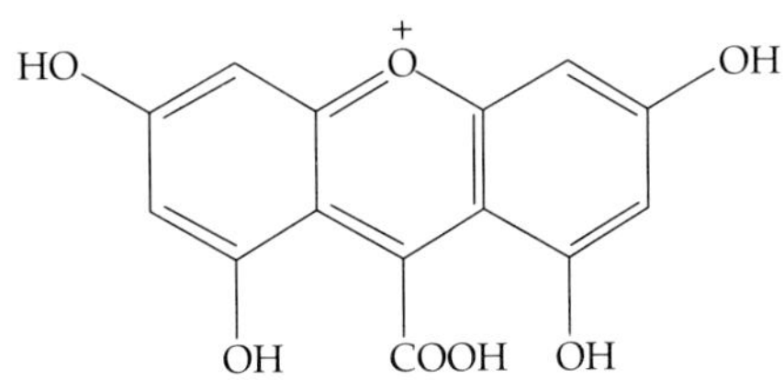

Cepaic acid

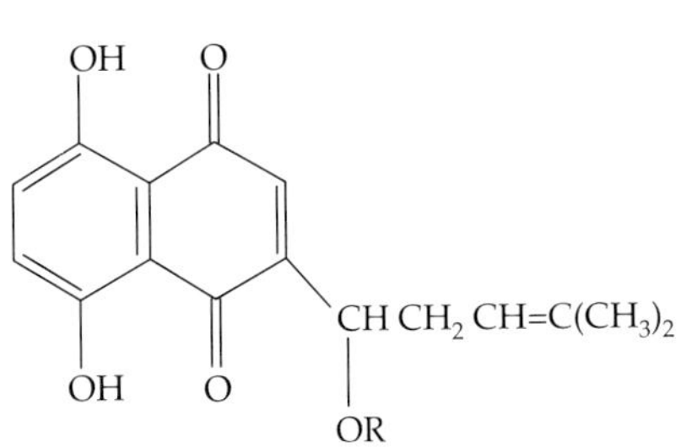

	R
Alkannin isovalerate	Isovaleryl
Alkannin angelate	Angeloyl

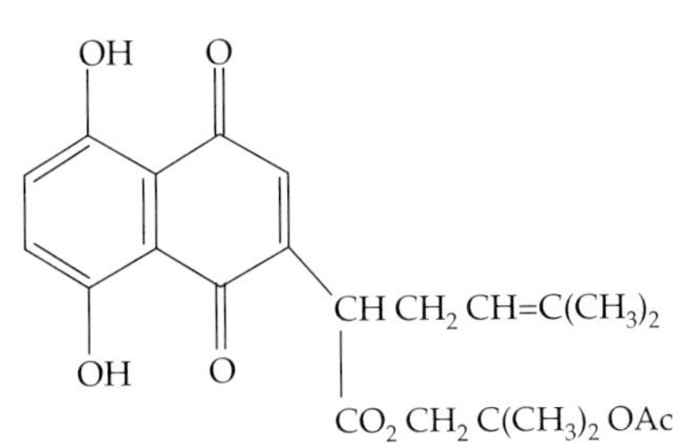

3-methyl-3-acetoxy-butanoic ester of alkannin

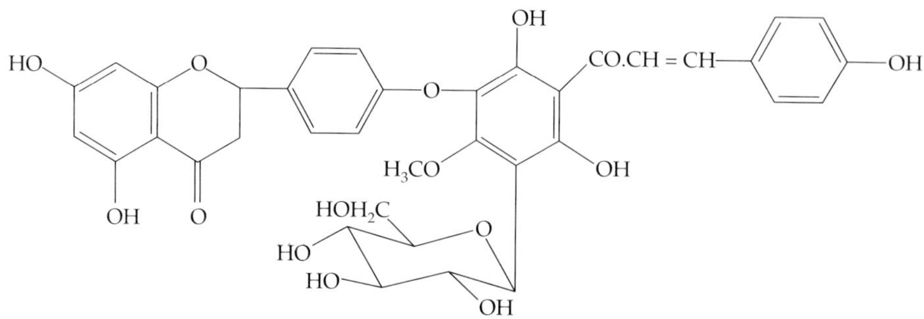

Occidentoside

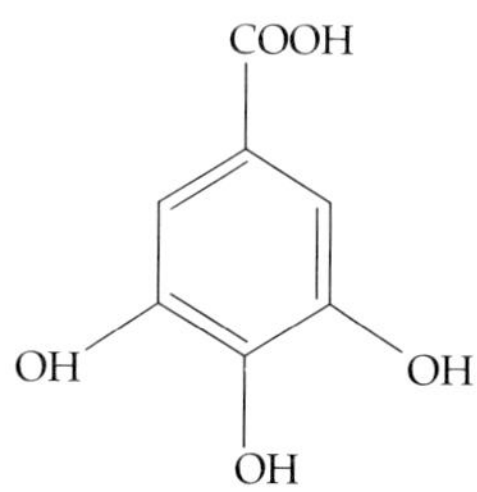

Gallic acid

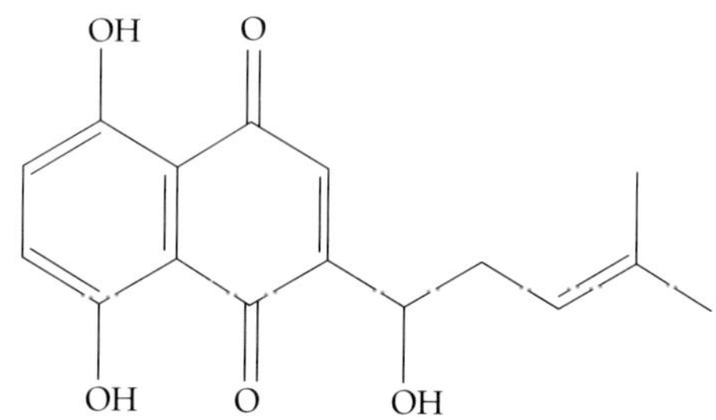

Shikonin

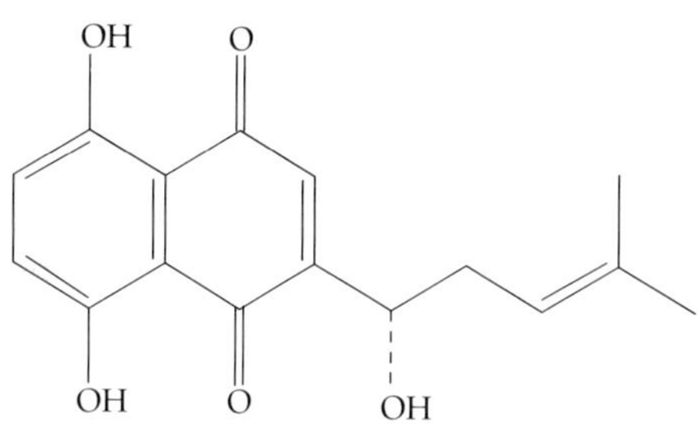

Alkannin

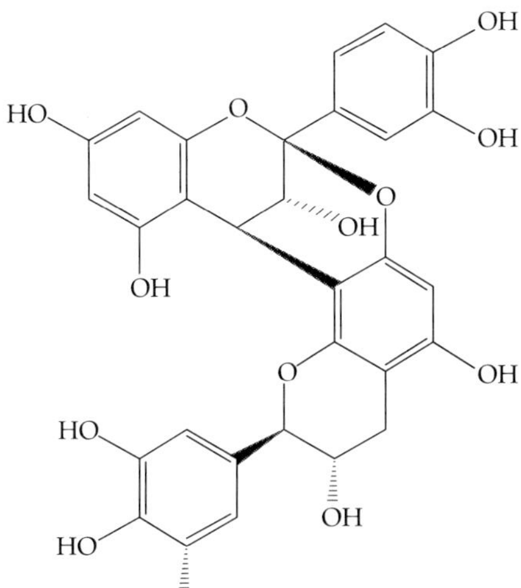

Procyanidin A1

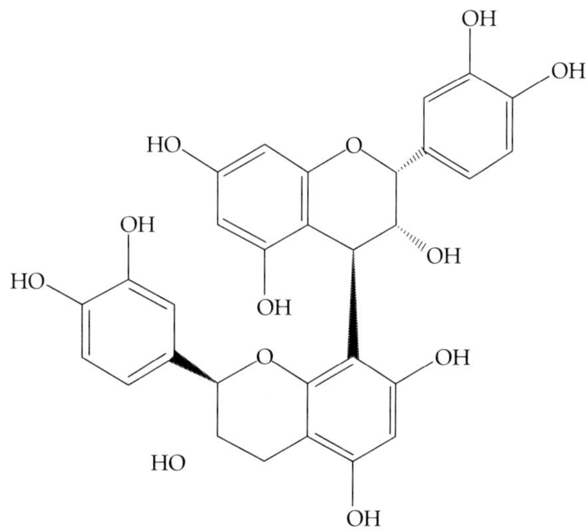

Procyanidin B1

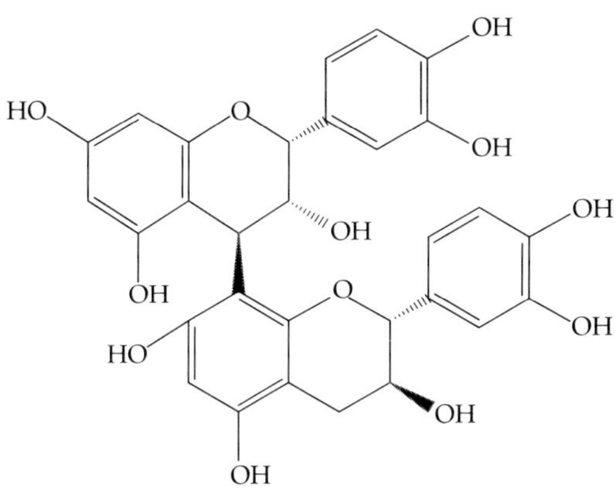

Procyanidin B2

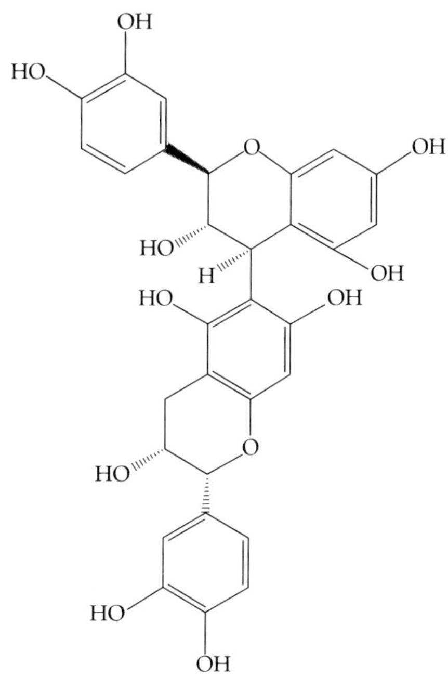

Procyanidin B7

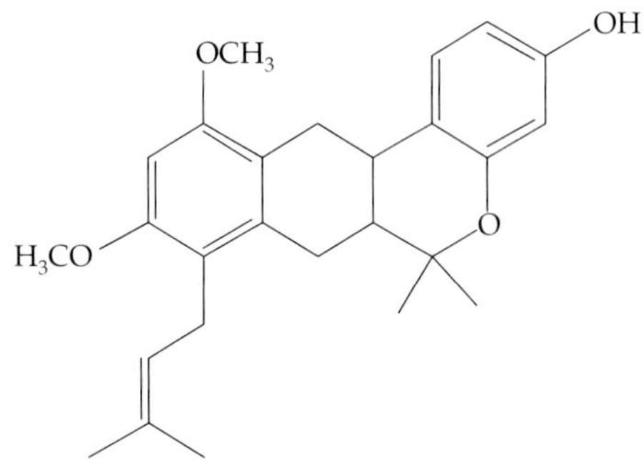

Heterophylol

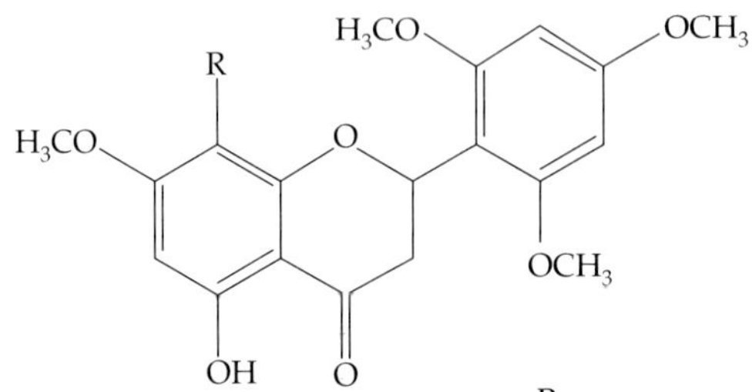

	R
Heteroflavanone A	H
Heteroflavanone B	Prenyl

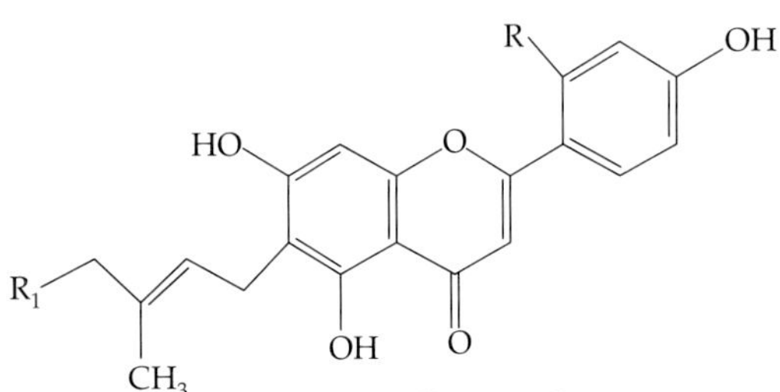

	R	R_1
Artocarmin A	H	OH
Artocarmin B	H	p-coumarate
Artocarmin C	H	Ferulate
Artocarmin D	OH	OH

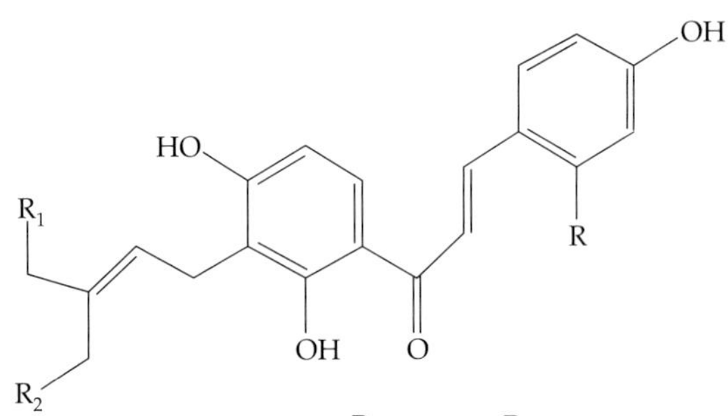

	R	R_1	R_2
Artocarmitin A	H	OH	H
Artocarmitin B	H	Ferulate	H
Artocarmitin C	H	OH	Sinapate

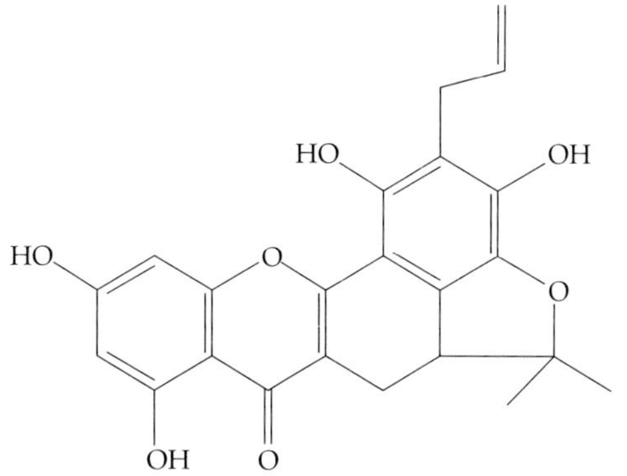

Artonin J

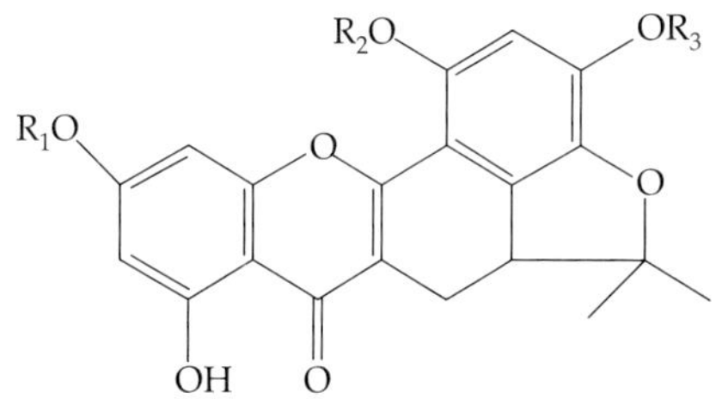

	R_1	R_2	R_3
Artonin K	CH_3	H	H
Artonin L	CH_3	CH_3	H

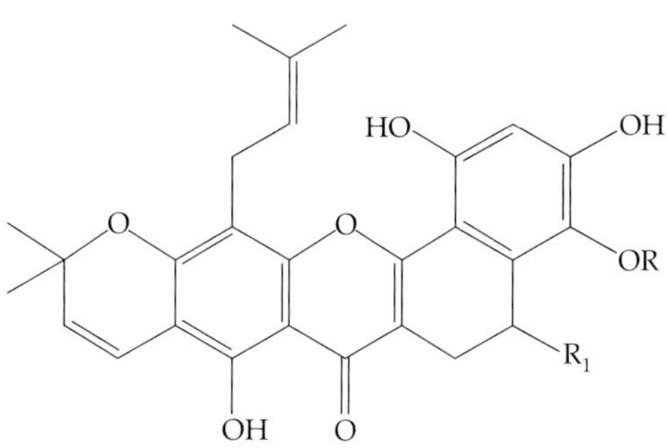

	R	R_1
Artonin A	H	$= CMe_2$
Artonin B	H	$- C\ (Me) = CH_2$

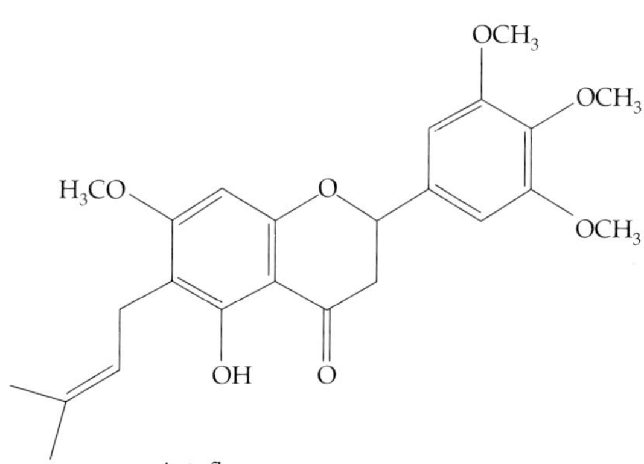

Artoflavanone

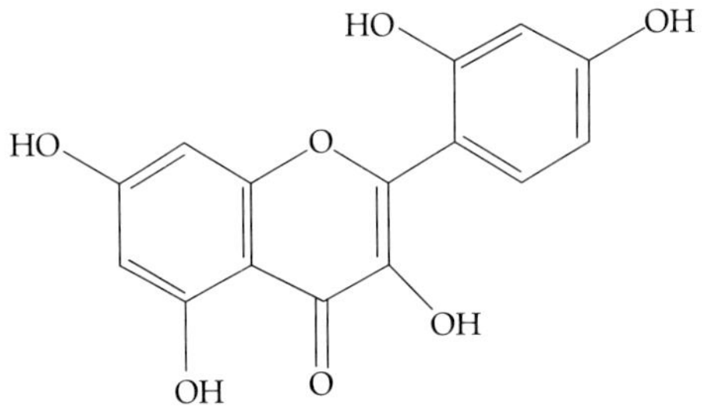

Morin

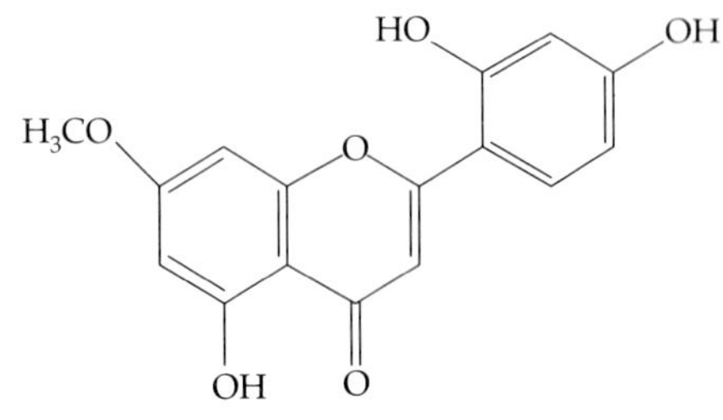

Artocarpanone

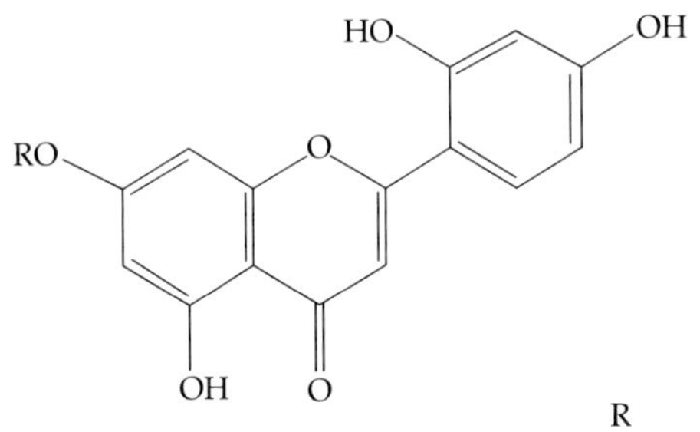

	R
Nor-artocarpetin	H
Artocarpetin	CH_3

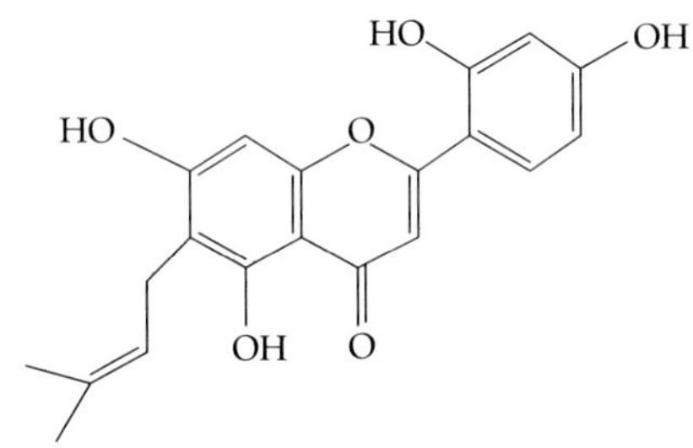

Artocarpasin

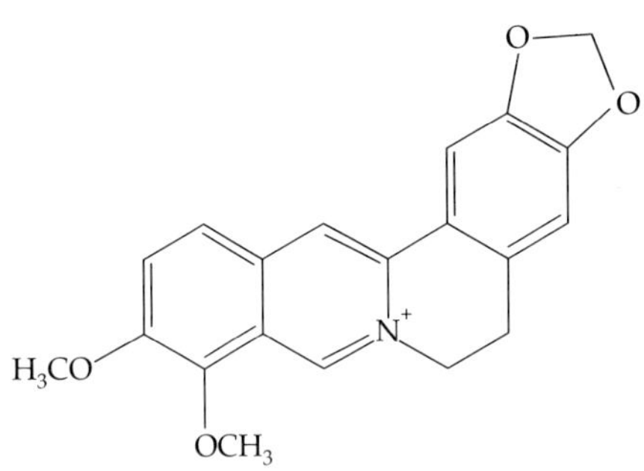

Berberine

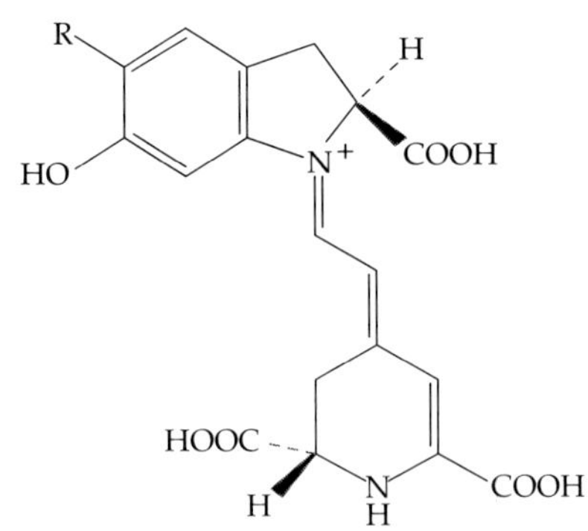

	R
Betanin	O-glucose
Betanidin	OH
Prebetanin	O-glucose-6-sulphate

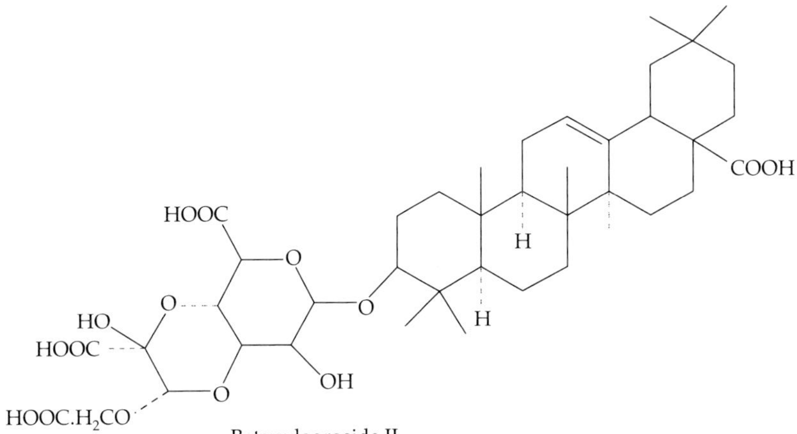

Betavulgaroside II

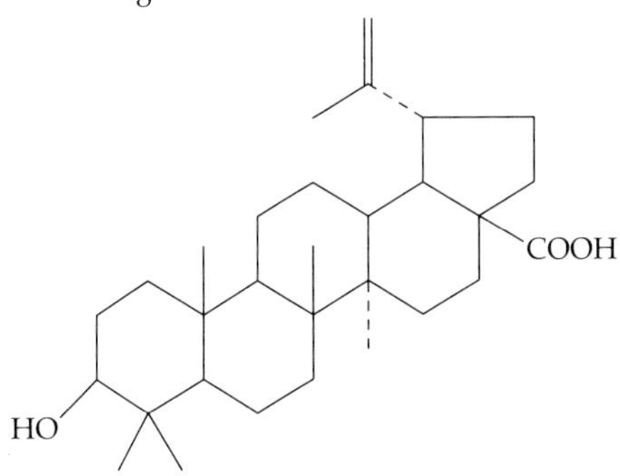

Betulinic acid

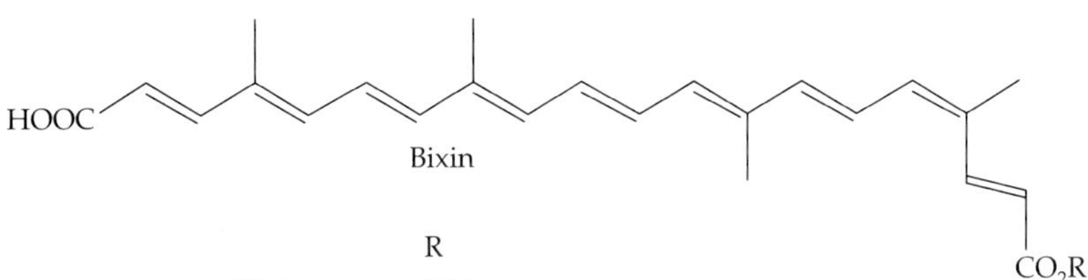

Bixin

	R
Bixin	CH_3
Nor Bixin	H

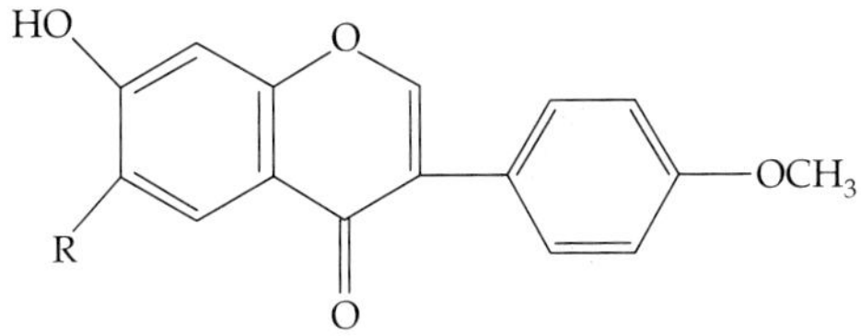

	R
Formononetin	H
Afromosin	OCH_3

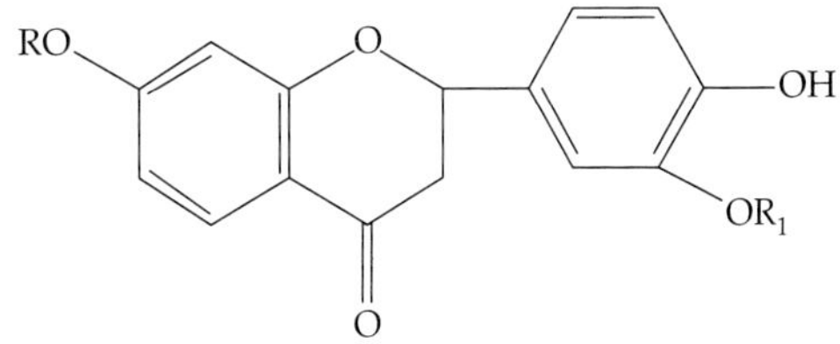

	R	R_1
Butin	H	H
Butrin	Glucosyl	Glucosyl
Isocoreopsin	Glucosyl	H
Isomonosperoside	H	Glucosyl

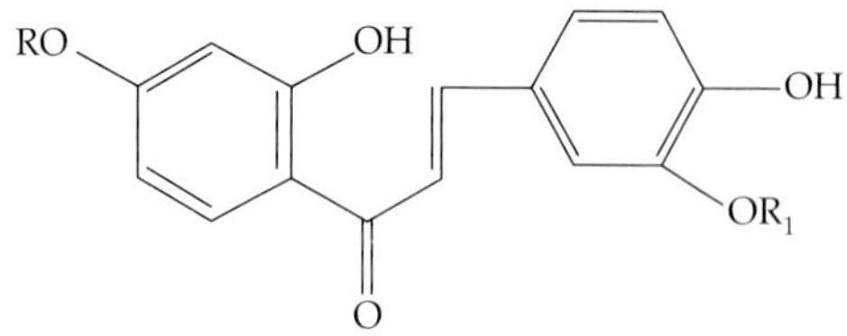

	R	R_1
Butein	H	H
Isobutrin	Glucosyl	Glucosyl
Coreopsin	Glucos	H
Monospermoside	H	Glucosyl

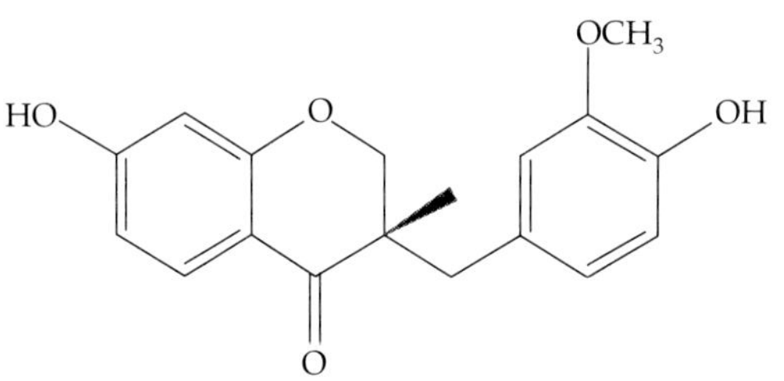

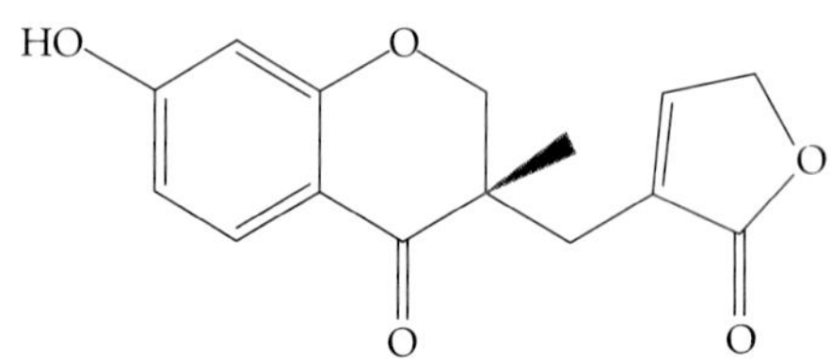

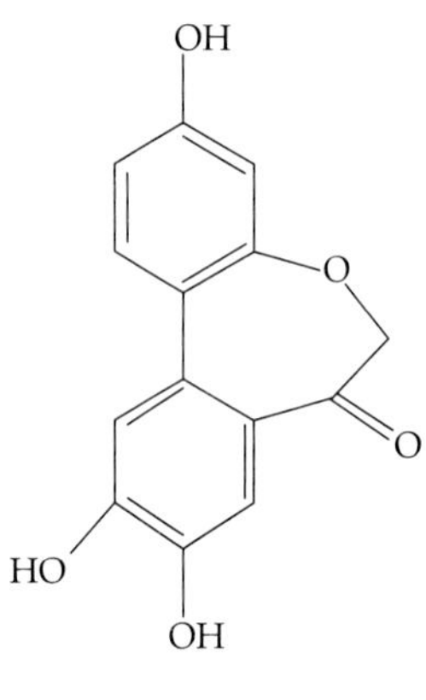

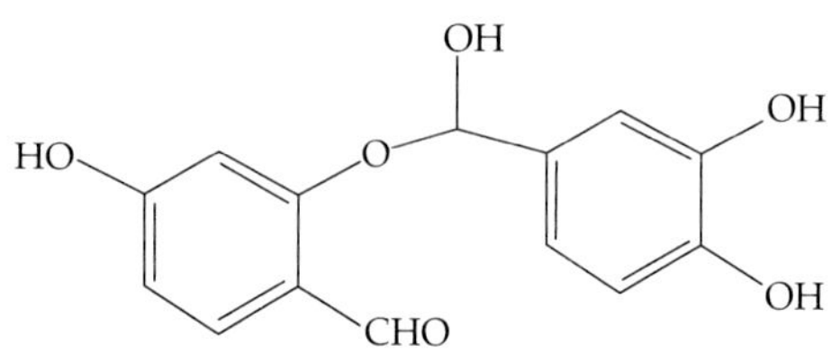

Caesalpiniaphenols A-D

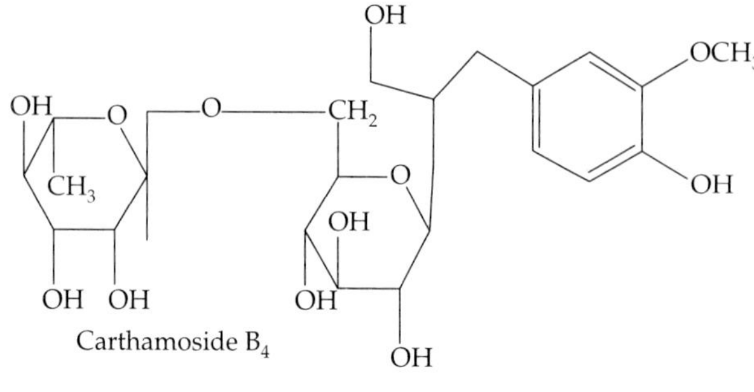

Carthamoside B_4

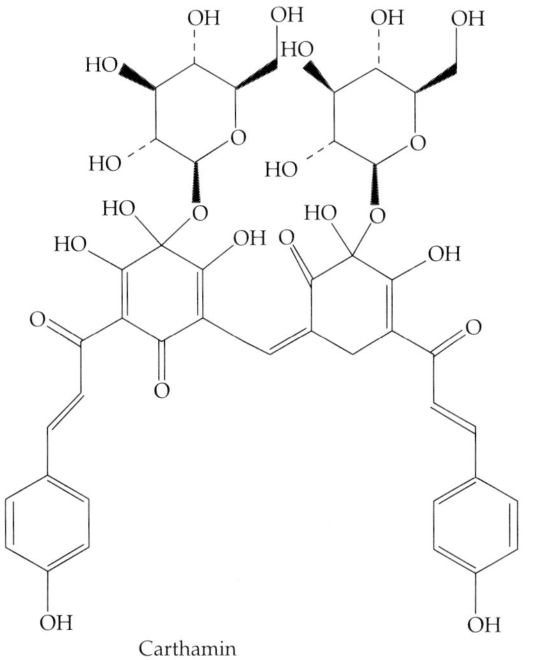

Carthamin

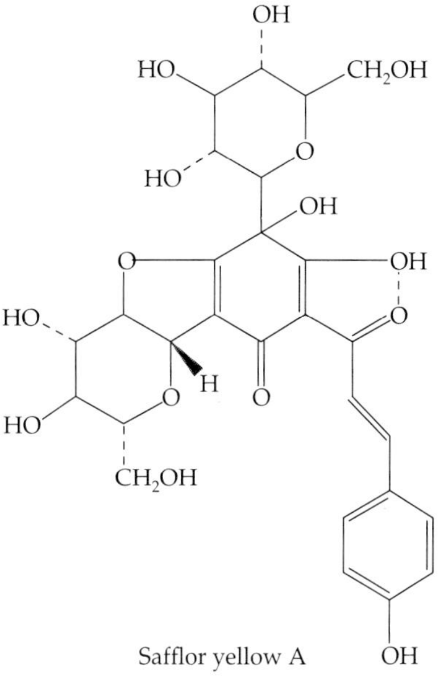

Safflor yellow A

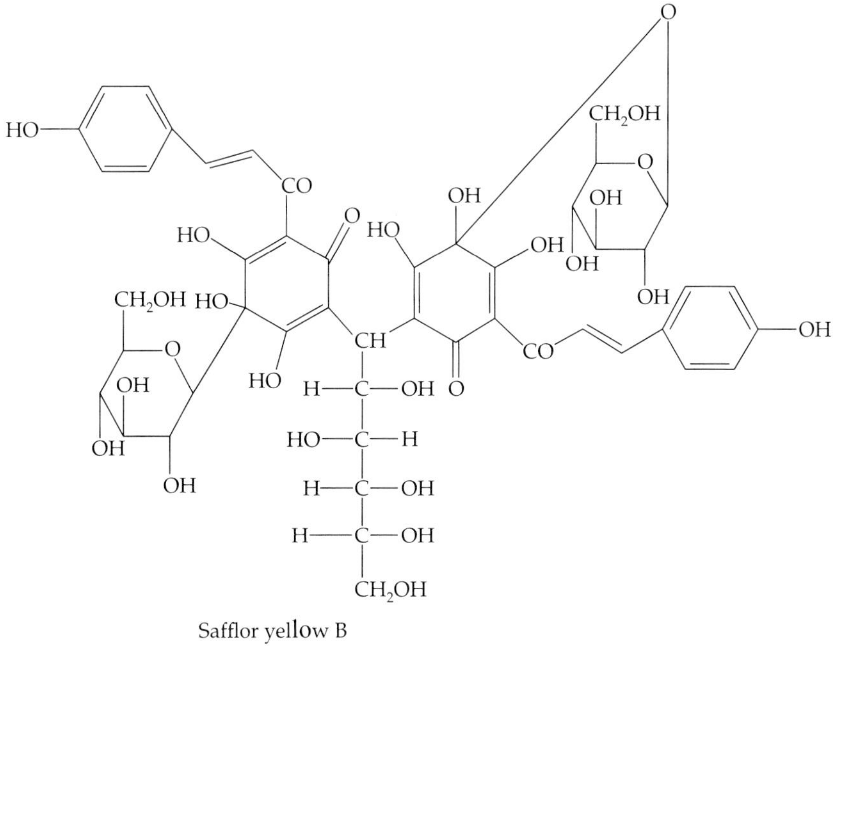

Safflor yellow B

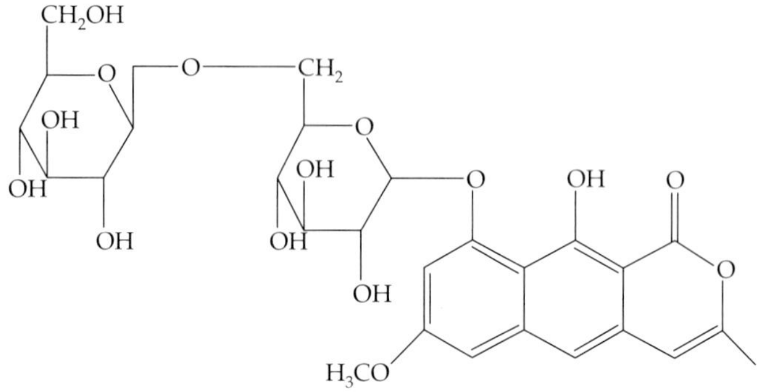

9-[(β-D-glucopyranosyl-(1 → 6)-O-β-D-glucopyranosyl)oxy]-10-hydroxy-7-methoxy-3-methyl -1H-naphtho[2, 3-c] pyran-1-one

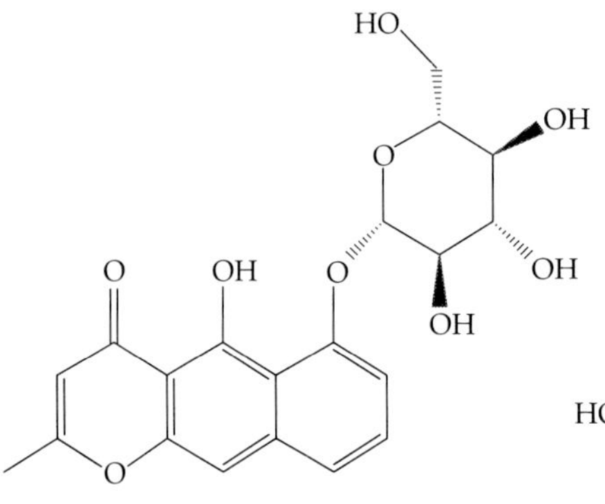

Cassiaside

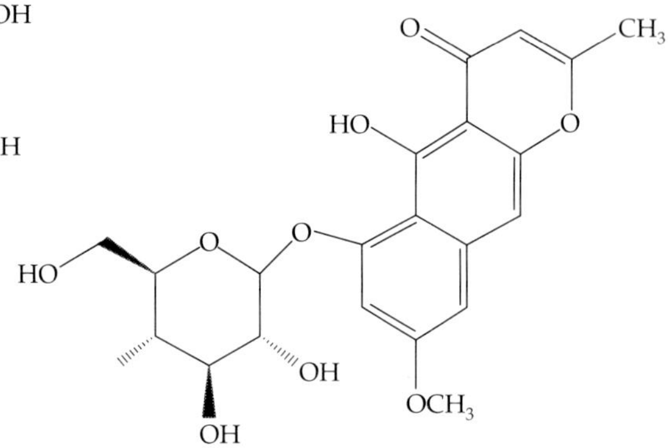

Rubrofusarin-6-β-glucoside

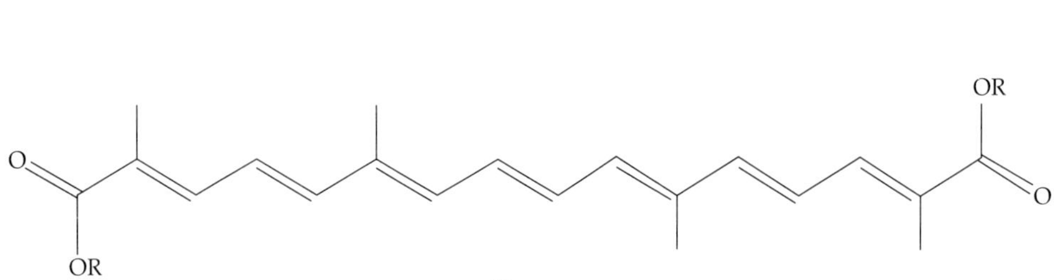

	R
Crocetin	H
Crocin	gentiobiose

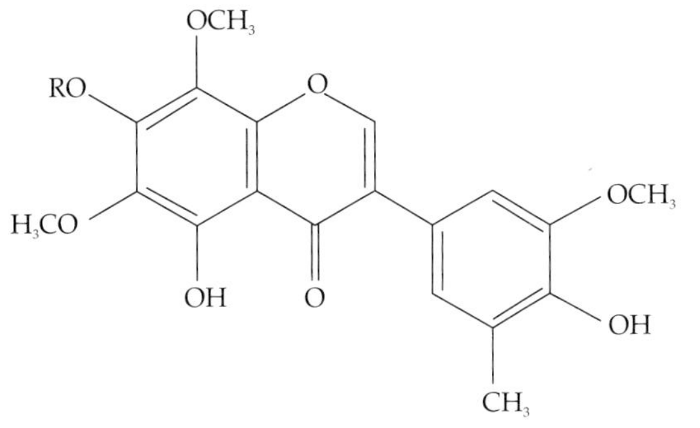

6, 8, 3′-trimethoxy-5, 4-dihydroxy-5′-methyl-7-(rhamnoglucosyl) isoflavone, R=rhamnoglucosyl

6, 8, 3′-trimethoxy-5, 4′-dihydroxy-5′-methyl-7(β-D-glucosyl) isoflavone, R=glucosyl

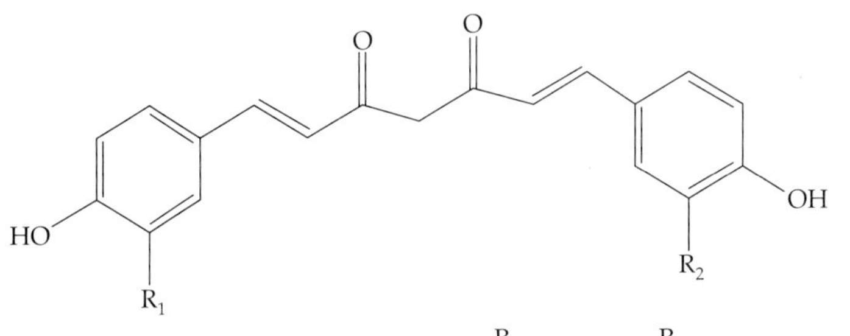

	R_1	R_2
Curcumin	OCH_3	OCH_3
Demethoxy curcumin	H	OCH_3
Bis Demethoxy curcumin	H	H

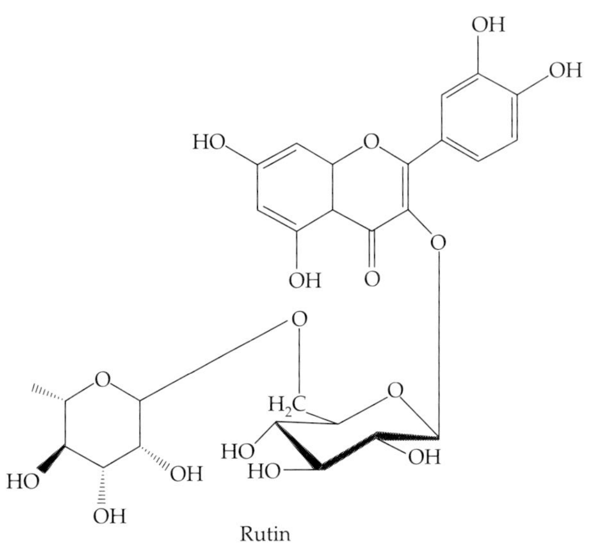

Rutin

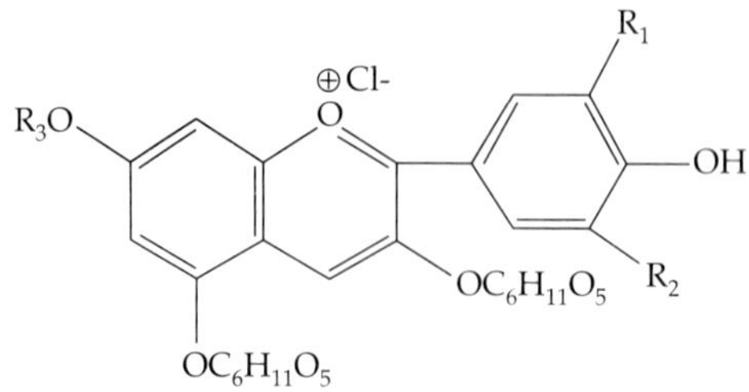

	R_1	R_2	R_3
Pelargonin chloride	H	H	H
Cyanin chloride	OH	H	H
Delphin chloride	OH	OH	H
Peonin chloride	OCH_3	H	H
Malvin chloride	OCH_3	OCH_3	H
Hirusutin chloride	OCH_3	OCH_3	OCH_3

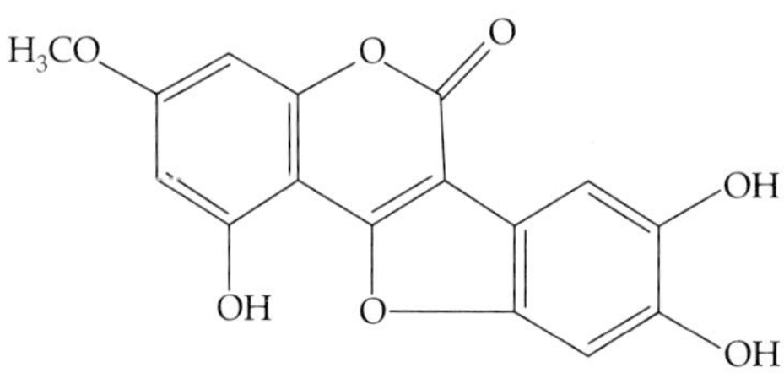

Wedelolactone

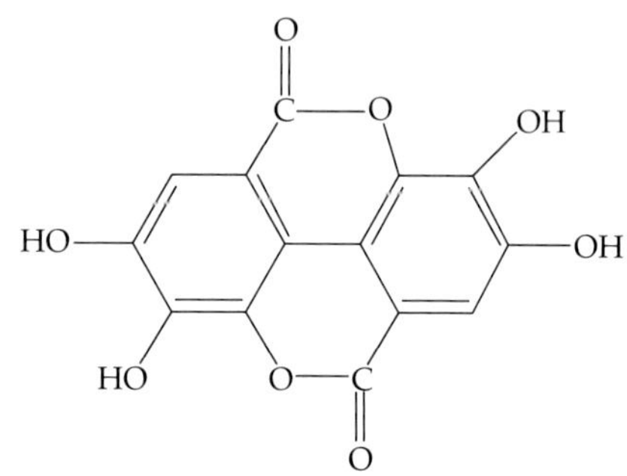

Ellagic acid

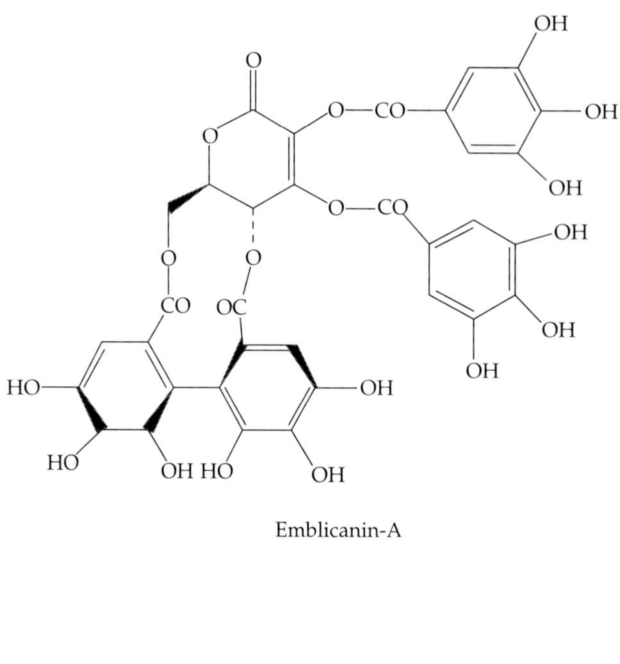

Emblicanin-A

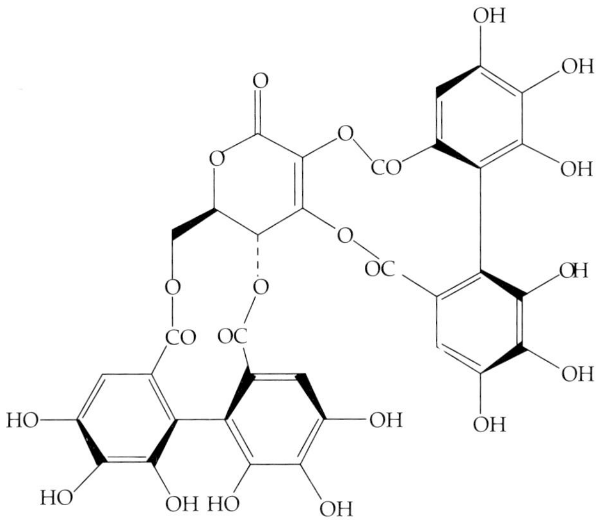

Emblicanin-B

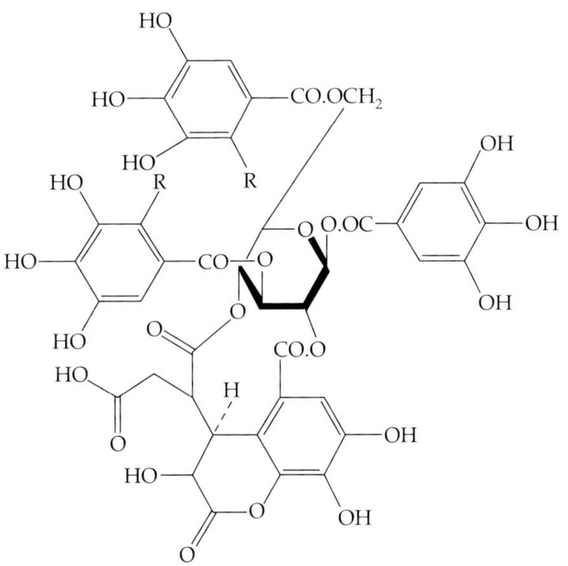

Chebulinic acid
R = H
Chebulagic acid
RR = Bond

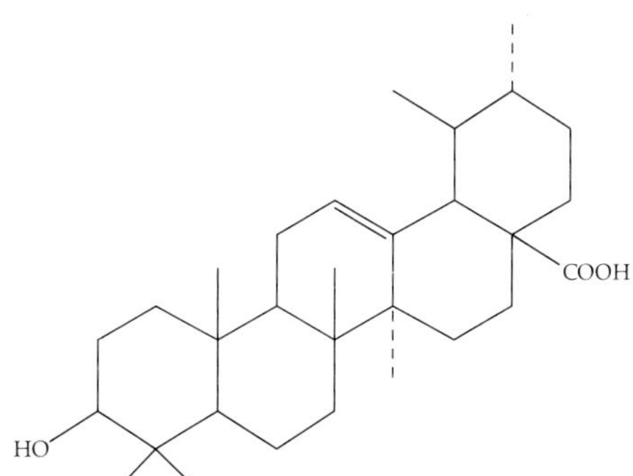

Ursolic acid

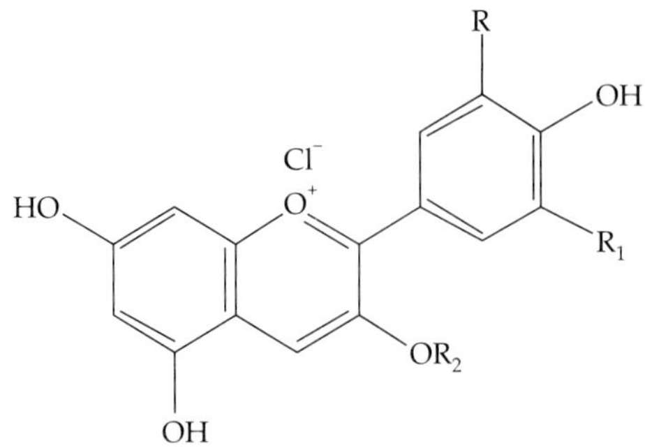

	R	R_1	R_2
Palargonidin-3-O-α-L-rhamnoside	H	H	Rhamnosyl
Delphinidin-3-O-α-L-rhamnoside	OH	OH	Rhamnosyl
Peonidin-3-O-rutinoside	OCH_3	H	Rutinosyl
Peonidin-3-O-glucoside	OCH_3	H	Glucosyl
Cyanidin-3-O-rutinoside	OH	H	Rutinosyl

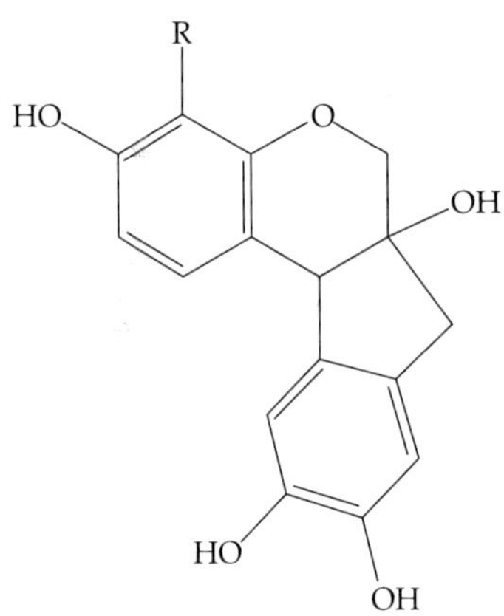

	R
Haematoxylin	OH
Brazilin	H

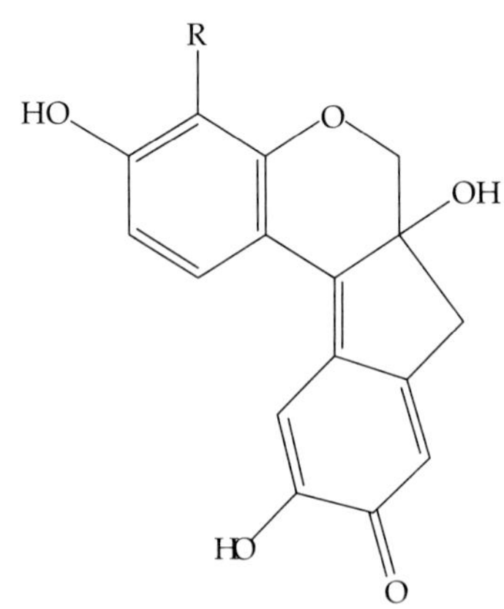

	R
Haematein	OH
Brazilein	H

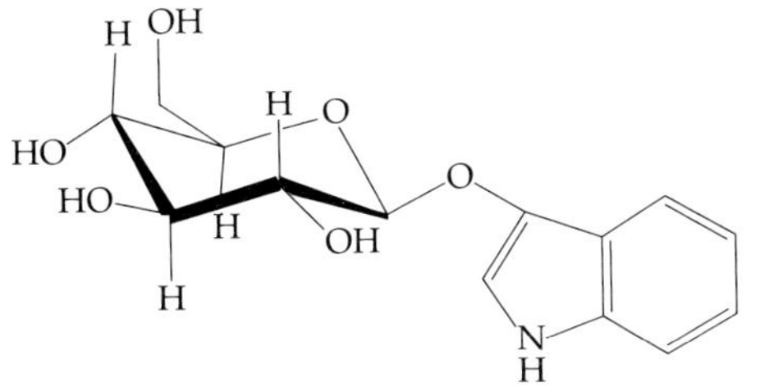

Indican

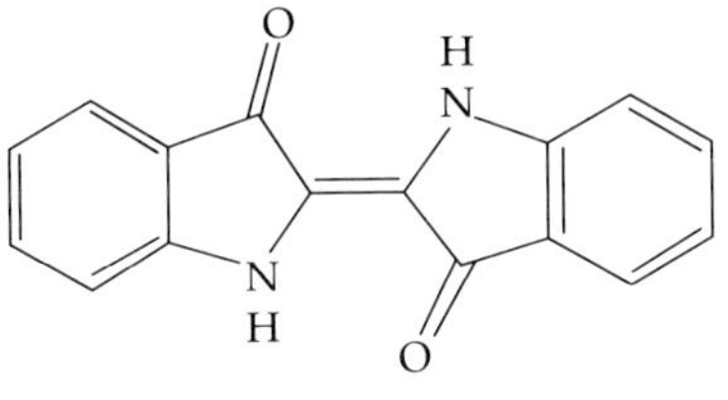

Indigotin

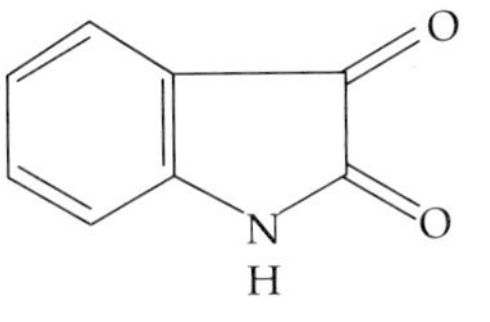

Isatin

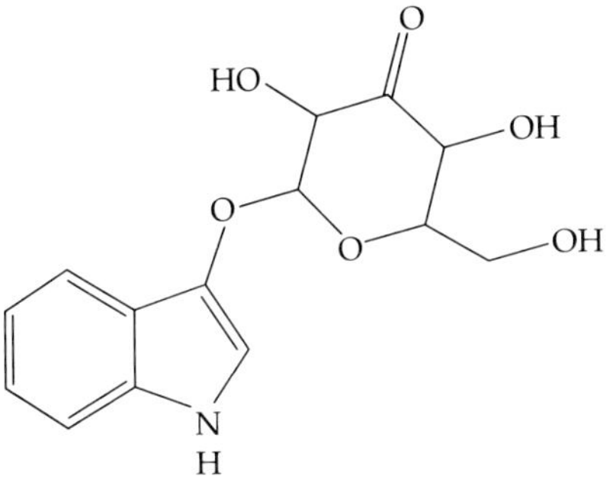

Isatan B

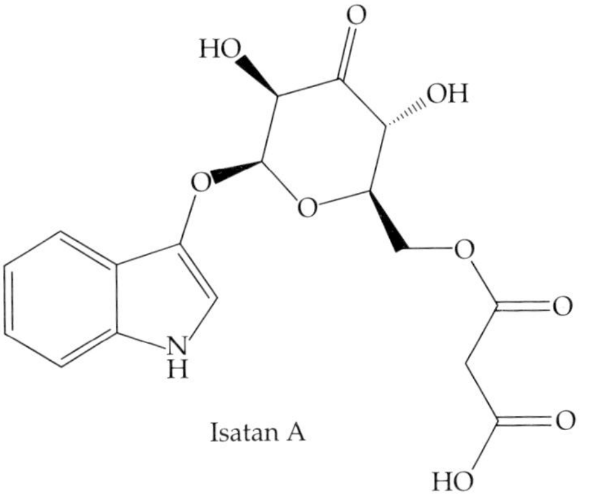

Isatan A

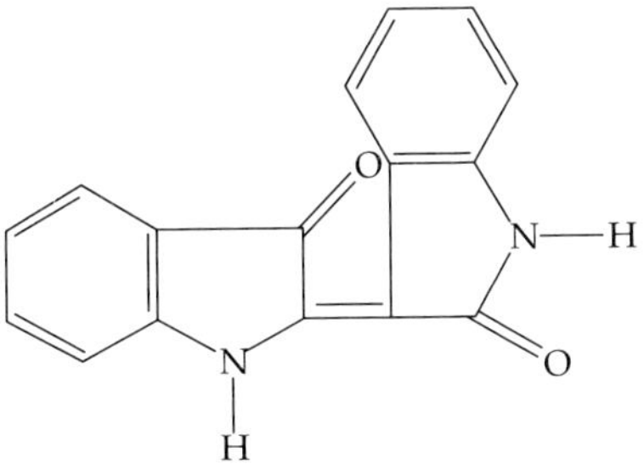

Indirubin

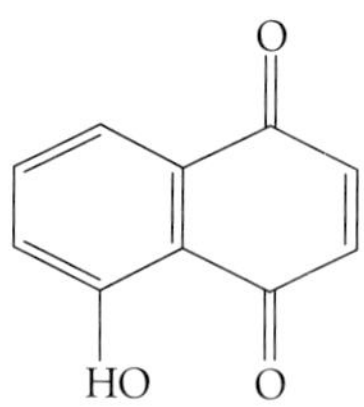

Juglone

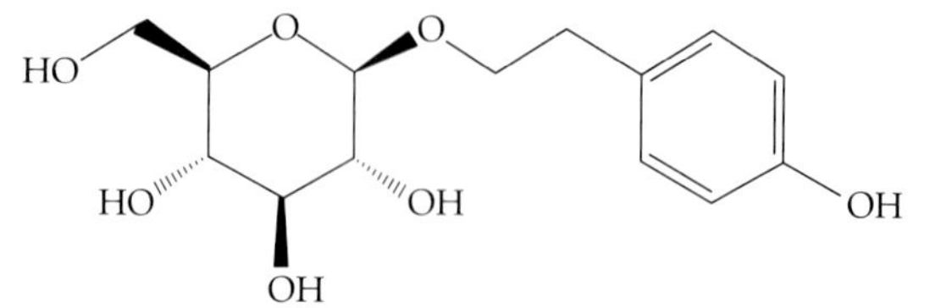

Salidroside

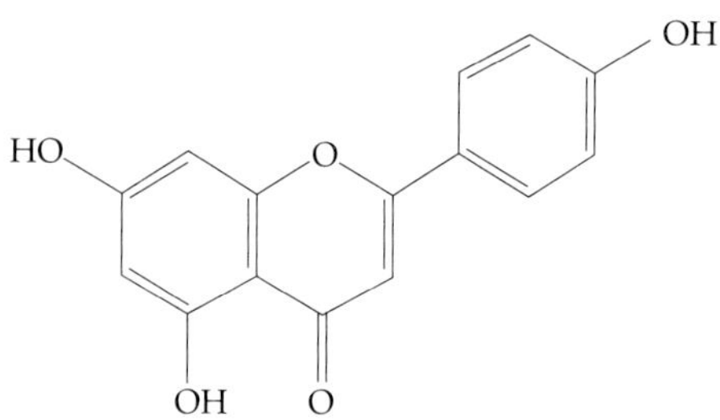

Apigenin

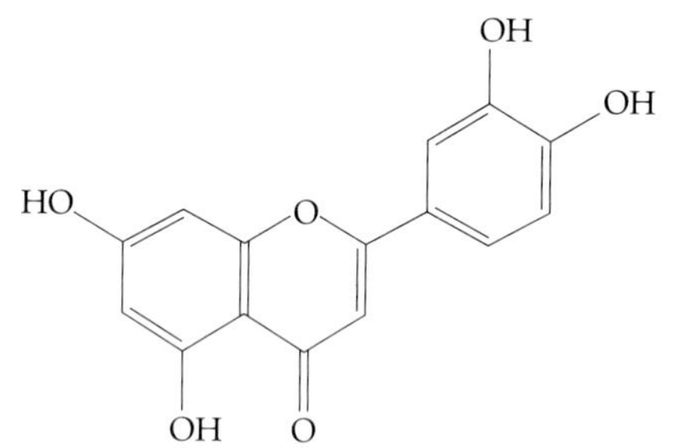

Luteolin

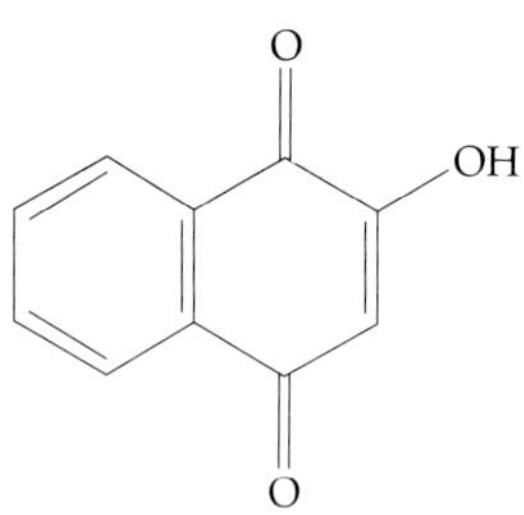

Lawsone

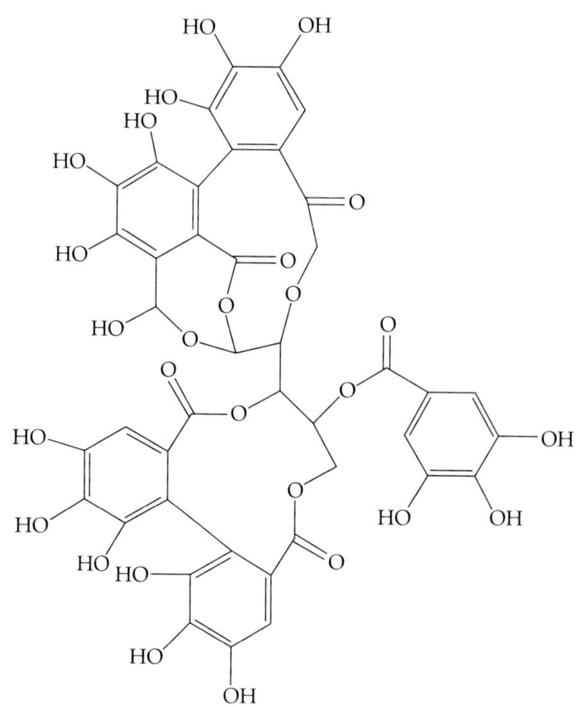

Casuarinin

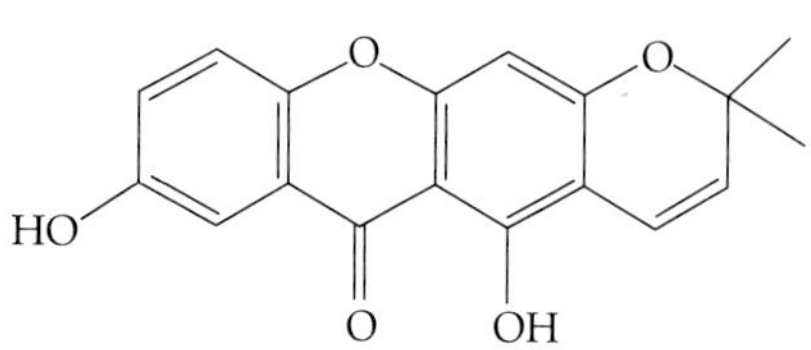

Osajaxanthone

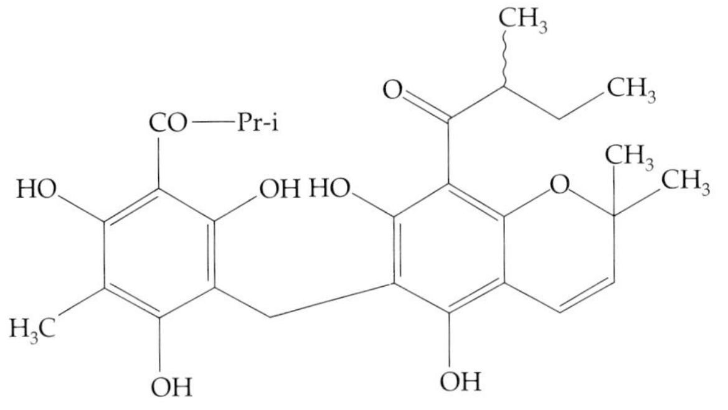

Mallotophilippen A

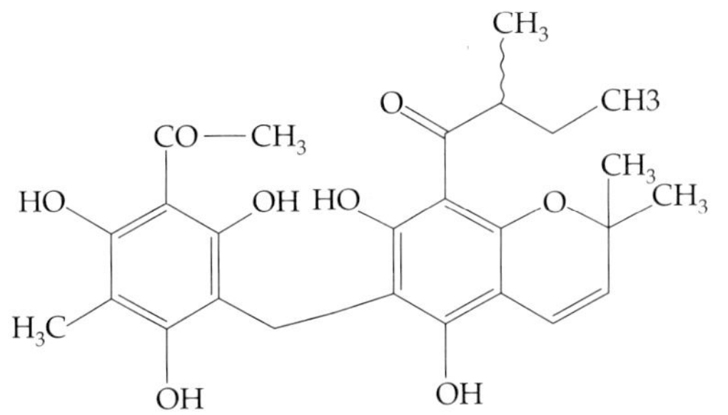

Mallotohilippen B

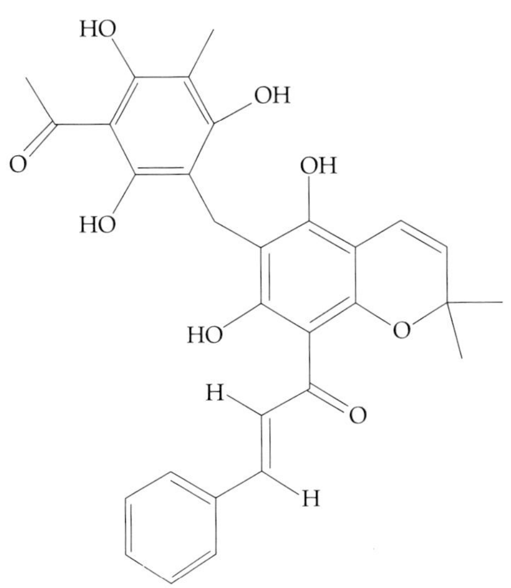

Rottlerin

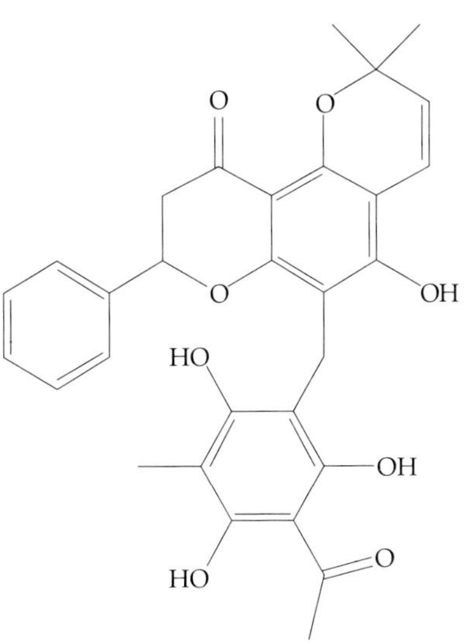

Isorottlerin

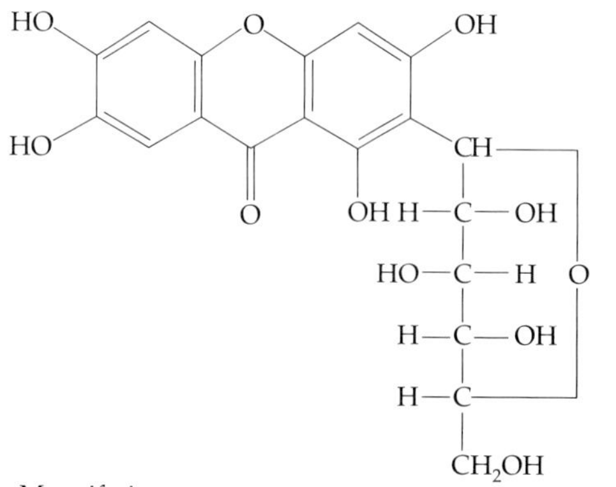

Mangiferin
(1,3,6,7-tetrahydroxyxanthone-2-glucopyranoside)

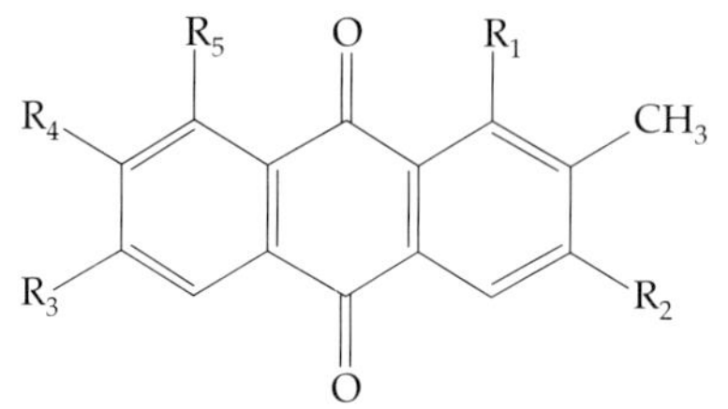

	R_1	R_2	R_3	R_4	R_5
Morenone-1	H	OH	OCH_3	H	OCH_3
Morenone-2	OH	H	H	OCH_3	OH

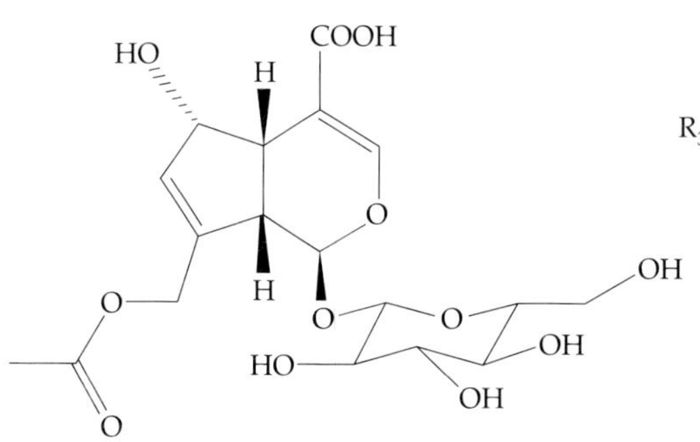

Asperulosidic acid

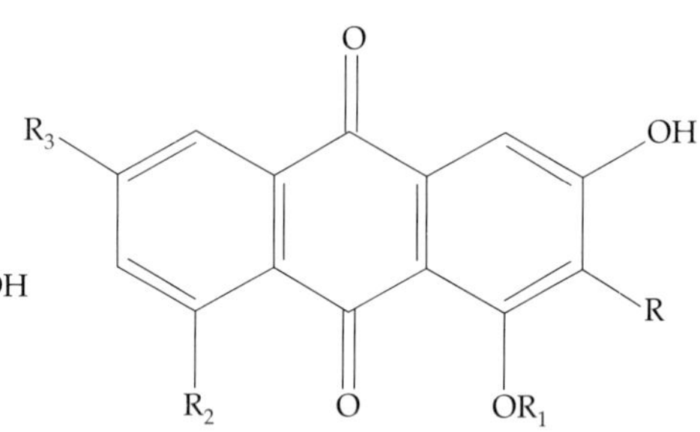

	R	R_1	R_2	R_3
Rubiadin	CH_3	H	H	H
Lucidin	CH_2OH	H	H	H
nor Damnaca-nthal	CHO	H	H	H
Munjistin	COOH	H	H	H
Rubiadin-α-methyl ether	CH_3	CH_3	H	H
Damnacanthol	CH_2OH	CH_3	H	H
Damnacanthal	CHO	CH_3	H	H
Xanthopurpurin	H	H	H	H
Xanthopurpurin-α methyl ether	H	CH_3	H	H

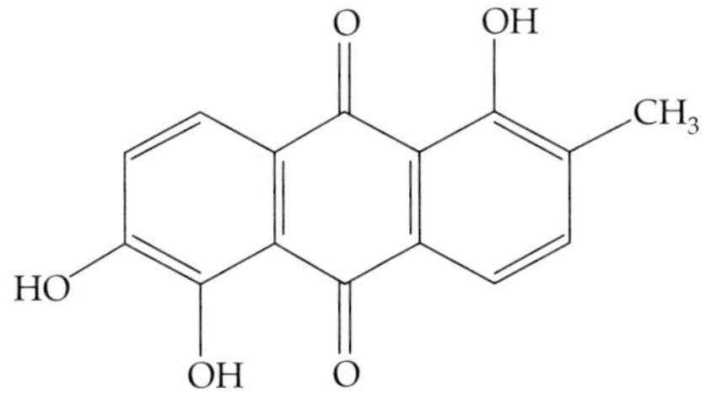

Morindone

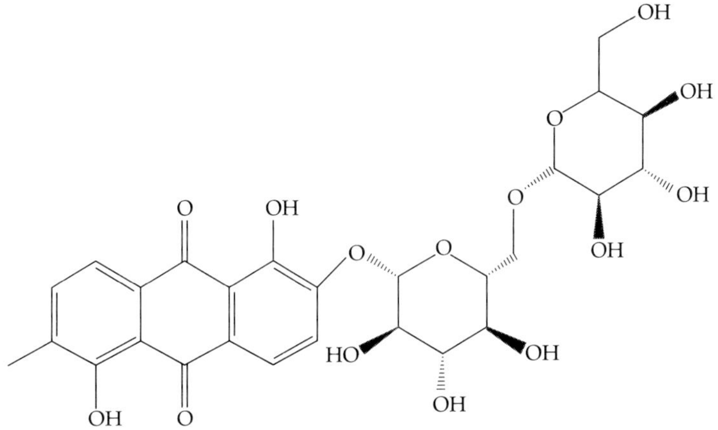

Morindin

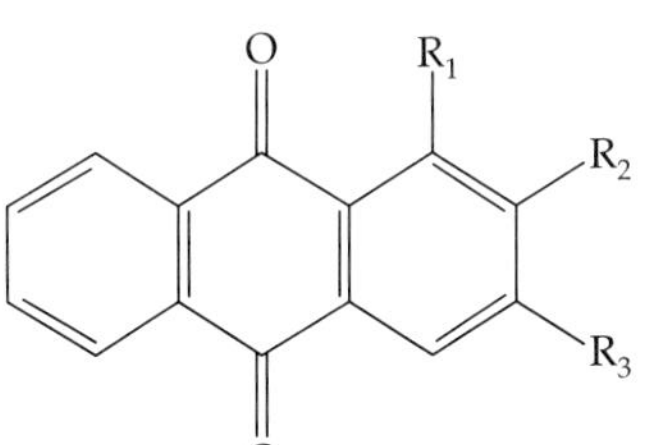

	R_1	R_2	R_3
1,2,3-trimethoxyanthraquinone	OCH_3	OCH_3	OCH_3
1,3,-dimethoxy-2-hydroxyanthraquinone	OCH_3	OH	OCH_3
1,2,-dimethoxyanthraquinone	OCH_3	OCH_3	H
1-methoxy-2-hydroxyanthraquinone	OCH_3	OH	H
1,2-dihydroxyanthraquinone (Alizarin)	OH	OH	H

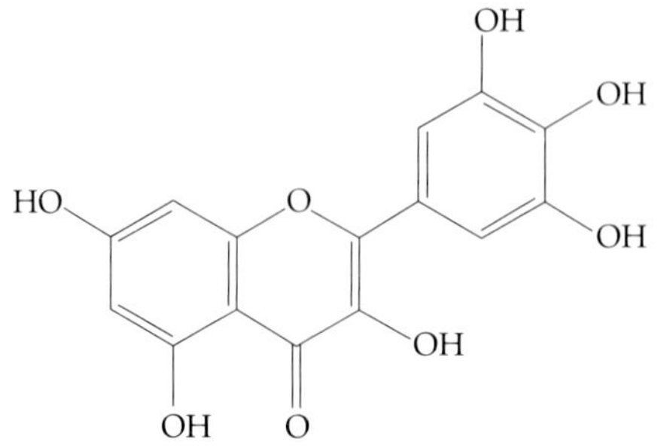

Myricetin

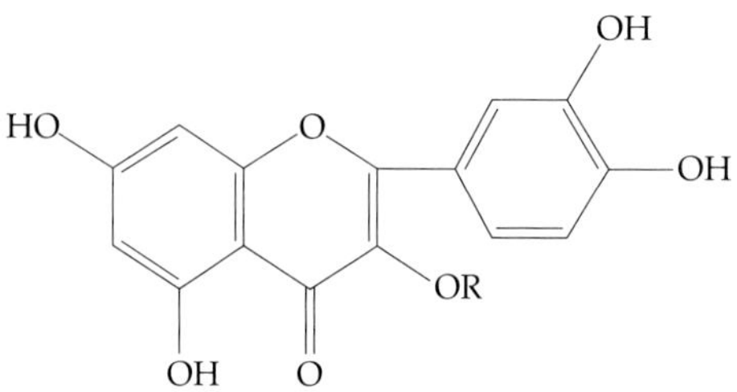

	R
Quercetin	H
Rutin	Rutinosyl
Quercitrin	Rhamnosyl
Quercimeritrin	Glucosyl

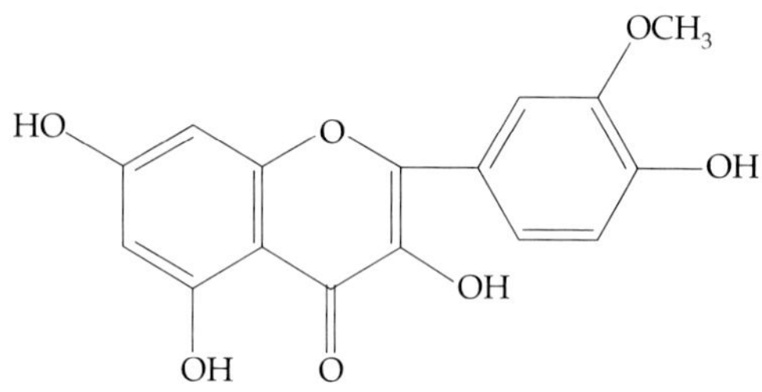

Isorhamnetin

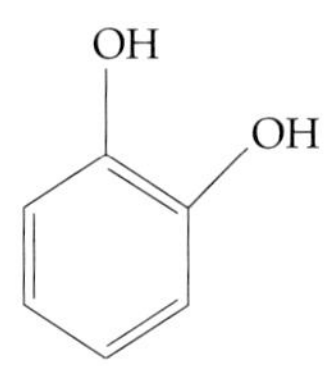

Pyrocatechol

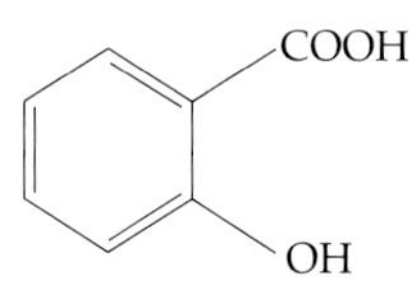

Salicylic acid

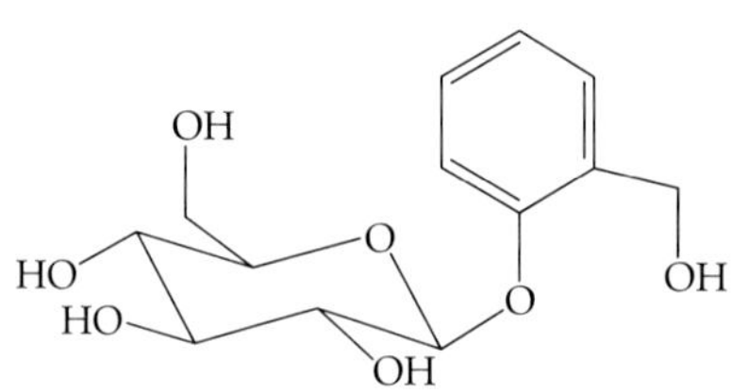

Salicin

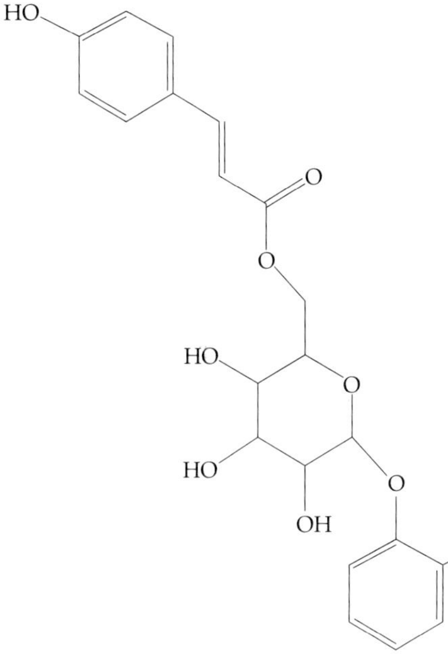

Trichocarposide

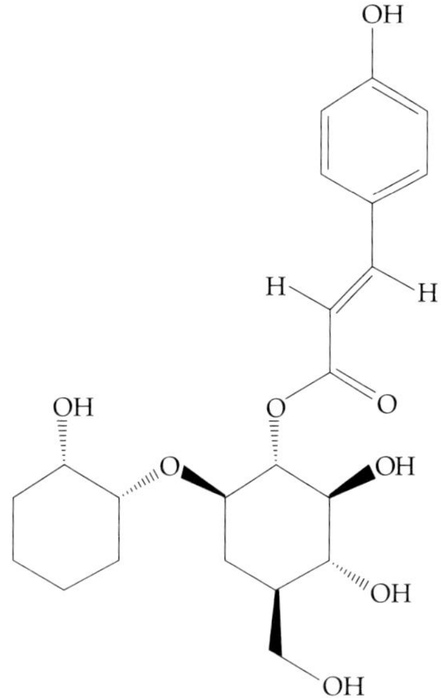

Grandidentatin

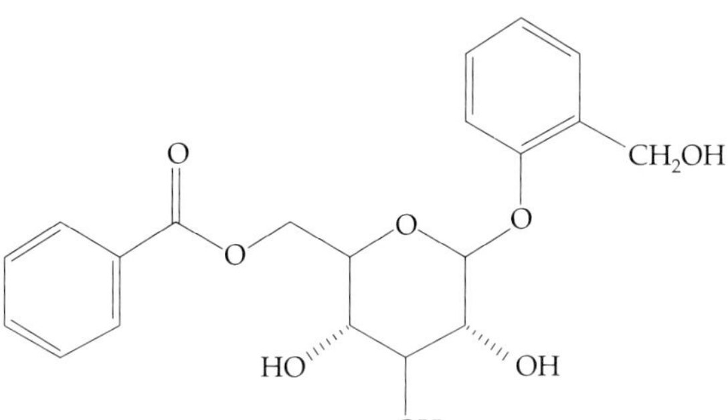

Populoside

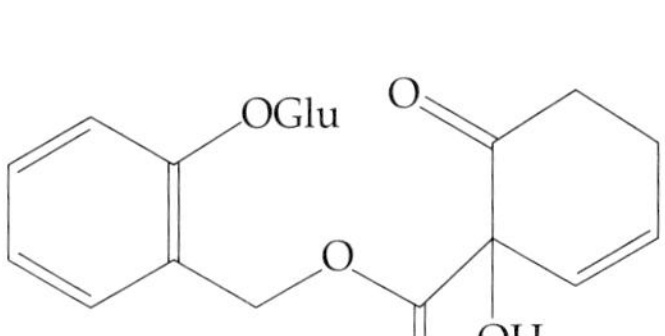

Salicortin

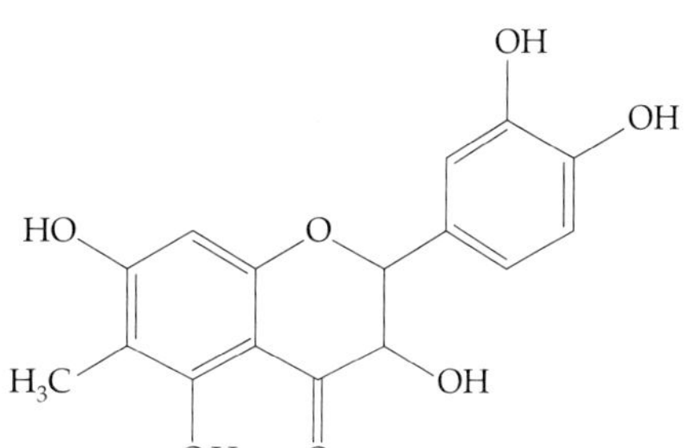

6-methyldihydroquercetin

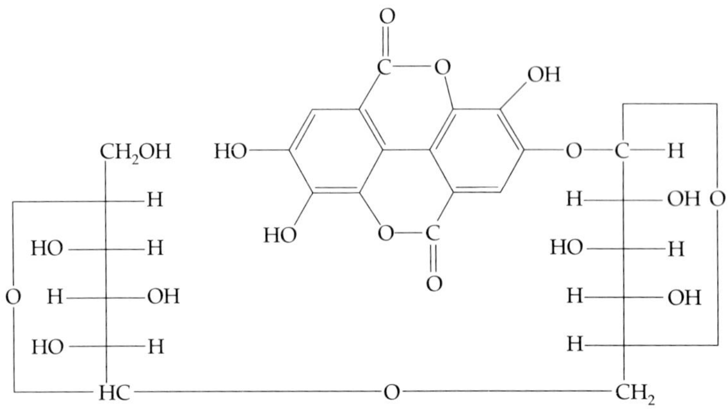

Amritoside

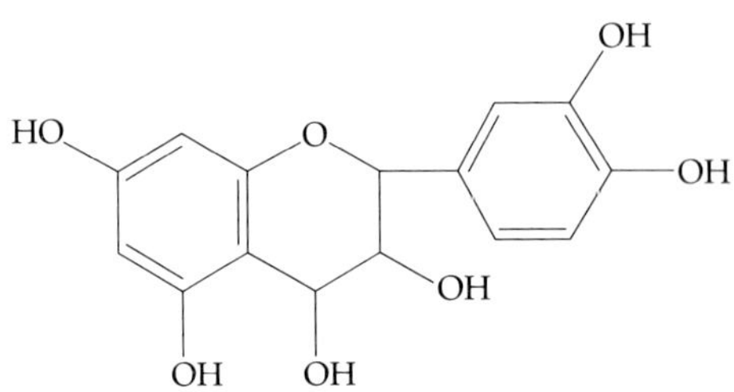

Leucocyanidin
5,7,3′4′-tetrahydroxyflavan-3,4-diol

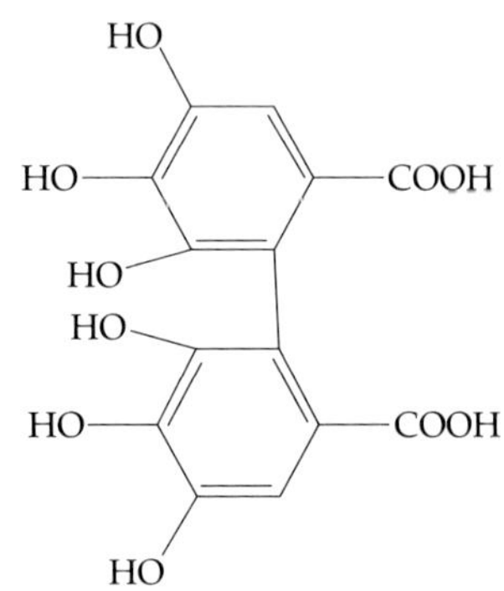

Hexahydroxydiphenic acid

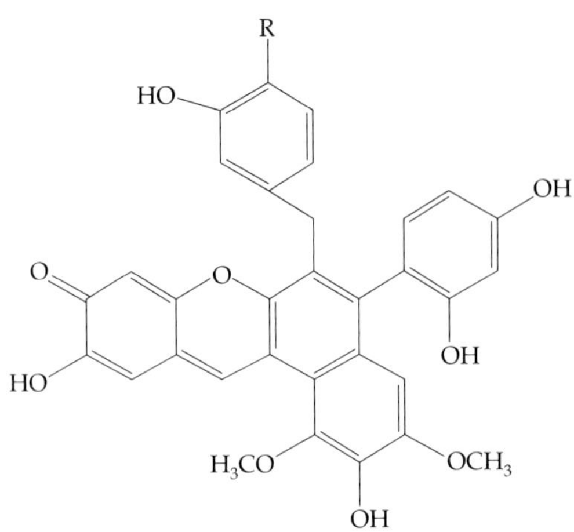

	R
Santalin A	OH
Santalin B	OCH_3

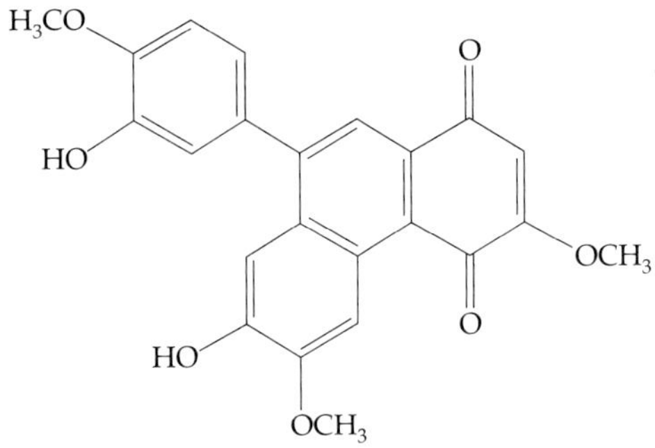

Pterolinus K

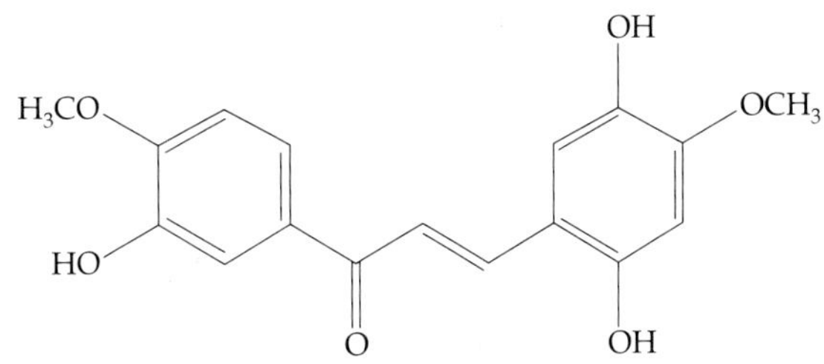

Peterolinus L

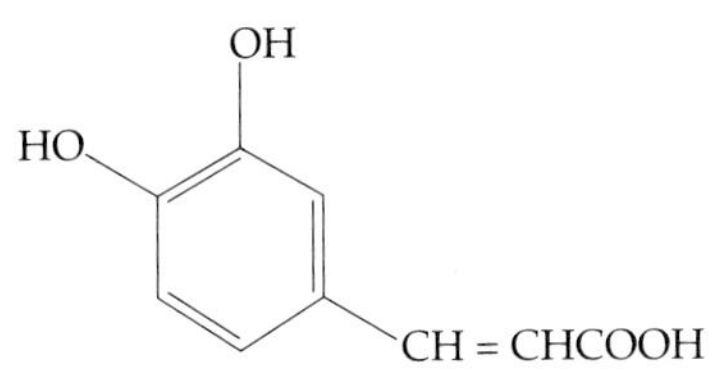

Caffeic acid

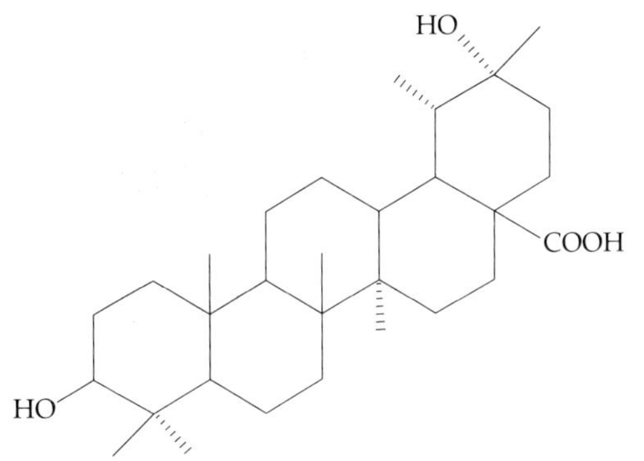

Punicanolic acid

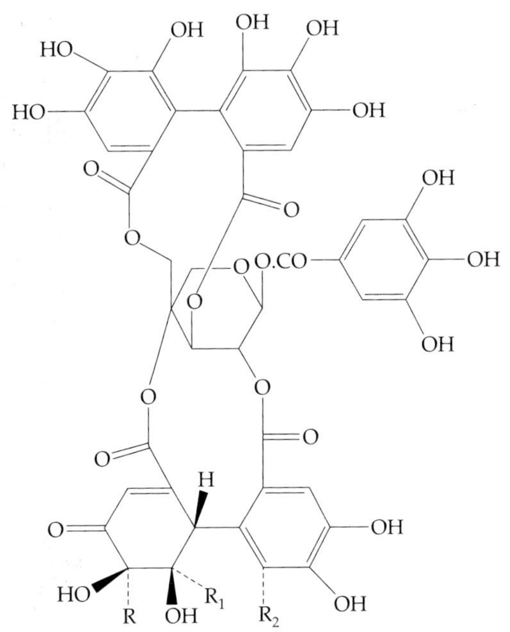

$RR_2 = O, R_1 = OH; R = OH, R_1 R_2 = O$

Granatin B

(Equivalent mixture of epoxy derivative)

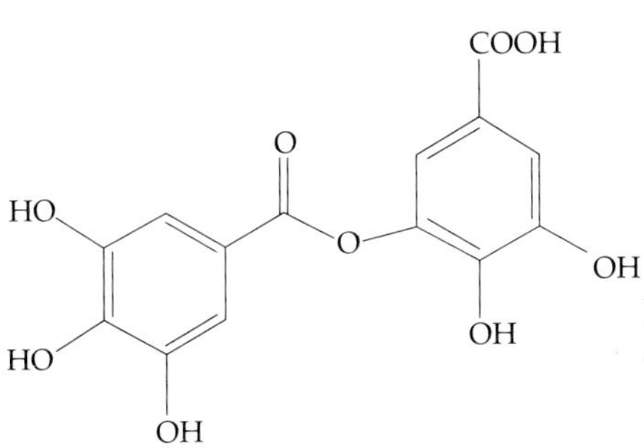

Digallic acid

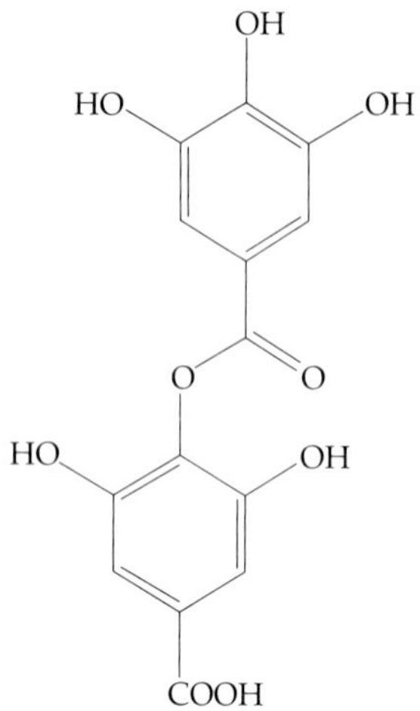

p-Digalloyl acid

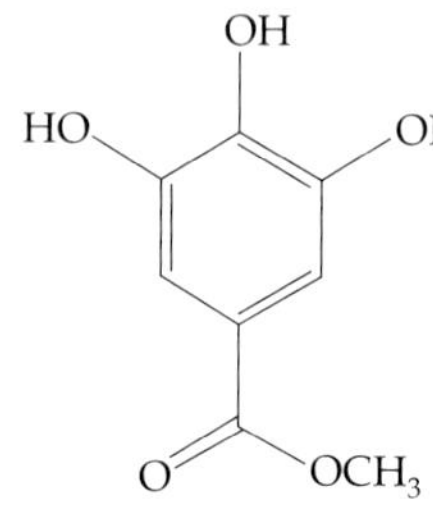

Methyl gallate

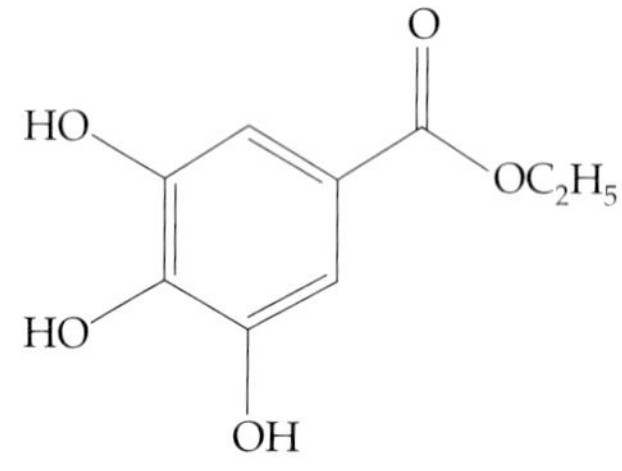

Ethyl gallate

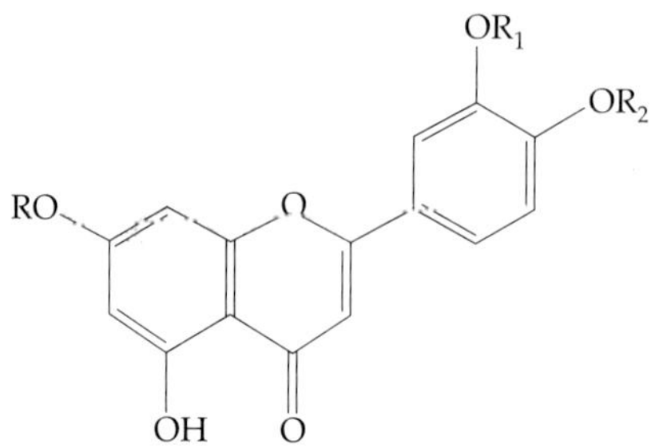

	R	R_1	R_2
Luteolin7-O-glucoside	β-D-glucopyranosyl	H	H
Luteolin4′-O-glucoside	H	H	β-D-glucopyranosyl
Luteolin3′,7-O-diglucoside	β-D-glucopyranosyl	β-D-glucopyranosyl	H
Luteolin	H	H	H

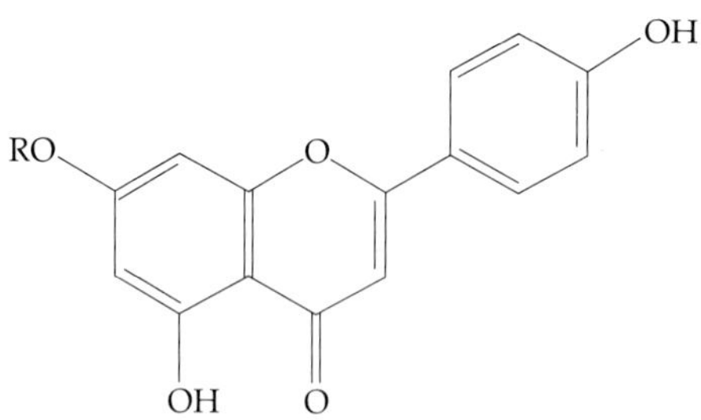

	R
Apigenin	H
Apigenin 7-O-glucoside	Glucosyl

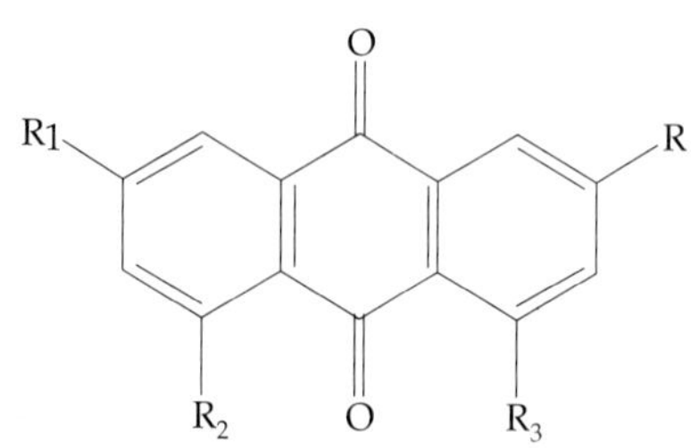

	R	R_1	R_2	R_3
Chrysophanol	CH_3	H	OH	OH
Aloe-emodin	CH_2OH	H	OH	OH
Rhein	COOH	H	OH	OH
5, 7-Dihydroxy-2-mehtylanthraquinone	CH_3	OH	OH	H

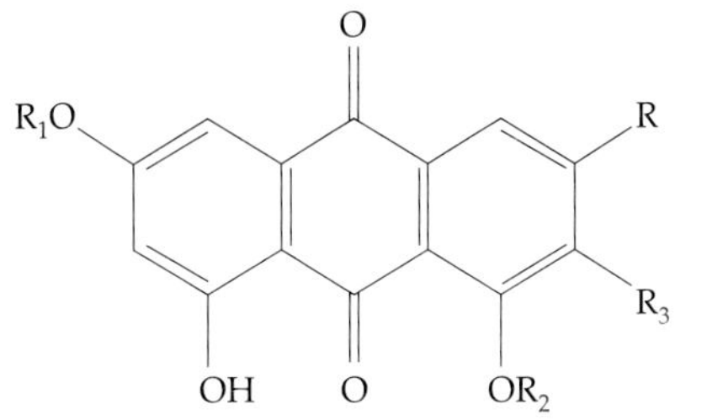

	R	R_1	R_2	R_3
Physcion	CH_3	CH_3	H	H
Emodin	CH_3	H	H	H

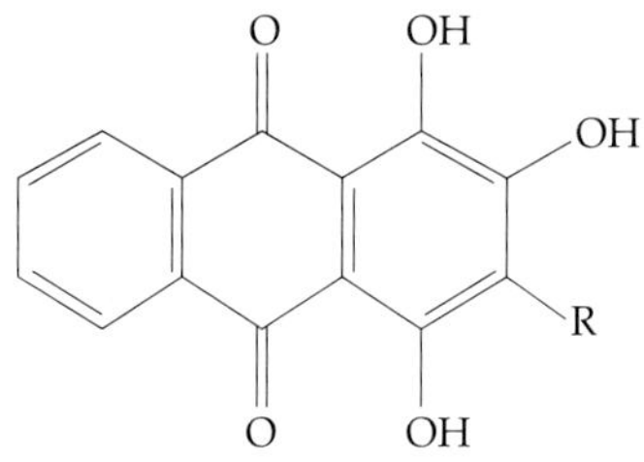

	R
Purpurin	H
Pseudo Purpurin	COOH

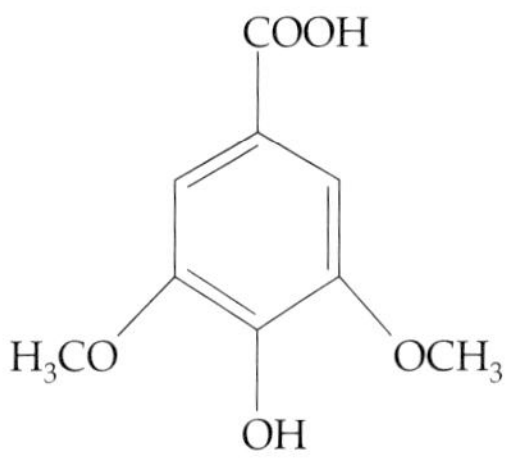

Syringic acid

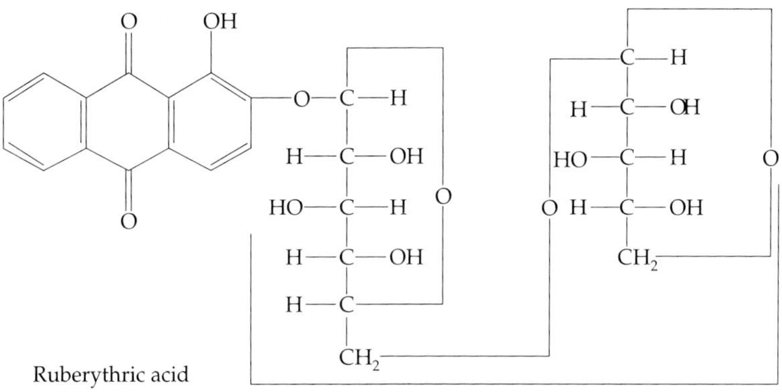

Ruberythric acid

Primeverose

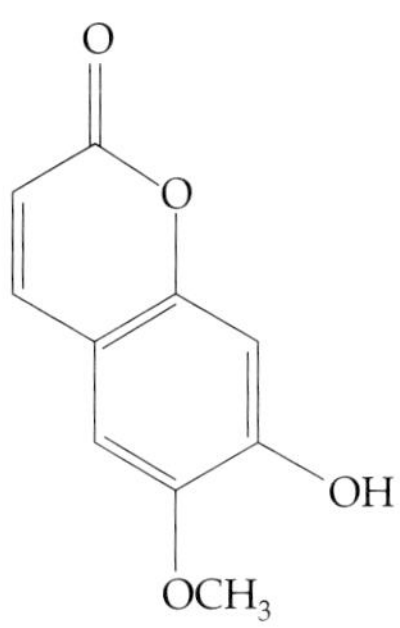

Scopoletol

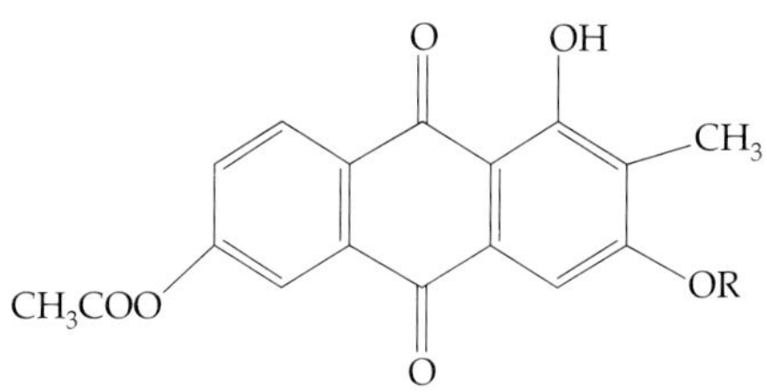

Rubiacordopne A, R = -[α-L-rhamnosyl]

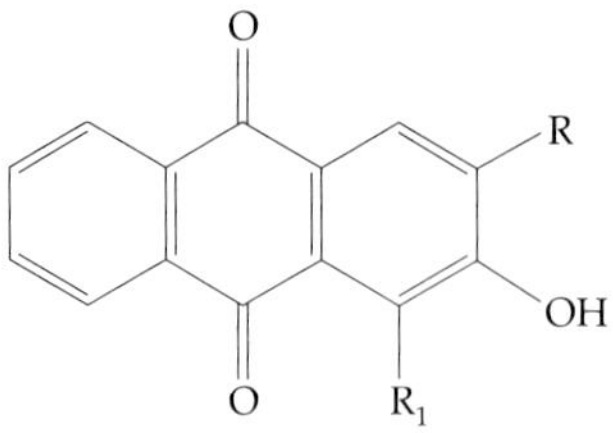

	R	R1
Alizarin	H	OH
Alizarin-α-methylether	H	OCH_3
3-Hydroxy anthraquinone	H	H

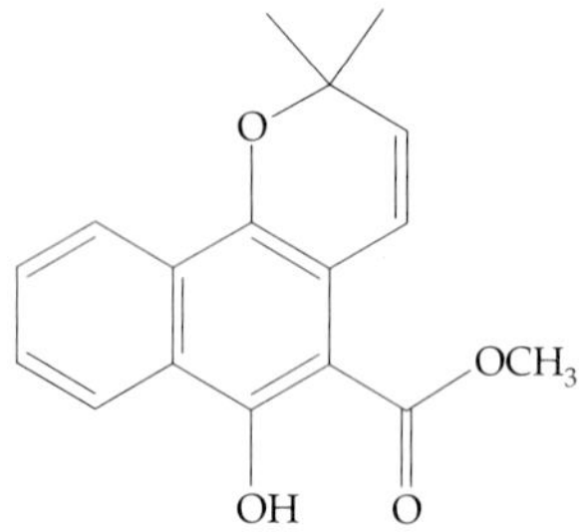

Mollugin

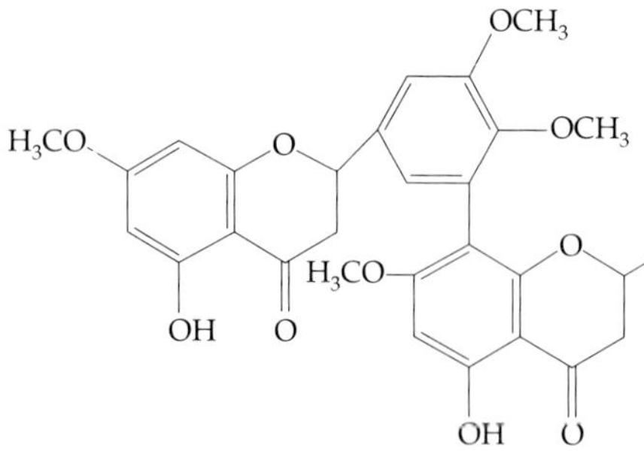

Anacarduflavanone

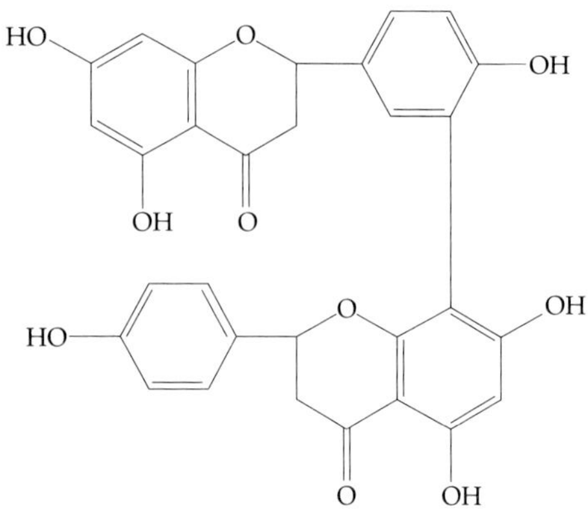

Tetrahydroamentoflavone

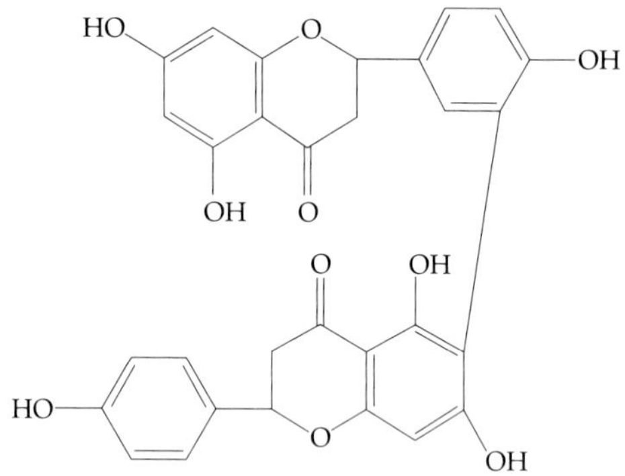

Tetrahydrorobustaflavone

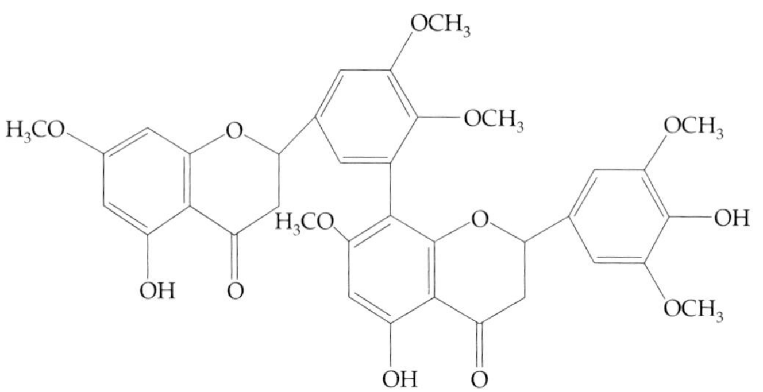

Nallaflavanone

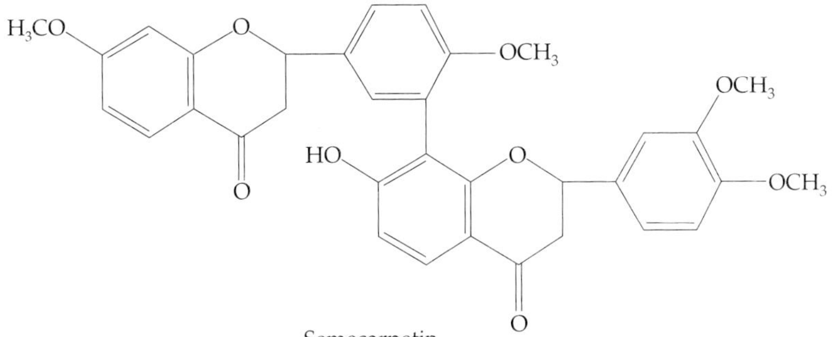

Semecarpetin

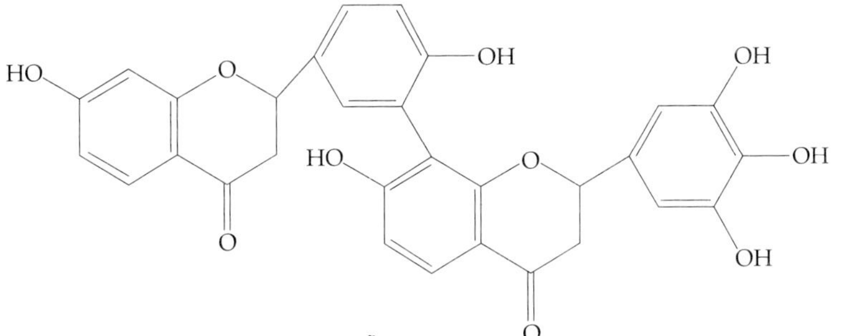

Semecarpuflavanone

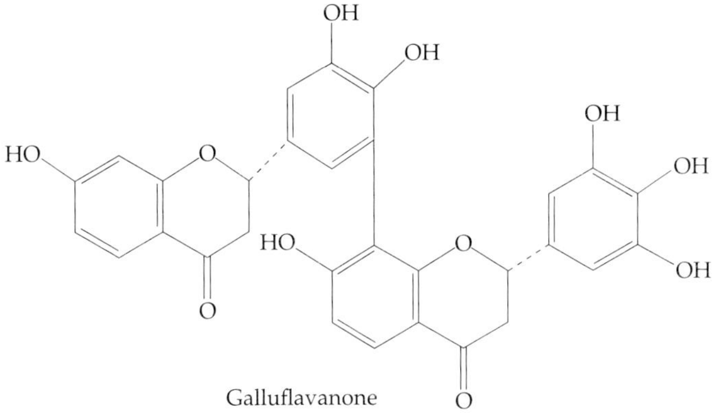

Galluflavanone

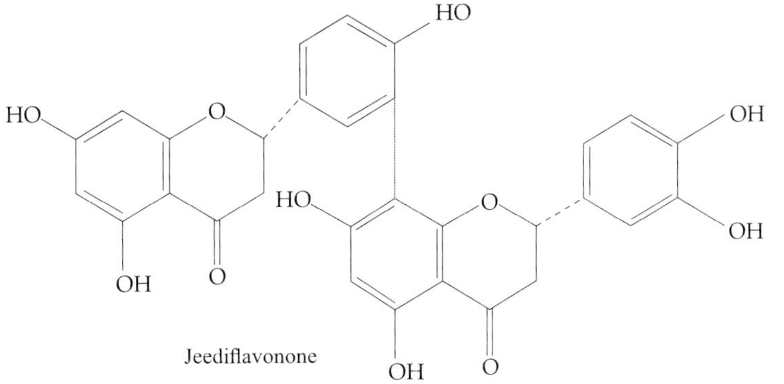

Jeediflavonone

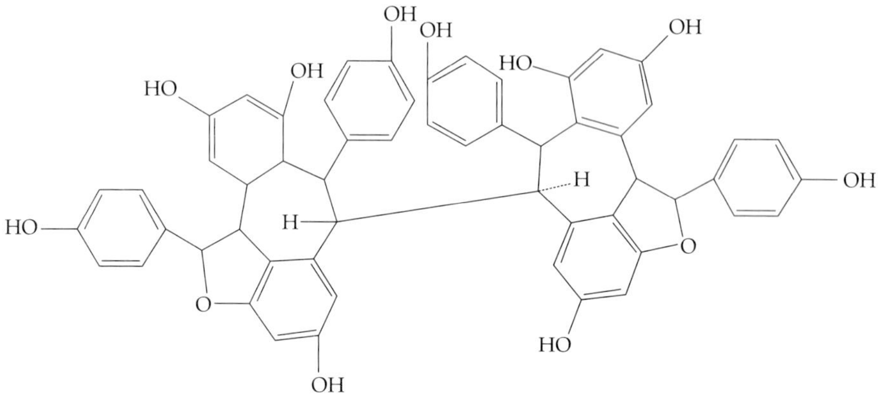

Hopeaphenol

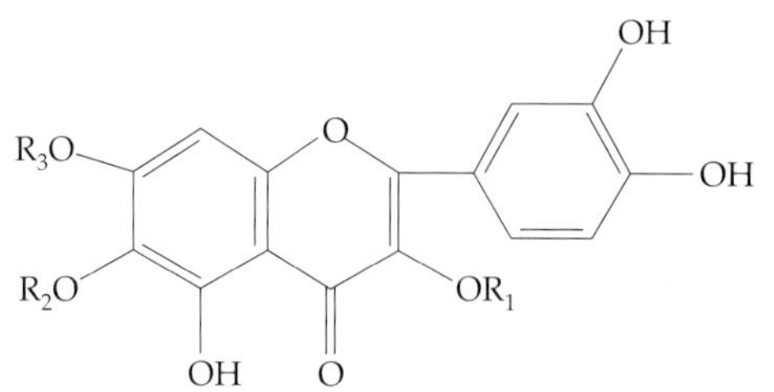

	R_1	R_2	R_3
Quercetagetin	H	H	H
Patuletin	H	CH_3	H
Quercetagitrin (Quercetagetin-7 -O-glucoside)	H	H	Glucosyl

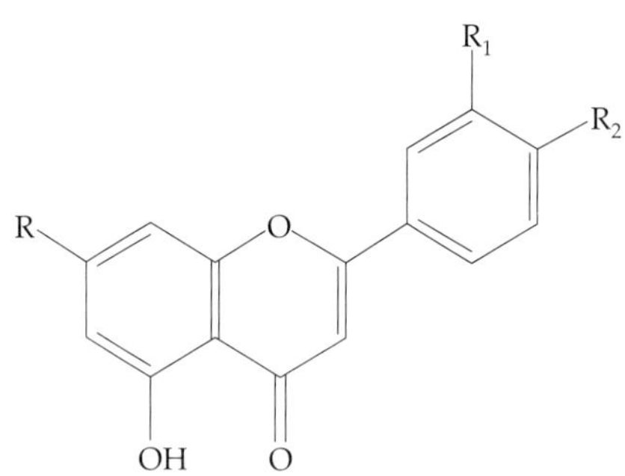

	R	R_1	R_2
Apigenin	OH	H	OH
4′-methoxyapigenin	OH	H	OH
7-methoxyluteolin	OCH_3	OH	OH

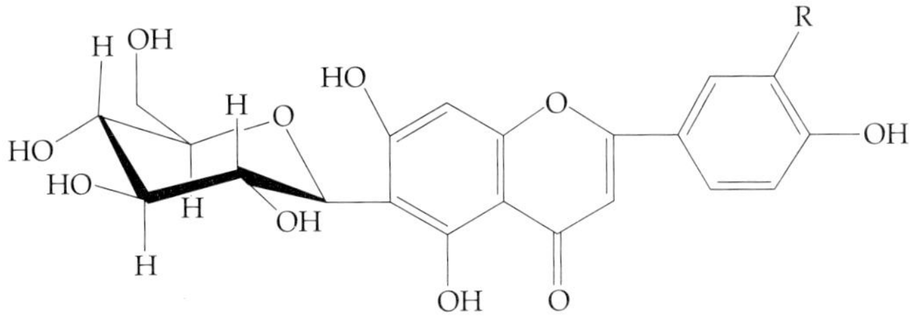

	R
Isoorientin (Homoorientin)	OH
Isovitexin	H

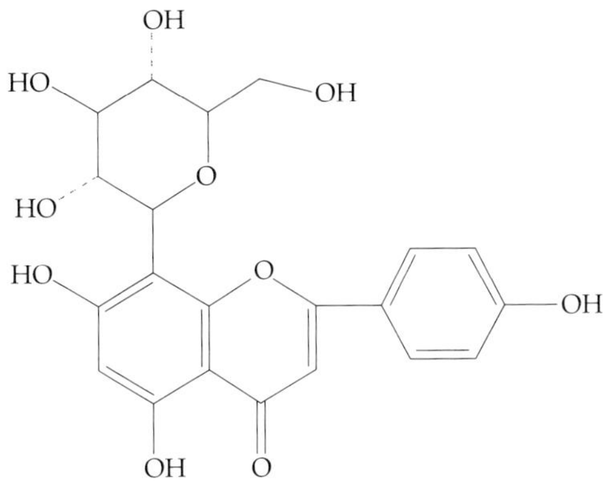

Vitexin (Apigenin-8-C-glucoside)

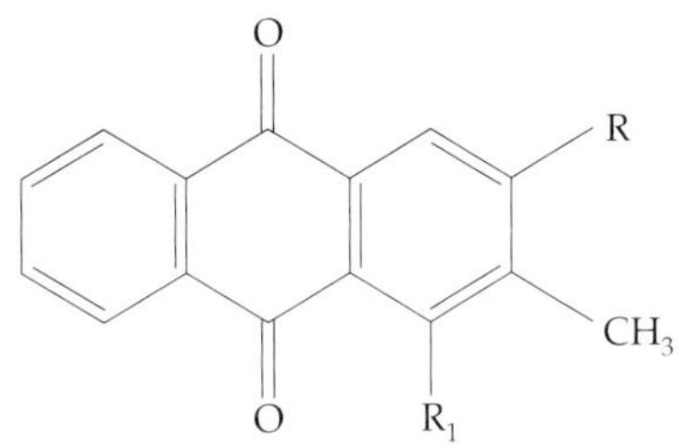

	R	R_1
Tectoquinone	H	H
2-Hydroxy-3-methylanthraquinone	OH	H
4-hydroxy-3-methylanthraquinone	H	OH

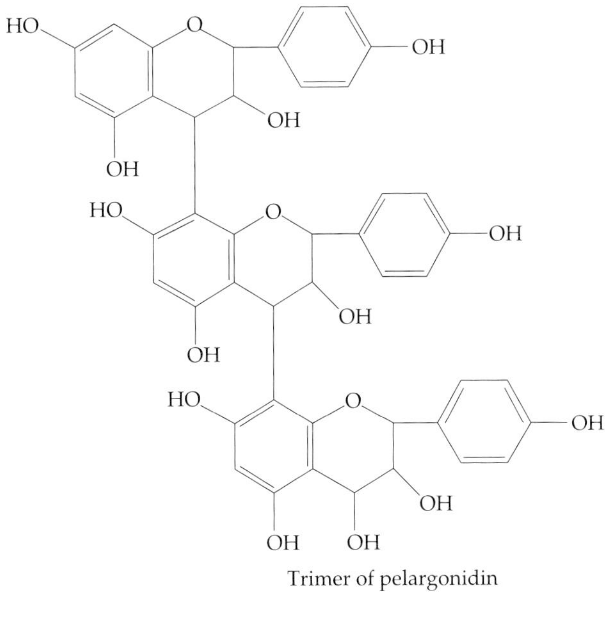

Trimer of pelargonidin

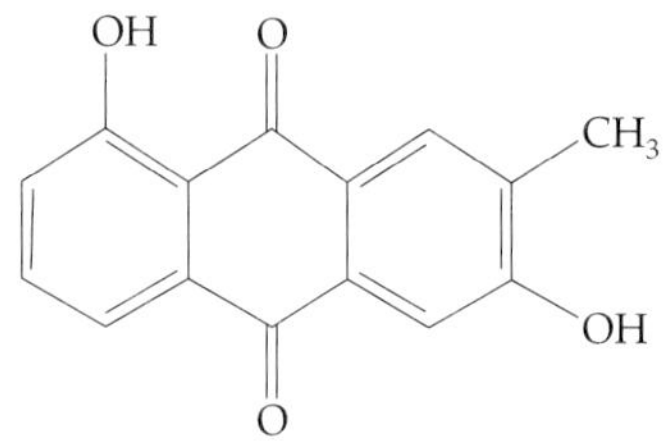

Tectone

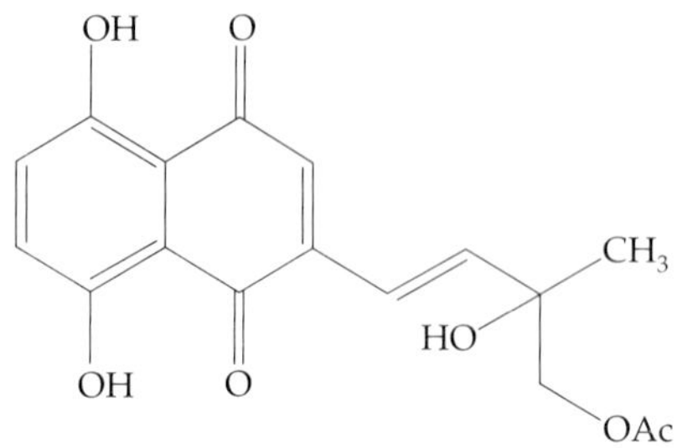

Naphthotectone

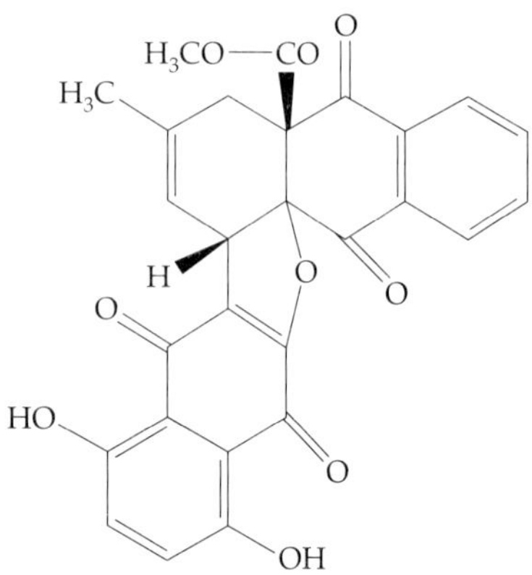

Anthratectone

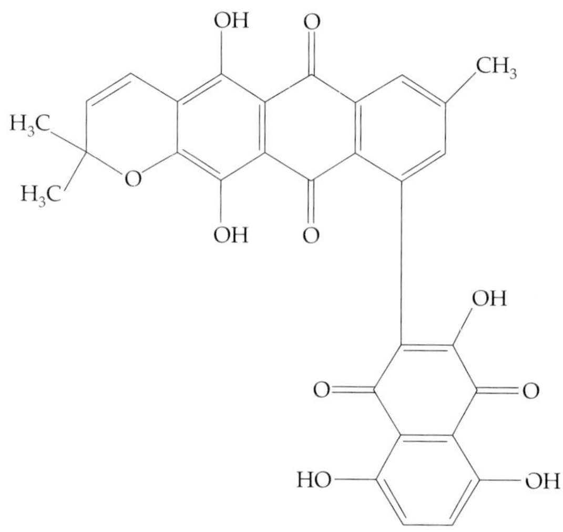

Tectograndone

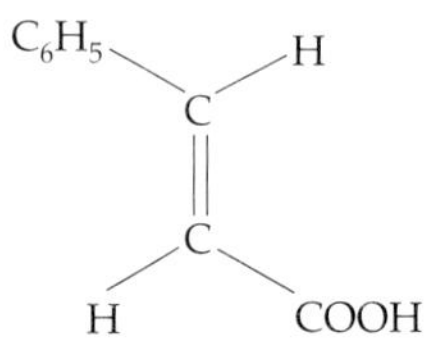

Trans-cinnamic acid

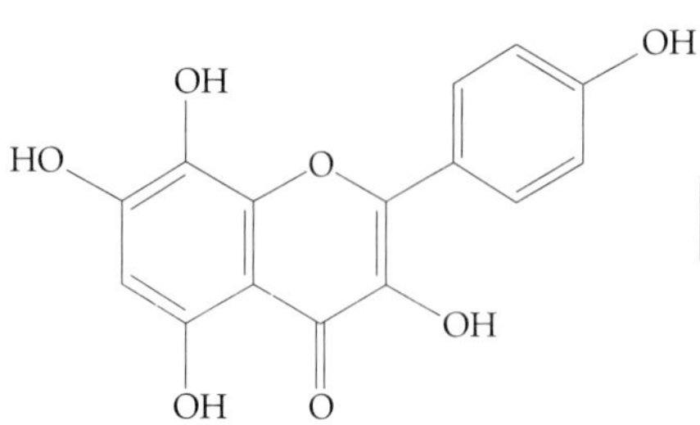

Herbacetin

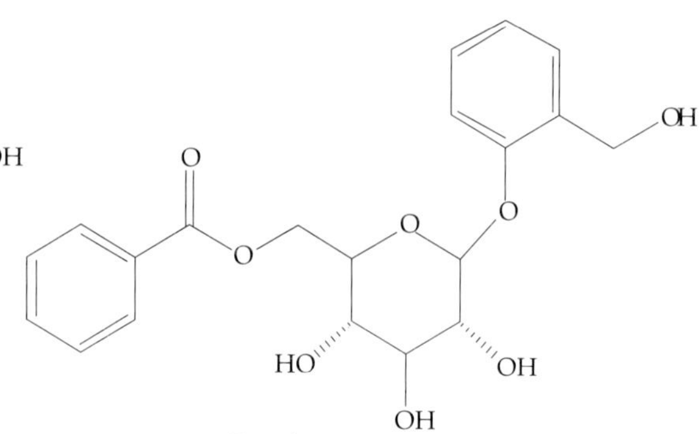

Populin

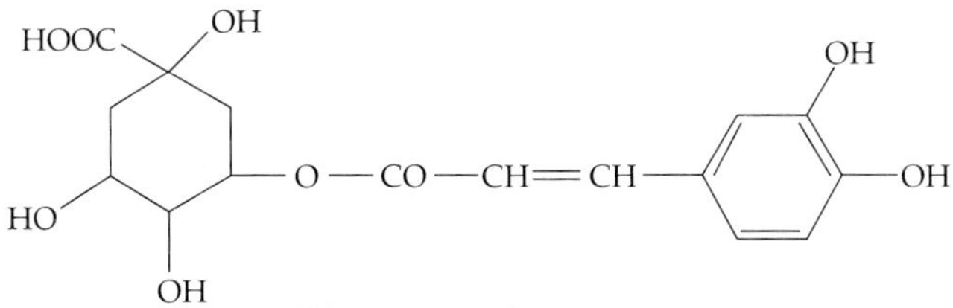

Chlorogenic acid

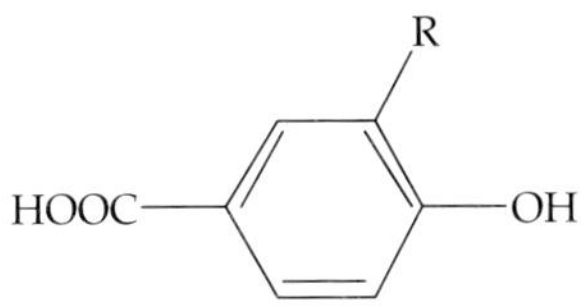

	R
p-Hydroxybenzoic acid	H
Protocatechuic acid	OH
Vanillic acid	OCH_3

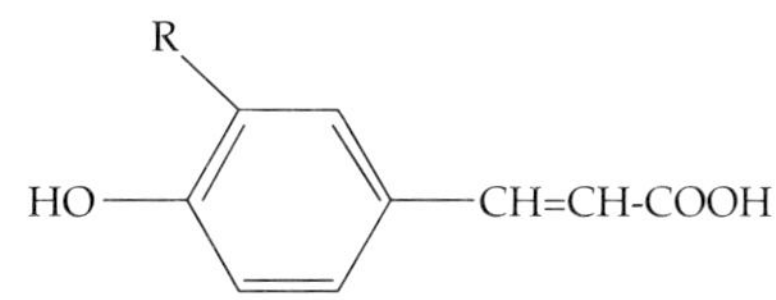

	R
p-Coumaric acid	H
Caffeic acid	OH
Ferulic acid	OCH_3

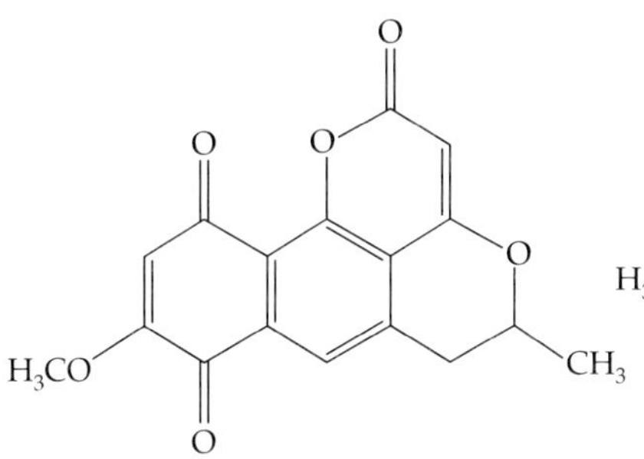

Ventilation A

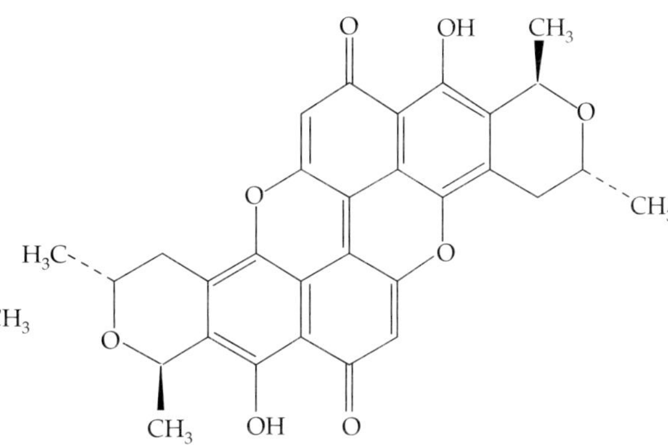

Ventilation B

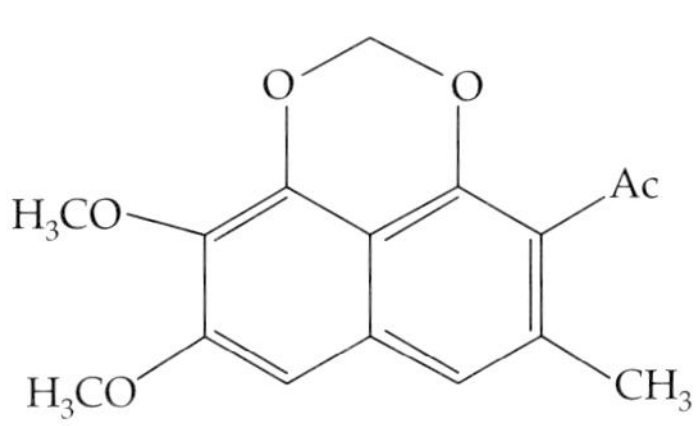

Ventilaginone

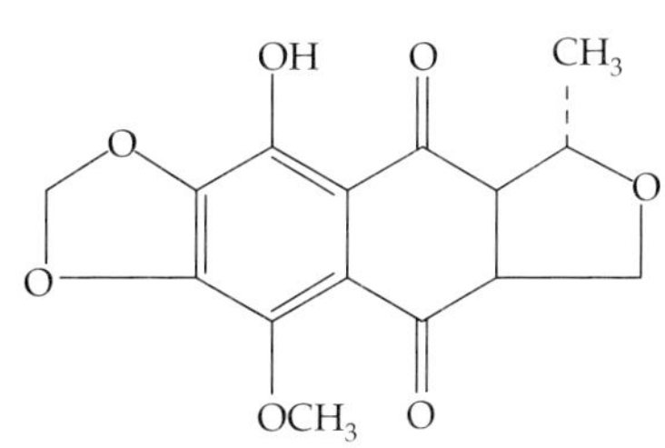

Ventilone C

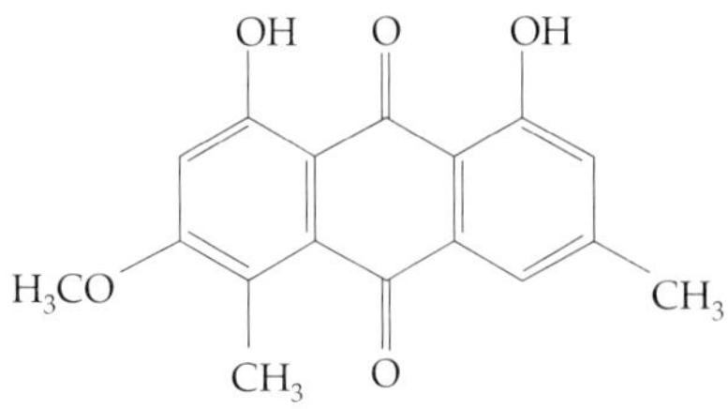

Xanthorin 5-Me ehter

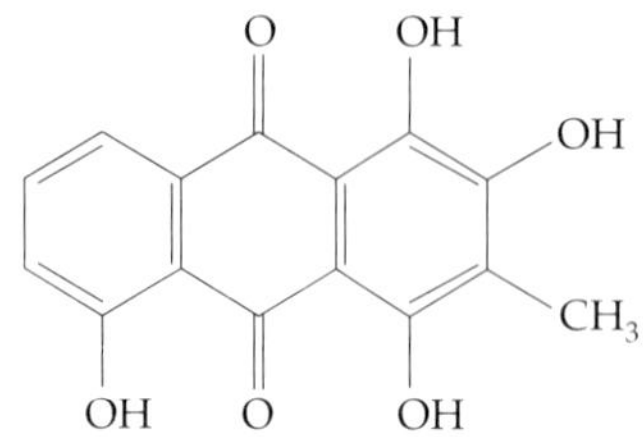

2-Hydroxyislandicin

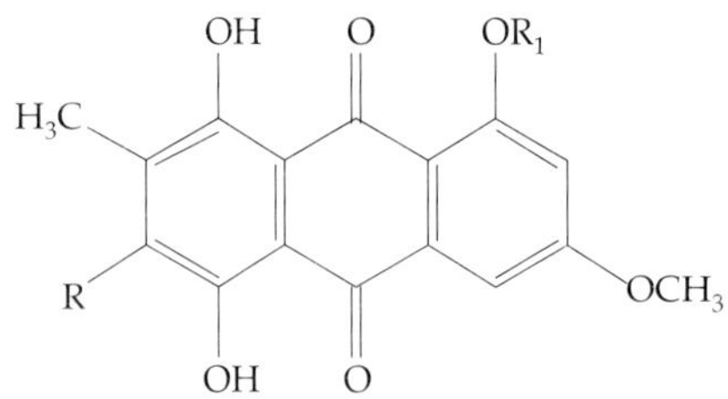

	R	R_1
Ventinone A	CH_3	H
Ventinone B	H	CH_3

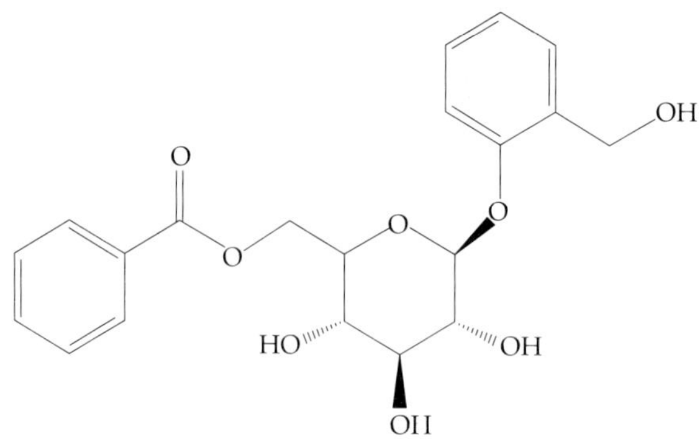

Populin

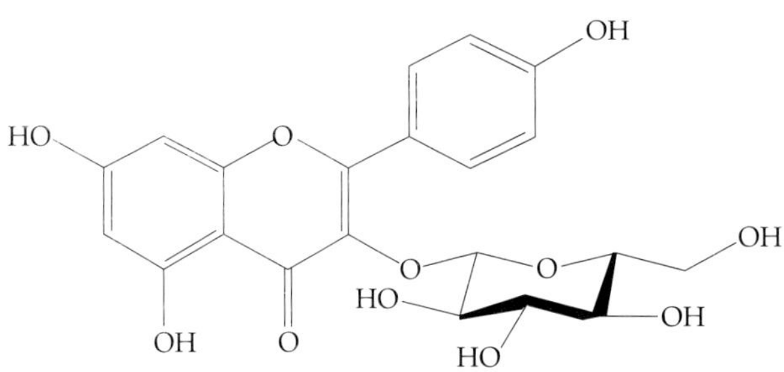

Astragalin (Kaempferol 3-O-β-D-glucopyranoside)

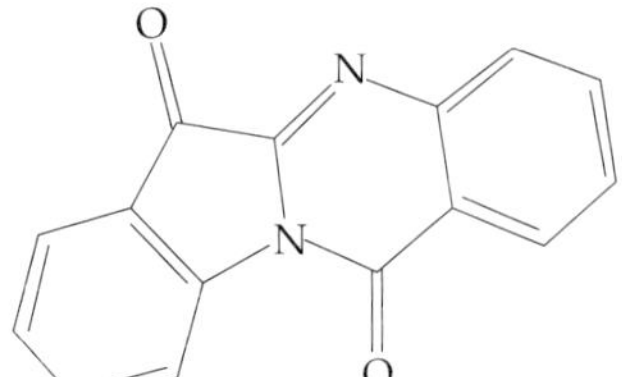

Tryptanthrin

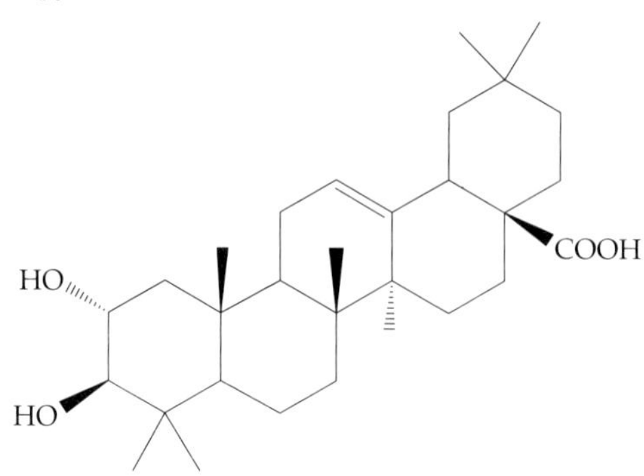

Maslinic acid

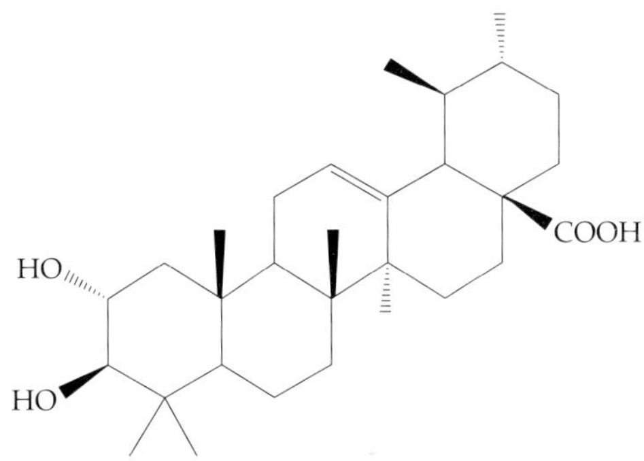

2α-hydroxyursolic acid

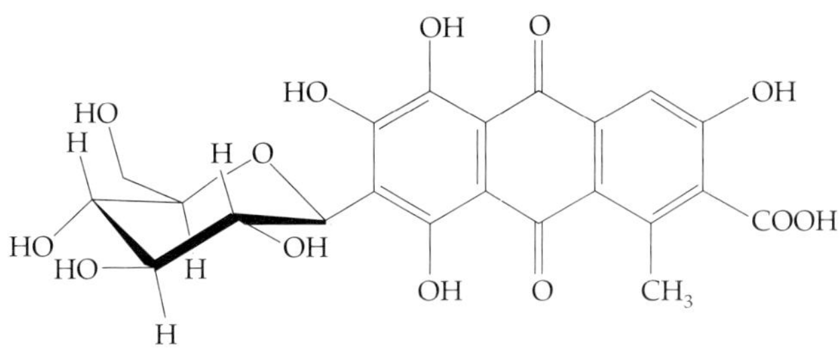

Carminic acid

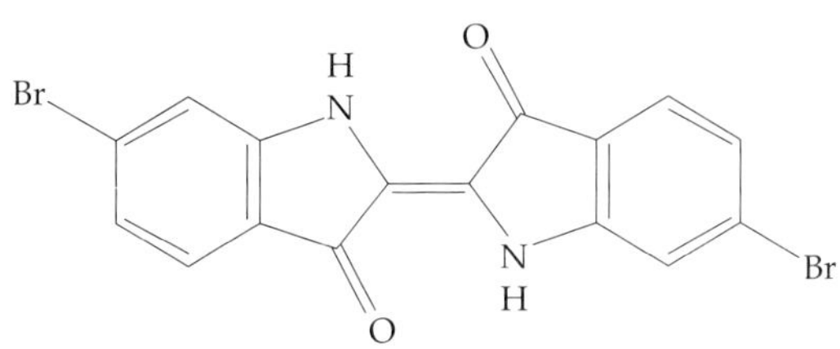

6,6′-dibromoindigotin

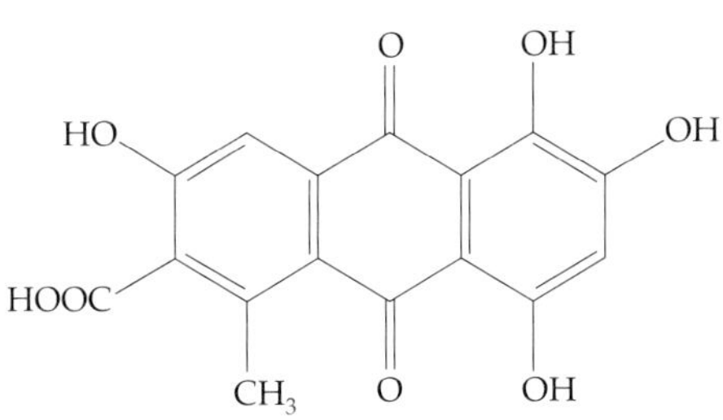

Kermesic acid

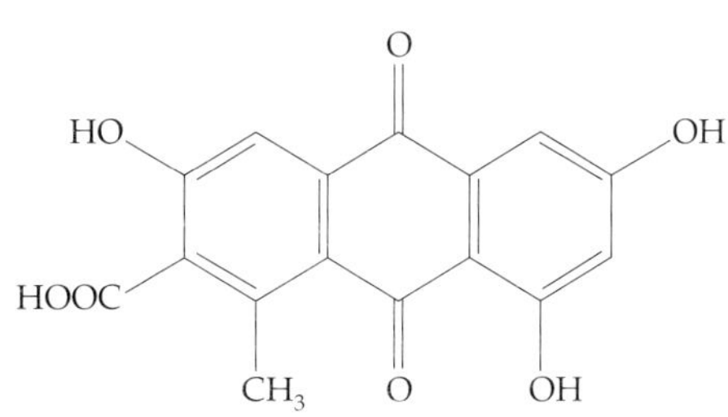

Flavokermesic acid

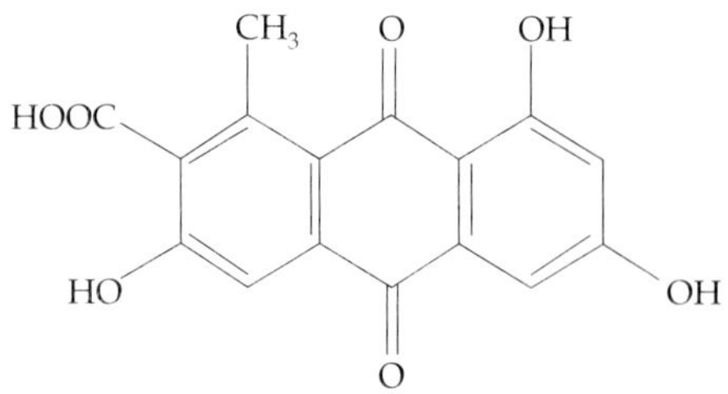

Laccaic acid D

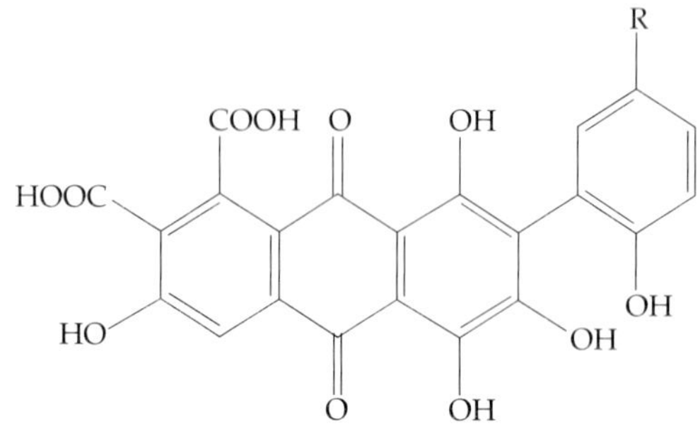

	R
Laccaic acid A	$CH_2CH_2NH\ COCH_3$
Laccaic acid B	CH_2CH_2OH
Laccaic acid C	$CH_2CH(NH_2)COOH$
Laccaic acid E	$CH_2CH_2NH_2$

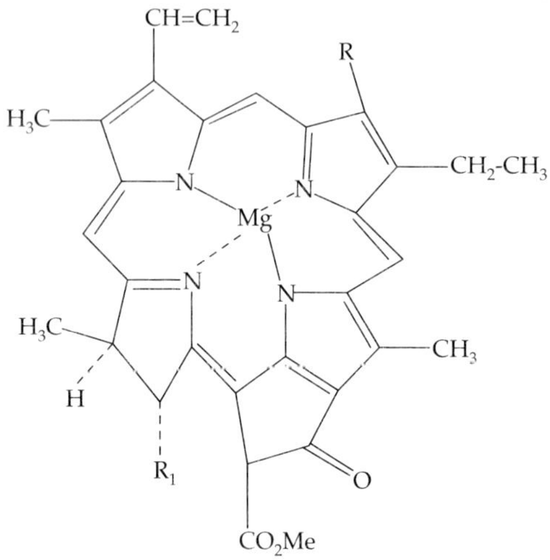

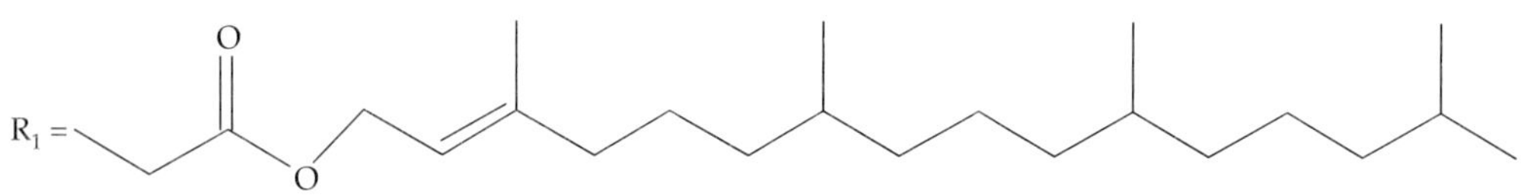

	R
Chlorophyll a	CH_3
Chlorophyll b	CHO

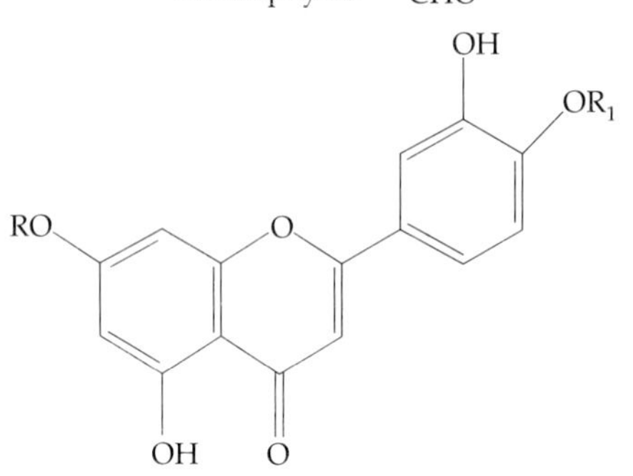

	R	R_1
Luteolin	H	H
Luteolin4′-O-rutinoside	H	β-D-glucopyranosyl-(6′ →1)-α-L-rhamnopyranoside

References

1. Singh, R. B. (2010). Diversity of *Abrus precatorius* plant with chemical and medicinal properties like foods, dye and sugar from seeds. *Journal Chemtracks*, 12(2), 279-282.
2. Yadava, R. N., Reddy, V. M.S. (2002). A new biologically active flavonol glycoside from the seeds of *Abrus precatorius*. *Journal of Asian Natural Products Research*, 4(2), 103-107.
3. Ma, Chao-Mei., Nakamura, N., Hattori, M. (1998). Saponins and C-glycosyl flavones from the seeds of *Abrus precatorius*. *Chemical and Pharmaceutical Bulletin*, 46(6), 982-987.
4. Markham, K.R., Wallace, J.W., Babu, Y.N., Murty, V.K., Rao, M.G. (1988). 8-C-Glucosylscutellarein 6,7-dimethyl ether and its 2″-O-apioside from *Abrus precatorius*. *Phytochemistry*, 28(1), 299-301.
5. Bhardwaj, D. K., Bisht, M. S., Mehta, C. K. (1980). Flavonoids from *Abrus precatorius*. *Phytochemistry*, 19(9), 2040-1.
6. Jain, R., Patni, V., Arora, D.K. (2007). Isolation and identification of flavonoid (quercetin) from *Acacia catechu*. *Journal of Phytological Research*, 20(1), 43-45.
7. Shen, D., Wu, Q., Wang, M., Yang, Y., Lavoie, E.J., Simon, J.E. (2006). Determination of the Predominant Catechins in *Acacia catechu* by Liquid- Chromatography/ Electrospray Ionization- Mass Spectrometry. *Journal of Agricultural and Food Chemistry*, 54(9), 3219-3224.
8. Kartik, M.G., Sharma, O.P. (1964). An improved method for the estimation of catechin in Katha and Khair wood samples. *ISI Bulletin*, 16(8), 325-327.
9. Azad, A.K., Ogiyama, K., Sassa, T. (2001). Isolation of (+)-catechin and a new polyphenolic compound in Bengal catechu. *Journal of Wood Science*, 47(5), 406-409.
10. Narang, K.S., Ralhan, N.K., Kaushal, A.N., Parmar, M.L., Sodhi, J.S., Pant, T.K., Singhal, V.K., Teotia, R.P.S. (1985) (+)-Catechin. Indian, IN 156326 A1 19850622.
11. Rao, P. R., Seshadri, T. R. (1948). (-) Epicatechol from *Acacia catechu*. *Journal of Scientific & Industrial Research*, 7B, 59.
12. Srivastava, S.K., Gupta, Rakesh K. (1983). Flavonoids from *Adina cordifolia*. *Indian Journal of Chemistry*, Section B, 22B(10), 1064.
13. Srivastava, S.K., Gupta, H.O. (1983). A new flavanone from *Adina cordifolia*. *Planta Medica*, 48(1), 58-9.
14. Srivastava, S.K., Gupta, R.K., Gupta, H.O., Mishra, K. M. (1981). Flavanones from *Adina cordifolia*. *Indian Journal of Pharmaceutical Sciences*, 43(4), 153-4.
15. Pal, B.C., Achari, B., Yoshikawa, K., Arihara, S. (1995). Saponins from *Albizia lebbeck*. *Phytochemistry*, 38(5), 1287-91.
16. Huu Tung, N., Du, Guang-Jian., Wang, Chong-Zhi., Yuan, Chun-Su., Shoyama, Y. (2013). Naphthoquinone Components from *Alkanna tinctoria*. *Phytotherapy Research*, 27(1), 66-70.

17. De Leo, P., Miceli, A., Ronzini, L., Sanasi, L., Sgarra, R. (1996). Chemical characterization of alkannin esterified derivatives. *Agro-Food-Industry Hi-Tech*, 7(1), 23-25.
18. Papageorgiou, V. P., Mellidis, A. S., Sagredos, A. N. (1980). Study on the antibiotic fraction of *Alkanna tinctoria. Chimika Chronika*, 9(1), 57-63.
19. Papageorgiou, V.P. (1979). 5, 8-Dihydroxynaphthoquinone derivatives. Ger. Offen., DE 2829744 A1 19790719.
20. Papageorgiou, V.P. (1977). A new pigment of *Alkanna tinctoria* having naphthaquinone structure. *Planta Medica*, 31(4), 390-4.
21. Papageorgiou, V.P.(1978). Wound healing properties of naphthaquinone pigments from *Alkanna tinctoria. Experientia*, 34(11), 1499-501.
22. Arung E. T., Wijaya, K. I., Shimizu, K., Kondo, R. (2011). Tyrosinase inhibitory effect of quercetin 4′-O-β-D-glucopyranoside from dried skin of red onion (*Allium cepa). Natural Product Research*, 25(3), 256-63.
23. Ito, Y., Sugimoto, N., Akiyama, T.,Yamazaki, T., Tanamoto, K. (2009). Cepaic acid, a novel yellow xanthylium pigment from the dried outer scales of the yellow onion *Allium cepa. Tetrahedron Letters*, 50(28), 4084-4086.
24. Sood, A.R., Joshi, K.K. (1974). Flavonoid components of *Allium cepa. Journal of the Indian Chemical Society*, 51(11), 975.
25. Murthy, S.S.N., Anjaneyulu, A.S.R., Row, L.R., Pelter, A., Ward, R.S. (1982). Chemical examination of *Anacardium occidentale*. Isolation and structure determination of a novel biflavonoid-Cglycoside. *Planta Medica*, 45(1), 3-10.
26. Denny, S., Gupta, S., Gavani, U., Paarakh, P.M., Mishra, A. (2011). Isolation of Phytoconstituents from the leaves of *Anogiessus latifolia. Asian Journal of Pharmacy & Life Science*, 1(1), 45-51.
27. Ma, Ying-Tsun., Hsu, Feng-Lin.,Lan, Shu-Jan J., Chen, Chieh-Fu. (1996). *Journal of the Chinese Chemical Society (Taipei)*, 43(1), 77-81.
28. Arora, A., Gupta, D., Rastogi, D., Gulrajani, M. (2012). Naphthoquinone colorants from *Arnebia nobilis. Coloration Technology*, 128(5), 350-355.
29. Nguyen, N.T., Nguyen, M.H.K., Nguyen, H.X., Bui, N.K.N., Nguyen, M.T.T. (2012). Tyrosinase Inhibitors from the Wood of *Artocarpus heterophyllus. Journal of Natural Products*, 75(11), 1951-1955.
30. Zheng, Zong-Ping., Cheng, Ka-Wing., To, James Tsz-Kin., Li, Haitao., Wang, Mingfu. (2008). Isolation of tyrosinase inhibitors from *Artocarpus heterophyllus* and use of its extract as anti browning agent. *Molecular Nutrition & Food Research*, 52(12), 1530-1538.
31. Lin, C.N., Lu, C.M. (1993). Heterophylol, a phenolic compound with novel skeleton from *Artocarpus heterophyllus. Tetrahedron Letters*, 34(51), 8249-50.
32. Lu, C.M., Lin, C.N. (1993). The bioactive principles of Formosan *Artocarpus* species. Part 5. Two 2′,4′,6′-trioxygenated flavanones from *Artocarpus heterophyllus. Phytochemistry*, 33(4), 909-11.

33. Aida, M., Shinomiya, K., Hano, Y., Nomura, T. (1993). Constituents of the Moraceae plants. Part 16. Artonins J, K, and L, three new isoprenylated flavones from the root bark of *Artocarpus heterophyllus*. *Heterocycles*, 36(3), 575-83.
34. Hano, Y., Aida, M., Nomura, T. (1990). Constituents of the Moraceae plants. Part 4. Two new natural Diels-Alder-type adducts from the root bark of *Artocarpus heterophyllus.Journal of Natural Products*, 53(2), 391-5.
35. Hano, Y., Aida, M., Shiina, M., Nomura, T., Kawai, T., Ohe, H., Kagei, K. (1989). Artonins A and B, two new prenylflavones from the root bark of *Artocarpus heterophyllus*. *Heterocycles*, 29(8), 1447-53.
36. Dayal, R., Seshadri, T.R. (1974). Colorless components of the roots of *Artocarpus heterophyllus*. Isolation of a new compound, artoflavanone. *Indian Journal of Chemistry*, 12(8), 895-6.
37. Atta-ur-Rahman., Ansari, A.A. (1983). Alkaloids of *Berberis aristata* - isolation of aromoline and oxyberberine. *Journal of the Chemical Society of Pakistan*, 5(4), 283-4.
38. Blasko, G., Murugesan, N., Freyer, A.J., Shamma, M., Ansari, A.A., Atta-ur-Rahman. (1982). Karachine: an unusual protoberberine alkaloid. *Journal of the American Chemical Society*, 104(7), 2039-41.
39. Yoshikawa, M., Murakami, T., Kodoya, M., Yamahara, J., Matsuda, H. (1998). Medicinal foodstuffs. XV. Sugar beet. (2): Structures of beta vulgarosides V, VI, VII, VIII, IX, and X from the roots and leaves of sugar beet *(Beta vulgaris)*. *Chemical & Pharmaceutical Bulletin*; 46(11), 1758-1763.
40. Yoshikawa, M., Murakami, T., Kadoya, M., Matsuda, H., Yamahara, J., Muraoka,O., Murakami, N. (1995). *Beta vulgarosides* I, II, III, IV, and V, hypoglycemic glucuronide saponins from the roots and leaves of *Beta vulgaris* (sugar beet). *Heterocycles*, 41(8), 1621-6.
41. Kujala, T.S., Loponen, J.M., Klika, K.D., Pihlaja, K. (2000). Phenolics and betacyanins in red beetroot (*Beta vulgaris*) root: distribution and effect of cold storage on the content of total phenolics and three individual compounds. *Journal of Agricultural and Food Chemistry*, 48(11), 5338-42.
42. Wada, Shun-ichi., Tanaka, Reiko. (2005). Betulinic acid and its derivatives, potent DNA topoisomerase II inhibitors, from the bark of *Bischofia javanica*. *Chemistry &Biodiversity*, 2(5), 689-94.
43. Scotter, M. (2009). The chemistry and analysis of annatto food colouring: A review. Food Additives & Contaminants, Part A*: Chemistry, Analysis, Control, Exposure & Risk Assessment*, 26(8), 1123-1145.
44. Oberoi, S., Ledwani, L. (2010). Isolation and characterization of new plant pigment along with three known compounds from *Butea monosperma* petals. *Archives of Applied Science Research*, 2(4), 68-71.
45. Yadava, R.N., Tiwari, L. (2007). New antifungal flavone glycoside from *Butea monosperma*. *Journal of Enzyme Inhibition and Medicinal Chemistry*, 22(4), 497-500.
46. Chokchaisiri, R., Suaisom, C., Sriphota, S., Chindaduang, A., Chuprajob, T., Suksamrarn, A. (2009). Bioactive flavonoids of the flowers of *Butea monosperma*. *Chemical & Pharmaceutical Bulletin*, 57(4), 428-32.

47. Namratha, V., Rao, G.V. (2009). Isolation and structure determination of a new flavonone di-glucoside from *Butea monosperma. International Journal of Chemical Sciences,* 7(2), 1107-1111.
48. Gupta, S.R., Ravindranath, B., Seshadri, T.R. (1970). The Glucosides of *Butea monosperma. Phytochemistry,* 9, 2231-2325.
49. Cuong, T.D., Hung, T.M., Kim, J.C., Kim, E.H., Woo, M.H., Choi, J.S., Lee, J.H., Min, B.S. (2012). Phenolic compounds from *Caesalpinia sappan* heartwood and their anti-inflammatory activity. *Journal of Natural Products,* 75(12), 2069-2075.
50. Min, B.S., Cuong, T.D., Hung, T.M., Min, B.K., Shin, B.S., Woo, M.H. (2012). Compounds from the heartwood of *Caesalpinia sappan* and their anti-inflammatory activity. *Bio-organic & Medicinal Chemistry Letters,* 22(24), 7436-7439.
51. Shu, Shi-Hui., Deng, An-Jun., Li, Zhi-Hong., Qin, Hai-Lin. (2011).Two novel biphenyl dimers from the heartwood of *Caesalpinia sappan. Fitoterapia,* 82(5), 762-766.
52. Cheng, Y., Wang, Y., Li, Y., Ge, Z. (2011). *Carthamus tinctorius* bioactive fraction containing hydroxyl safflor yellow A and 6-hydroxykaempferol-3,6,7-triglucoside, and preparation method and medical application thereof. *Faming Zhuanli Shenqing,* CN 102119984 A 20110713.
53. Zhou, Yu-Zhi., Qiao, Li., Chen, Huan., Li, Rui-Feng., Hua, Hui-Ming., Pei, Yue-Hu. (2008).New aromatic glucosides from *Carthamus tinctorius. Journal of Asian Natural Products Research,* 10(9), 817-821.
54. Zhou, Yu-Zhi., Chen, Huan., Qiao, Li., Lu, Xuan., Hua, Hui-Ming., Pei, Yue-Hu. (2008).Five new aromatic glycosides from *Carthamus tinctorius. Helvetica Chimica Acta,* 91(7), 1277-1285.
55. Zhou, Yu-Zhi., Chen, Huan., Qiao, Li., Xu, Nan., Cao, Jia-Qing., Pei, Yue-Hu. (2008). Two new compounds from *Carthamus tinctorius. Journal of Asian Natural Products Research,* 10(5), 429-433.
56. Takahashi, Y., Saito, K., Yanagiya, M., Ikura, M., Hikichi, K., Matsumoto, T., Wada, M. (1984). Chemical constitution of safflor yellow B, a quinochalcone C-glycoside from the flower petals of *Carthamus tinctorius. Tetrahedron Letters,* 25(23), 2471-4.
57. Takahashi, Y., Miyasaka, N., Tasaka, S., Miura, I., Urano, S., Ikura, M., Hikichi,K., Matsumoto, T., Wada, M. (1982). Constitution of two coloring matters in the flower petals of *Carthamus tinctorius.Tetrahedron Letters,* 23(49), 5163-6.
58. Wong, S.M., Wong, M.M., Seligmann, O., Wagner, H. (1989).Drugs for liver therapy. Part 17. New antihepatotoxic naphtho-pyrone glycosides from the seeds of *Cassia tora. Planta Medica,* 55(3), 276-80.
59. Wong, S.M., Wong, M.M., Seligmann, O., Wagner, H. (1988).Drugs for liver therapy. Part 16. Anthraquinone glycosides from the seeds of *Cassia tora. Phytochemistry,* 28(1), 211-14.

60. Singla, R.K., Varadaraj Bhat, G. (2011). Crocin: An overview. *Indo Global Journal of Pharmaceutical Sciences*, 1(4), 281-286.
61. Vignolini, P., Heimler, D., Pinelli, P., Ieri, F., Sciullo, A., Romani, A. (2008). Characterization of by-products of saffron (*Crocus sativus*) production. *Natural Product Communications*, 3(12), 1959-1962.
62. Hadizadeh, F., Khalili, N., Hosseinzadeh, H., Khair-Aldine, R. (2003). Kaempferol from saffron petals. *Iranian Journal of Pharmaceutical Research*, 2(4), 251-252.
63. Jitoe-Masuda, A., Fujimoto, A., Masuda, T. (2013). Curcumin: from chemistry to chemistry-based functions. *Current Pharmaceutical Design*, 19(11), 2084-2092.
64. Ittadwar, A.M. (2004). The process for isolation of anticancer compound, 6,8,3′-trimethoxy-5,4′-dihydroxy-5′-methyl-7-rhamnoglucosyl)isoflavone, from the rhizomes of *Curcuma longa*. Indian, IN 192940 A1 20040612.
65. Li, W., Wang, S., Feng, J., Xiao, Y., Xue, X., Zhang, H., Wang, Y., Liang, X. (2009). Structure elucidation and NMR assignments for curcuminoids from the rhizomes of *Curcuma longa*. *Magnetic Resonance in Chemistry* : MRC, 47(10), 902-8.
66. Rasmussen, H B., Christensen, S B., Kvist, L P., Karazmi, A. (2000). A simple and efficient separation of the curcumins, the antiprotozoal constituents of *Curcuma longa*. *Planta Medica*, 66(4), 396-8.
67. Veigas, J.M., Divya, P., Neelwarne, B. (2012). Identification of previously unreported pigments among carotenoids and anthocyanins in floral petals of *Delonix regia*. *Food Research International*, 47(1), 116-123.
68. Adje, F.A., Lozano, Y.F., Le Gerneve, C., Lozano, P.R., Meudec, E., Adima, A.A., Gaydou, E.M. (2012). Phenolic acid and flavonol water extracts of *Delonix regia* red flowers. *Industrial Crops and Products*, 37(1), 303-310.
69. Patel, M. B., Mishra, S.H. (2006). A simple spectrofluorometric estimation of wedelolactone in methanol extract of *Eclipta alba*. *Pharmacognosy Magazine*, 2(7), 165-168.
70. Abdel-Kader, M.S., Bahler, B.D., et al. (1998). DNA-Damaging Steroidal Alkaloids from *Eclipta alba* from the Suriname Rainforest. *Journal of Natural Products*, 61(10), 1202-1208.
71. Alam, P., Gupta, J., Firdouse, S., Sultana, A., Sultana, S. (2012). A HPTLC method for qualitative and quantitative estimation of gallic acid from *Emblica officinalis* in poly herbal formulation. *Research Journal of Pharmaceutical, Biological and Chemical Sciences*, 3(4), 1218-1223.
72. Zhang, L., Zhao, W., Guo, Y., Tu, G., Lin, S., Xin, L. (2003). Studies on chemical constituents in fruits of Tibetan Medicine *Phyllanthus emblica*. *Zhongguo Zhongyao Zazhi*, 28(10), 940-943.
73. Ghosal, S., Tripathi, V.K., Chauhan, S. (1996). Active constituents of *Emblica officinalis*: Part 1. The chemistry and antioxidative effects of two new hydrolysable tannins, emblicanin A and B. *Indian Journal of Chemistry*, Section B, 35B(9), 941-948.

74. Verma, S.C., Nigam, S., Jain, C.L., Pant, P., Padhi, M.M. (2011). Microwave-assisted extraction of gallic acid in leaves of *Eucalyptus* x hybrida and its quantitative determination by HPTLC. *Chemica Sinica*, 2(2), 268-277.
75. Dayal, R., Onial, P. (2010). A process for the isolation of ursolic acid from *Eucalyptus* hybrid (leaves). *Indian Pat. Appl.*, IN 2009DE00361 A 20100903.
76. Dayal, R., Dobhal, P.C. (1991). Chemical constituents of *Eucalyptus* hybrid. *Journal of the Indian Chemical Society*, 68(8), 478-9.
77. Dayal, R. (1987).Occurrence of ursolic acid and related compounds in *Eucalyptus* hybrid leaves. *Current Science*, 56(13), 670-1.
78. Verma, S.C., Jain, C.L., Kumari, A., Padhi, M.M., Devalla, R.B. (2013). Microwave-assisted extraction and rapid isolation of ursolic acid from the leaves of *Eucalyptus* × hybrida and its quantification using HPLC-diode array technique. *Journal of Separation Science*, 36(7), 1255-62.
79. Subramanian, P.M., Misra, G. S. (1977). Chemical constituents of *Ficus bengalensis. Indian Journal of Chemistry*, Section B, 15(8), 762-3.
80. Masuda, H., Ohtani, K., Mizutani, K., Ogawa, S., Kasai, R., Tanaka, O. (1991). Chemical study on *Haematoxylon campechianum*: A sweet principle and new dibenz[b,d]oxocin derivatives. *Chemical & Pharmaceutical Bulletin*, 39(6), 1382-4.
81. Cooksey C. (2010). Hematoxylin and related compounds: An annotated bibliography concerning their origin, properties, chemistry, and certain applications. *Biotechnic & Histochemistry: Official Publication of the Biological Stain Commission*, 85(1), 65-82.
82. Andreotti, A., Bonaduce, I., Colombini, Maria P., Ribechini, E. (2004). Characterisation of natural indigo and shellfish purple by mass spectrometric techniques. *Rapid Communications in Mass Spectrometry*, 18(11), 1213-1220.
83. Li, L., Yang, G., Dong, T., Chen, Z. (1996). Studies on the chemical constituents of dyers woad (*Isatis tinctoria*). *Zhongcaoyao*, 27(7), 389-391.
84. Oberthur, C., Schneider, B., Graf, H., Hamburger, M. (2004). The elusive indigo precursors in woad (*Isatis tinctoria*)-Identification of the major indigo precursor, isatan A, and a structure revision of isatan B. *Chemistry & Biodiversity*, 1(1), 174-82.
85. Berkeley, C. (1961). Indigotin content of woad. *Nature*, 191, 1414-15.
86. Liu, C., Tai, Z., Feng, S., Fang, Y., Cai, Le., Ding, Z. (2012). Chemical constituents from the seed coat of *Juglans regia. Zhongguo Zhongyao Zazhi*, 37(10), 1417-1421.
87. Thakur, A. (2011). Juglone: A therapeutic phytochemical from *Juglans regia. Journal of Medicinal Plants Research*, 5(22), 5324-5330.
88. Jin, Z., Qu, Z. (2011). Study on the hydrolyzable tannin components of *Juglans regia. Zhongchengyao*, 33(4), 715-717.
89. Zohourian, T. H., Quitain, A.T., Sasaki, M., Goto, M. (2012). Extraction of Bioactive Compounds from Leaves of *Lawsonia inermis* by Green Pressurized Fluids. *Separation Science and Technology*, 47(7), 1006-1013.

90. Nguyen, T.B., Pham, T.K., Tran, T.O., Phan, V.K. (2008). Isolation of 3-O-β-L-Rhamnopyranosyl 5,7,4-trihydroxyflavone from *Lawsonia inermis. Tap Chi Duoc Hoc*, 48(9), 20-21.
91. Mikhaeil, B.R., Badria, F.A., Maatooq, G.T., Amer, M.M. A. (2004). Antioxidant and immunomodulatory constituents of henna leaves. *Zeitschrift fuer Naturforschung, C.Journal of Biosciences*, 59(7/8), 468-476.
92. Mahmoud, Z.F., Salam, N. A. A., Khafagy, S. M. (1980). Constituents of henna leaves (*Lawsonia inermis*) growing in Egypt. *Fitoterapia*, 51(3), 153-5.
93. Jain, A.C., Khanna, V.K., Seshadri, T.R. (1969). Synthesis of Osajazanthone. *Tetrahedron*, 25, 2787-2793.
94. Sharma, J., Varmal, R. (2011). A review on endangered plant of *Mallotus philippensis. Pharmacologyonline* (3, Newsletters), 1256-1265.
95. Daikonya, A., Katsuki, S., Wu, Jin-Bin., Kitanaka, Susumu, (2002). Anti-allergic agents from natural sources (4): Anti-allergic activity of new phloroglucinol derivatives from *Mallotus philippensis. Chemical & Pharmaceutical Bulletin*, 50(12), 1566-1569.
96. Yu, X., Liu, C., Li, G., Xu, Z., Wang, D. (2013). Chemical constituents of bark of *Mangifera indica. Zhongguo Xiandai Yingyong Yaoxue*, 30(3), 257-260.
97. Nunez Selles, A. J., Velez Castro, H.T., Aguero-Aguero, J., Gonzalez-Gonzalez, J., Naddeo, F., De Simone, F., Rastrelli, L. (2002). Isolation and quantitative analysis of phenolic antioxidants, free sugars, and polyols from mango (*Mangifera indica*) stem bark aqueous decoction used in Cuba as a nutritional supplement. *Journal of Agricultural and Food Chemistry*, 50(4), 762-766.
98. Sharma, S. K., Ali, M. (1995). Chemical constituents of stem bark of *Mangifera indica*. (cultivar Desi). *Journal of the Indian Chemical Society*, 72(5), 339-42.
99. Khan, M.A., Nizami, S.S., Khan, M. N.I., Azeem, S.W. (1992). Biflavone from *Mangifera indica. Pakistan Journal of Pharmaceutical Sciences*, 5(2), 155-9.
100. Bhatia, V.K., Seshadri, T.R. (1968). Synthesis of mangiferin. *Tetrahedron Letters*, 14, 1741-1742.
101. Mohideen, S., Balasubramaniyan, M. P., Krishnan, S. N., Ch Narasimha Raju, B. H., Delhi Raju, N. (2012). Development and validation of HPLC method for quantitative estimation of ursolic acid in leaves extract of *Memecylon umbellatum. International Journal of Pharmacy and Pharmaceutical Sciences*, 4(2), 503-505.
102. Joshi, H., Joshi, A.B., Gururaj, M.P., Sati, H. (2009). Flavonoids of *Memecylon umbellatum.Asian Journal of Chemistry*, 21(2), 1639-1641.
103. Ee, G. C. L., Wen, Y. P., Sukari, M. A., Go, R. (2011). Anthraquinones from *Morinda citrifolia* roots. *Asian Journal of Chemistry*, 23(1), 77-80.
104. Kamiya, K., Hamabe, W., Tokuyama, S., Hirano, K., Satake, T., Kumamoto-Yonezawa, Y., Yoshida, H., Mizushina, Y. (2009). Inhibitory effect of anthraquinones isolated from the Noni (*Morinda citrifolia)* root on animal A-, B- and Y-families ofDNA polymerases and human cancer cell proliferation. *Food Chemistry*, 118(3), 725-730.

105. Kamiya, K., Hamabe, W., Tokuyama, S., Satake, T. (2009). New anthraquinone glycosides from the roots of *Morinda citrifolia*. *Fitoterapia*, 80(3), 196-9.

106. Kamiya, K., Hamabe, W., Harada, S., Murakami, R., Tokuyama, S., Satake, T. (2008). Chemical constituents of *Morinda citrifolia* roots exhibit hypoglycemic effects in streptozotocin-induced diabetic mice. *Biological & Pharmaceutical Bulletin*, 31(5), 935-8.

107. Deng, Y., Chin, Young-Won., Chai, H., Keller, W.J., Kinghorn. A.D. (2007). Anthraquinones with quinone reductase-inducing activity and benzophenones from *Morinda citrifolia* roots. *Journal of Natural Products*, 70(12), 2049-52.

108. Jain, R.K., Srivastava, S.D. (1992). Two new anthraquinones in the roots of *Morinda citrifolia*. *Proceedings of the National Academy of Sciences, India, Section A: Physical Sciences*, 62(1), 11-13.

109. Balakrishna, S., Seshadri, T.R., Venkatramani, B. (1961). Special chemical components of commercial woods and related plant material, part X-Heartwood of *Morinda citrifolia*. *Journal of Scientific and Industrial Research*, 20B, 331-333.

110. Balakrishna, S., Seshadri, T.R., Venkatramani, B. (1960). Special chemical components of commercial woods and related plant material: Part IX-Morindonin, A new glycoside of morindone, *Journal of Scientific and & Industrial. Research*, 19B(11), 433-436.

111. Murti, V.V.S., Neelakantan, S., Seshadri, T.R., Venkatramni, B.(1959). Special chemical constituents of commercial woods and related plant material: Part VIII-Heartwood of *Morinda tinctoria*. *Journal of scientific and Industrial Research*, 18 B(9), 367-370.

112. Ramamoorthy, S., Mudgal, G., Rajesh, D., Nawaz, K.F., Vijayakumar V., Rajasekaran, R.C. (2009). Characterisation of novel pH indicator of natural dye *Oldenlandia umbellate*. *Natural Product Research*, 23(13), 1210-7.

113. Yadava, R. N., Khan, Shirin. (2013). Isolation and characterisation of a new allelochemical from *Parthenium hysterophorus*. *International Journal of Pharmaceutical Sciences and Research*, 4(1), 311-315.

114. Madan, H., Sharma, S. (2011). Hentriacontanyl eicosanoate ester from *Parthenium hysterophorus* and its antimicrobial activity. *Journal of the Indian Chemical Society*, 88(7), 1033-1036.

115. Naeem, I., Taskeen, A., Mubeen, H., Maimoona, A. (2010). Characterization of flavonols present in barks and needles of *Pinus wallichiana* and *Pinus roxburghii*. *Asian Journal of Chemistry*, 22(1), 41-44.

116. Huang, J., Hou, P., Wu, L., Gao, H. (2012). Isolation and identification of chemical constituents from whole plant of *Polygonum hydropiper*.*Shenyang Yaoke Daxue Xuebao*, 29(1), 22-25.

117. Yagi, A., Uemura, T., Okamura, N., Haraguchi, H., Imoto, T., Hashimoto, K. (1994). Antioxidative sulfated flavonoids in leaves of *Polygonum hydropiper*. *Phytochemistry*, 35(4), 885-7.

118. Haraguchi, H., Hashimoto, K., Yagi, A. (1992). Antioxidative substances in leaves of *Polygonum hydropiper*. *Journal of Agricultural and Food Chemistry*, 40(8), 1349-51.
119. Peng, Z.F., Strack, D., Baumert, A., Subramaniam, R., Goh, N. K., Chia, T.F., Tan, S.N., Chia, L.S. (2003). Antioxidant flavonoids from leaves of *Polygonum hydropiper*. *Phytochemistry*, 62(2), 219-28.
120. Pearl, I.A., Darling, S.F. (1971). Barks of the family Salicaceae. XXVI. Hot water extractives of the bark and leaves of *Populus deltoides*. *Canadian Journal of Chemistry*, 49(1), 49-55.
121. Nair, A.G.R., Subramanian, S.S. (1964). Variation in the chemical components of the stem-bark of *Psidium guajava*. *Indian Journal of Pharmacy*, 26, 140-1.
122. Seshadri, T.R., Vasishta. (1965). Polyphenols of the stem bark of *Psidium guava*. *Phytochemistry*, 4, 317-326.
123. Wu, Shou-Fang., Hwang, Tsong-Long., Chen, Shu-Li., Wu, Chin-Chung., Ohkoshi, Emika., Lee, Kuo-Hsiung., Chang, Fang-Rong., Wu, Yang-Chang. (2011). Bioactive components from the heartwood of *Pterocarpus santalinus*. *Bio-organic & Medicinal Chemistry Letters*, 21(18), 5630-5632.
124. Kesari, A.N., Gupta, R.K., Watal, G. (2004). Two aurone glycosides from heartwood of *Pterocarpus santalinus*. *Phytochemistry*, 65(23), 3125-3129
125. Krishnaveni, K.S., Rao, J.V., Srinivasa. (2000). A new isoflavone glucoside from *Pterocarpus santalinus*. *Journal of Asian Natural Products Research*, 2(3), 219-223.
126. Arnone, A., Camarda, L., Merlini, L., Nasini, G. (1975). Structures of the red sandalwood pigments santalins A and B.*Journal of the Chemical Society, Perkin Transactions 1: Organic and Bio-Organic Chemistry* (1972-1999) (2), 186-93.
127. Singh, S.S., Srivastava, A., Saxena, R., Pandey, S.C., Sharma, V. (2005). Red sandal *Pterocarpus santallius*: Chemistry, biological activities and use- a review. *Journal of Medicinal and Aromatic Plant Sciences*, 27(2), 303-308.
128. Naz, S., Siddiqi, R., Ahmad S., Rasool, S A., Sayeed, S A. (2007). Antibacterial activity directed isolation of compounds from *Punica granatum*. *Journal of Food Science*, 72(9), M341-5.
129. Lansky, E.P., Harrison, G., Froom, P., Jiang, W.G. (2005). Pomegranate (*Punica granatum*) pure chemicals show possible synergistic inhibition of human PC-3 prostatecancer cell invasion across Matrigel. *Investigational New Drugs*, 23(2), 121-2.
130. Lal, C., Sharma, M.C., Shakyawar, D.B., Raja, A.S. M.,Sharma, K. K., Pareek, P.K. (2011). Natural dye constituents from rind of*Punica granatum* and its application on Pashmina fabrics.*Archives of Applied Sciences Research*, 3(3), 350-357.

130a Okuda, T., Hatano, T., Nitta, H., Fujii, R. (1980). Hydrolyzable tannins having enantiomeric dehydrohexahydroxydiphenoyl group., revised structure of terchebin and structure of granatin B. *Tetrahedron Letters*, 21(45), 4361-4.

131. Liu, Yan-Lin., Deng, Xu-Ming., Wang, Da-Cheng., Lei, Lian-Cheng., Han, Wen-Yu., Zhao, Rui-Li., Wang, Jiao-Yu. (2008). Studies on chemical constituents and antifungal activity of Turkish galls. *Linchan Huaxue Yu Gongye*, 28(4), 44-48.

132. Hamid, H., Kaur, G., Abdullah, S.T., Ali, M., Athar, M., Alam, M.S. (2005). Two new compounds from the galls of *Quercus infectoria* with nitric oxide and superoxide inhibiting ability. *Pharmaceutical Biology,* 43(4), 317-323.
133. Hwang, Jae-Kwan., Kong, Tae-Wook., Baek, Nam-In., Pyun, Yu-Ryang. (2000).A-Glycosidase inhibitory activity of hexagalloylglucose from the galls of *Quercus infectoria*.Planta Medica, 66(3), 273-274.
134. Gaspar, H., Moiteiro, C., Turkman, A., Coutinho, J., Carnide, V. (2009). Influence of soil fertility on dye flavonoids production in weld (*Reseda luteola*) accessions from Portugal. *Journal of Separation Science,* 32(23-24), 4234-40.
135. Villela. A., van der Klift Elbert, J.C., Mattheussens Eric, S.G.M., Derksen Goverdina, C.H., Zuilhof, H., van Beek Teris, A. (2011). Fast Chromatographic separation for the quantitation of the main flavones dyes in *Reseda luteola* (Weld). *Journal of Chromatography.* A, 1218 (47), 8544-50.
136. Liu, B., Yang, J., Wang, S. (2007). The chemical constituents in rhubarb rhizomes and roots derived from *Rheum emodi. Huaxi Yaoxue Zazhi,* 22(1), 33-35.
137. Singh, S.S., Pandey, S.C., Singh, R., Agarwal, S.K. (2005). 1,8- Dihydroxyanthraquinone derivatives from rhizomes of *Rheum emodi. Indian Journal of Chemistry,* Section B, 44B(7), 1494-1496.
138. Suresh Babu, K., Srinivas, P.V., Praveen, B., Hara Kishore, K., Suryanarayana Murty, U., Madhusudana Rao, J. (2003). Antimicrobial constituents from the rhizomes of *Rheum emodi.Phytochemistry,* 62(2), 203-207.
139. Agarwal, S.K., Singh, S.S., Verma, S., Kumar, S. (1999).Two new anthraquinone derivatives from *Rheum emodi. Indian Journal of Chemistry,*Section B,38B(6), 749-751.
140. Agarwal, S K., Singh, S S., Verma, S., Kumar, S. (2000). Antifungal activity of anthraquinone derivatives from *Rheum emodi. Journal of Ethnopharmacology,* 72(1-2), 43-6.
141. Sharma, N., Sharma, U.K., Gupta, A.P., Sinha, A.K. (2010). Simultaneous determination of epicatechin, syringic acid, quercetin-3-O-galactoside and quercitrin in the leaves of *Rhododendron* species by using a validated HPTLC method. *Journal of Food Composition and Analysis,* 23(3), 214-219.
142. Li, X., Liu, Z., Chen, Y., Wang, Li-Juan., Zheng, Yi-Nan., Sun, Guang-Zhi., Ruan, Chang-Chun. (2009). Rubiacordone A: a new anthraquinone glycoside from the roots of *Rubia cordifolia. Molecules,* 14(1), 566-572.
143. Akhtar, M. S., Ali, M.,Madhurima., Mir, S.R., Singh, O. (2006).New anthraquinones from *Rubia cordifolia* roots. *Indian Journal of Chemistry,* Section B, 45B(8), 1945-1950.
144. Abdullah, S.T., Ali, A., Hamid, H., Ali, M., Ansari, S.H., Alam, M. S. (2003). Two new anthraquinones from the roots of *Rubia cordifolia. Pharmazie,* 58(3), 216-217.
145 Chang, L.C., Chavez, D., Gills, J.J., Fong, H.H.S., Pezzuto, J. M., Kinghorn, A.D. (2000).Rubiasins A-C, new anthracene derivatives from the roots and stems of *Rubia cordifolia. Tetrahedron Letters,* 41(37), 7157-7162
146. Hua, H. M., Wang, S. X., Wu, L. J., Li, X., Zhu, T. R. (1992). Naphthoic acid esters from the root of *Rubia cordifolia.Yaoxue Xuebao,* 27(4), 279-82.

147. Qiao, Y.F., Wang, S.K., Wu, L.J., Li, X., Zhu, T.R. (1990). Antibacterial constituents from the roots of *Rubia cordifolia. Yaoxue Xuebao,* 25(11), 834-9.

148. Vidal-Tessier, A.M., Delaveau, P., Champion, B. (1987). New anthraquinones of *Rubia cordifolia* roots. *Annales Pharmaceutiques Francaises,* 45(3), 261-7.

149. Vidal-Tessier, A.M., Delaveau, P., Champion, B. (1986). New quinones of *Rubia cordifolia* roots. *Annales Pharmaceutiques Francaises,* 44(2), 117-22

150. Itokawa, H., Mihara, K., Takeya, K. (1983). Studies on a novel anthraquinone and its glycosides isolated from *Rubia cordifolia* and *T. akane. Chemical & Pharmaceutical Bulletin,* 31, (7), 2353-8.

151. Dosseh, C., Tessier, A.M., Delaveau, P. (1981). New quinones in *Rubia cordifolia* roots, III. *Planta Medica,* 43(4), 360-6.

152. Dosseh, C., Tessier, A. M., Delaveau, P. (1981). *Rubia cordifolia* roots. II. New quinones. *Planta Medica,* 43(2), 141-7.

153. Tessier, A.M., Delaveau, P., Champion, B. (1981). New anthraquinones in *Rubia cordifolia* root. *Planta Medica,* 41(4), 337-43.

154. Boldizsar, I., Laszlo-Bencsik, A., Szucs, Z., Danos, B. (2004). Examination of the anthraquinone composition in root-stock and root samples of *Rubia tinctorum* plants of different origins. *Acta Pharmaceutica Hungarica,* 74(3), 142-8.

155. Gil, R.R., Lin, Long-Ze., Cordell, G.A., Kumar, M.R., Ramesh, M., Reddy, B.M., Mohan, G,K., Rao, A. V. N. A. (1995). Anacardoside from the seeds of *Semecarpus anacardium.Phytochemistry,* 39(2), 405-7.

156. Murthy, S.S.N. (1992). Naturally occurring biflavonoid derivatives. 11. New biflavonoid from *Semecarpus anacardium. Chimica Acta Turcica,* 20(1), 33-7.

157. Murthy, S.S.N. (1988). Naturally occurring biflavonoid derivatives. Part 10. Semecarpetin, a biflavanone from *Semecarpus anacardium. Phytochemistry,* 27(9), 3020-2.

158. Murthy, S. S. N. (1986). Semecarpuflavanone-a new biflavanone from *Semecarpus anacardium.Proceedings of Indian Academy of Sciences, Chemical Sciences,* 97(1), 63-9.

159. Murthy, S. S. N. (1985). Naturally occurring biflavonoid derivatives: Part VII - galluflavanone, a new biflavonoid from *Semecarpus anacardium. Indian Journal of Chemistry,* Section B, 24B(4), 398-402.

160. Murthy, S. S. N. (1985). Naturally occurring biflavonoid derivatives. Part 6. Jeediflavanone: a biflavonoid from *Semecarpus anacardium. Phytochemistry,* 24(5), 1065-9.

161. Ishratullah, K., Ansari, W. H., Rahman, W., Okigawa, M., Kawano, N. (1977). Biflavanoids from *Semecarpus anacardium.Indian Journal of Chemistry,* Section B, 15B(7), 615-19.

162. Madhav, R., Seshadri, T.R., Subranmanian, G.B.V. (1967). Indentity of the Polyphenol of *Shorea* species with hopeaphenol. *Phytochemistry,* 6, 1155-1156.

163. Semwal, R.B., Semwal, D.K., Semwal, R., Singh, R., Rawat, M.S.M. (2011). Chemical constituents from the stem bark of *Symplocos paniculata* with antimicrobial, analgesic and anti-inflammatory activities. *Journal of Ethnopharmacology*, 135(1), 78-87.

164. Mir, Q.Y., Ali, M., Alam, P. (2009). Lignan derivatives from the stem bark of *Syzygium cuminii. Natural Product Research*, 23(5), 422-430.

165. Abdala, L.R., Seeligmann, P. (1995). Natural distribution of flavonoids in *Tagetes minuta. Biochemical Systematic and Ecology*, 23(5), 567-8.

166. Kumar, R., Dayal, R., Sharma, P. (2009). GC-MS analysis of essential oil from aerial parts of *Tagetes minuta. Indian Perfumer*, 53(2), 16-19.

167. Arya, C. (1999).Post infectional changes in flavonoids and phenolic acids by fungal pathogens in *Tamarindus indica.Research Journal of Chemistry and Environment*, 3(4), 27-28.

168. Bhatia, V.K., Gupta, S.R., Seshadri T.R. (1966). C-Glycosides of Tamarind leaves. *Phytochemistry*, 5, 177-181.

169. Lacret, R., Varela, R.M., Molinillo, J.M.G., Nogueiras, C., Macias, F.A. (2011). Anthratectone and Naphthotectone, two quinones from Bioactive Extracts of *Tectona grandis. Journal of Chemical Ecology*, 37(12), 1341-1348.

170. Nayeem, N., Karvekar, M.D. (2010). Isolation of phenolic compounds from the methanolic extract of *Tectona grandis. Research Journal of Pharmaceutical, Biological and Chemical Sciences*, 1(2), 221-225.

171. Shukla, N., Kumar, M., Akanksha., Ahmad, G., Rahuja, N., Singh, A.B., Srivastava, A.K., Rajendran, S.M., Maurya, R. (2010). Tectone, a new antihyperglycemic anthraquinone from *Tectona grandis* leaves. *Natural Product Communications*, 5(3), 427-430.

172. Aguinaldo, A.M., Ocampo, O.P.M., Bowden, B.F., Gray, A.I., Waterman, P.G. (1993). Tectograndone, an anthraquinone-naphthoquinone pigment from the leaves of *Tectona grandis. Phytochemistry*, 33(4), 933-5.

173. Agarwal, S. C., Sarngadharan, M. G., Seshadri, T.R. (1965). Coloring matter of teak leaves. Isolation and constitution of tectoleafquinone. *Tetrahedron Letters*, (30), 2623-6.

174. Hegazy, Mohamed-Elamir F., Abd el-Razek, M.H., Nagashima, F., Asakawa, Y., Pare, P.W. (2009). Rare prenylated flavonoids from *Tephrosia purpurea. Phytochemistry*, 70(11-12), 1474-7.

175. Yang, J., Sun, F., Li, Z., Du, G., Qin, H. (2008). Chemical constituents from *Terminalia chebula. Tianran Chanwu Yanjiu Yu Kaifa*, 20(3), 450-451.

176. Kumar, K. J., Bhatnagar, S. P. (2005). H.P.L.C. method for simultaneous estimation of tannic acid, gallic acid, chebulinic acid and ethyl gallate in *Terminalia chebula* fruits. *Pharma Review*, 3(15), 141-143.

177. Ding, G., Liu, Y., Song, M., Zou, D., Sheng, L. (2001). Polyphenols from *Terminalia chebula. Zhongguo Yaoke Daxue Xuebao*, 32(3), 193-196.

178. Datta, S.C., Murti, V.V.S., Seshadri, T.R.(1968). A new component of the flowers of *Thespesia populnea* (+). Gossypol *Current Science*, 37(5), 135.

179. Chowdhury, R. (2004). 5-Methylcoumarins from *Toona ciliata* stem bark and their chemotaxonomic significance. *Biochemical Systematics and Ecology*, 32(1), 103-105.

180. Singh, D.P., Plant, A.K. (1995). Constituents of *Toona ciliata. Fitoterapia*, 66(2), 192.

181. Nagasampagi, B. A., Sriraman, M. C., Yankov, L., Dev,. (1975). Siderin from *Cedrela toona. Phytochemistry*, 14(7), 1673.

182. Rao, J.U.M., Hanumaiah, T., Rao, B.K., Rao, K.V.J. (1986). Emodin glycosides from *Ventilago calyculata. Journal of Natural Products*, 49(2), 343-5.

183. Hanumaiah, T., Marshall, D.S., Rao, B.K., Rao, J.U.M., Rao, K.V.J., Thomson, R.H. (1985). Naphthoquinone-lactones and extended quinones from *Ventilago calyculata. Phytochemistry*, 24(11), 2669-72.

184. Hanumaiah, T., Marshall, D.S., Rao, B.K., Rao, C.P., Rao, G.S.R., Rao, J.U.M., Rao, K.V.J., Thomson, R.H. (1985). Benzisochromanquinones in *Ventilago* species. *Phytochemistry*, 24(10), 2373-8.

185. Hanumaiah, T., Rao, B.K., Rao, C.P., Rao, G.S.R., Rao, J.U.M., Rao, K.V.J., Marshall, D.S., Thomson, R.H. (1985). Naphthalenes and naphthoquinones from *Ventilago* species. *Phytochemistry*, 24(8), 1811-15.

186. Rao, B.K., Hanumaiah, T., Rao, J.U.M., Rao, K.V.J., Thomson, R.H. (1984). Two anthraquinones from *Ventilago calyculata. Phytochemistry*, 23(9), 2104-5.

187. Rao, B.K., Hanumaiah, T., Rao, C.P., Rao, G.S.R., Rao, K.V.J., Thomson, R.H. (1983). Anthraquinones in *Ventilago* species. *Phytochemistry*, 22(11), 2583-5.

188. Hanumaiah, T., Rao, G.S.R., Rao, C.P., Rao, K.V.J., Cowe, H., Cox, P.J., Howie, R.A., Marshall, D.S., Thomson, R.H. (1985). Isofuranonaphthoquinones from *Ventilago maderaspatana*. Crystal structure of ventilone-C. *Tetrahedron*, 41(3), 635-42.

189. Rao, G.S.R., Hanumaiah, T., Rao, K.V.J. (1980). New anthraquinones from the roots of *Ventilago madaraspatana.Indian Journal of Chemistry*, Section B, 19B(2), 97-100.

190. Yadava, R.N., Bhargava, B. (2012). New potential allelochemical from *Wedelia chinensis. Journal of the Institution of Chemists (India)*, 84(1), 15-21.

191. Muruganandam, A.V., Bhattacharya, S.K. (2000). Indole and flavonoid constituents of *Wrightia tinctoria, W. tomentosa and W. coccinea. Indian Journal of Chemistry*, Section B, 39B(2), 125-131.

192. Li, Wei-Jen., Lin, Rong-Jyh., Yeh, Yu-Ting., Chen, Hsin-Liang., Chen, Chung-Yi. (2013). Phenylalkanoids from *Zingiber officinale. Chemistry of Natural Compounds*, 49(3), 440-442.

193. Guo, S., Tang, Yu-Ping., Duan, Jin-Ao., Su, Shu-Lan., Qian, Da-Wei. (2009). Chemical constituents from the fruits of *Ziziphus jujube. Zhongguo Tianran Yaowu*, 7(2), 115-118.

194. Woo, W.S., Kang, S.S., Wagner, H., Seligmann, O., Chari, V. M. (1980). Structure of flavone-C-glycosides. Part 18. Acylated flavone-C-glycosides from the seeds of *Zizyphus jujube*. *Phytochemistry*, 19(12), 2791-3.

195. Cooksey, C.J. (2001). Tyrian purple: 6,6′-Dibromoindigo and related compounds. *Molecules*, 6(9), 736-769.

196. Wouters, J., Verhecken, A. (1991). High-performance liquid chromatography of blue and purple indigoid natural dyes. *Journal of the Society of Dyers and Colorists*, 107(7-8), 266-9.

197. Thomson, R.H.(1971). *Naturally Occurring Quinones*, second edition, Academic Press, London, New York p. 459, 457, 462-476.

198. Quye, A. (1999). *Chemistry Review*, 8(5), 27-31.

199. Harborne, J.B. (1999). Classes and functions of secondary products from plants in "*Chemicals from Plants: Perspectives on Plant Secondary Products*" edited by N.J. Walton and D.E. Brown, Imperial College Press, London, pp. 1-23.

200. Heath, M.C. (2002). Secondary metabolites and plant defence. *Physiol. Mol. Plant pathology*, 60, 273-274.

200a. Singh, R. (2009). Phytochemical examination of *Clematis roylei*. Doctor of Philosophy Thesis submitted to Forest Research Institute University, Dehradun, India, pp. 3-6

200b. Pichersky, E., Gershenzon, J. (2002). The formalities and function of plant volatiles., perfumes for pollinator attraction and defence. *Curr. Opin. Plant Biol.*, 5, 237-243.

201. Metcalf, R.L., Metcalf, E.R. (1992). *Plant Kairomones in Insect Ecology and Control*. Chapman and Hall, New York p. 1.

202. Southon, I.W., Buckingham, J. (1989). *Dictionary of Alkaloids*, vol. 1&2. Chapman and Hall, London.

203. Hill, R.A., Connolly, J.D. (1991).*Dictionary of Terpenoids*, Chapman and Hall, London.

204. Harborne, J.B. (1994). *The Flavonoids: Advances in Research Since 1986*. Chapman and Hall, London.

Part B: Colouring Constituents of Some Prominent Ancient Natural Dyes

In the ancient days, natural dyes were obtained from vegetable, animal, and mineral sources. The chemical constituents of some prominent natural dyes used in the past, before the discovery of synthetic dyes in 1856, belong mainly to the following classes of compounds:

1. Quinone-based dyes
   (i) Anthraquinones
   (ii) Naphthoquinones
2. Flavonoid-based dyes
   (i) Flavones and flavonols
   (ii) Anthocyanins
   (iii) Pyran colourants (Chromenes)
3. Carotenoid-based dyes
4. Indigoid-based dyes
5. Chlorophylls

The colouring constituents of some prominent ancient natural dyes belonging to the above classes of compounds are described below.

QUINONE-BASED DYES

(1) Anthraquinone dyes

(i) Anthraquinone dyes from plants: The anthraquinone-based chemical compounds, viz., alizarin, purpurin, ruberythric acid, munjistin, xanthopurpurin, pseudopurpurin, and a number of others compounds have been isolated and characterized from the dyes obtained from *Rubia cordifolia, Rubia tinctorum,* and other *Rubia* species. Purpurin and munjistin were found to be the main colouring compounds found in the roots dye of *Rubia cordifolia,* while the *Rubia tinctorum* roots dye contained alizarin and purpurin as the major constituents along with other compounds (Anon, 1972; Thomson, 1971). The number of other compounds isolated from the *Rubia* species by researchers all over the world are given in Part A of Chapter 5. The colouring constituents existed as glycosides and also in the free form. Natural dyes containing these chemical compounds when used for dyeing different fabrics imparted charming shades with very good colourfastness properties.

Alizarin (1, 2-dihydroxy anthraquinone) is a vital dye compound isolated from the roots of *Rubia tinctorum* and other *Rubia* species (Anon, 1972., Schweizer, 1964., Ref. 142, 154, ch. V A; Thomson, 1971). It was used for imparting red colour in ancient times. The use of this dye existed on the clothes found in Egyptian tombs, dating 6000 years ago. The dye (alizarin) occurs mainly as a glycoside in the roots, which was characterized as ruberythric acid, and its structure was confirmed by synthesis (Richter, 1936; Zemplen and Bongnar, 1939).

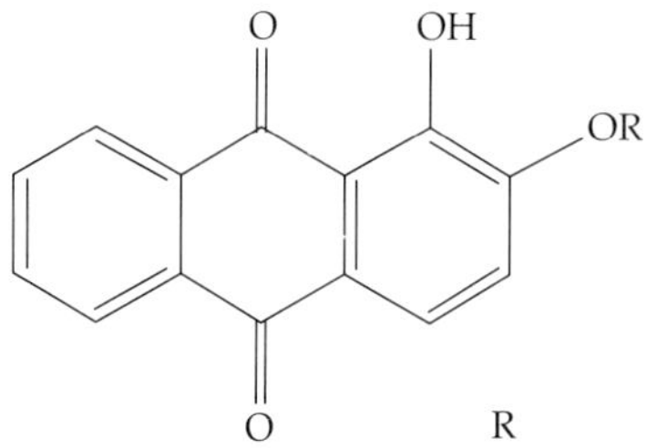

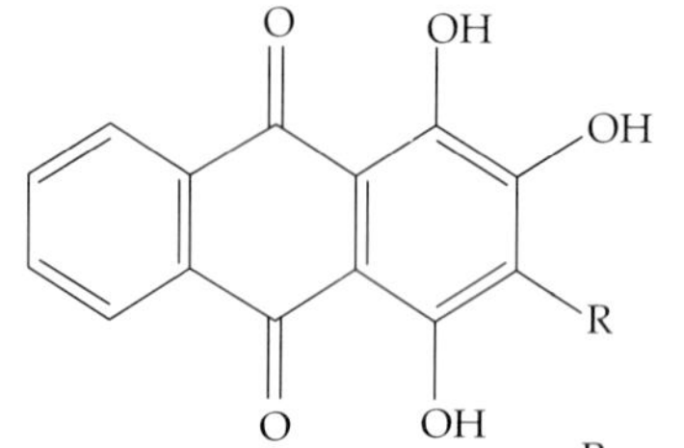

	R		R
Alizarin	H	Purpurin	H
Ruberythric acid	Primeverose	Pseudo Purpurin	COOH

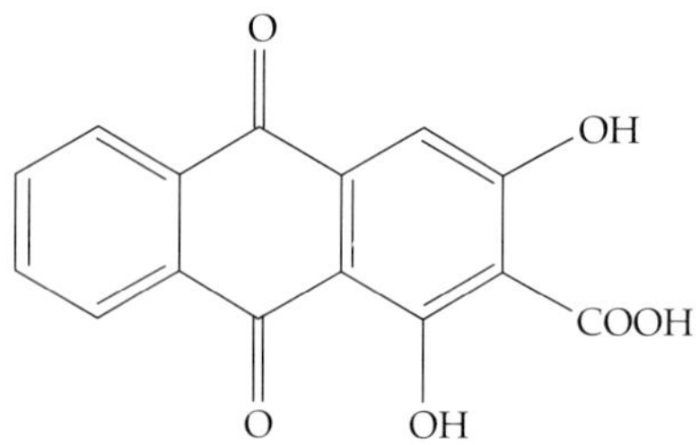

Munjistin

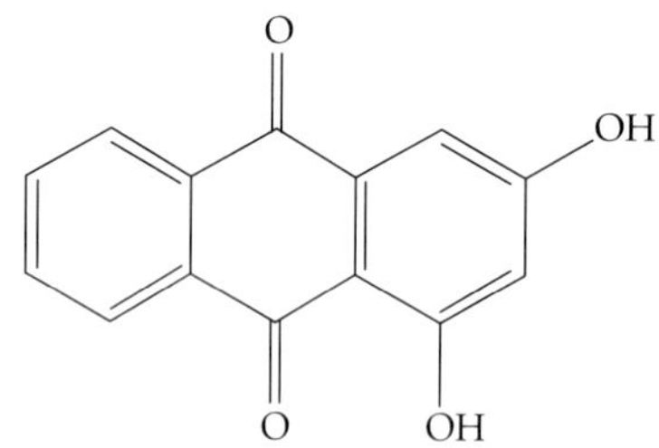

Xanthopurpurin (Purpuroxanthin)

A commercial viable process for the synthesis of alizarin was developed by BASF company (Germany) and almost simultaneously by Perkin in 1869 (Kroschwitz and Howe-Grant, 1993). The chemical modification of the process led to the reduction in the cost of synthetic alizarin considerably (from 100% to 2.5%) from the year 1870 to 1912, as a consequence of which, natural dyes containing alizarin were wiped out from the market. The natural alizarin in dyeing of cotton was first used in India, and later, a beautiful red colour (Turkey red) was developed by Turks using the dye (Haller, 1940; Venkataraman, 1952).

(ii) Anthraquinone dyes from animals

(a) Natural dyes isolated from insects are known as animal dyes. Lac is one of the most valuable gifts of nature to the mankind. In India, use of lac can be traced to the ancient times of *Mahabharata*, if not earlier. Lac dye is one of the most ancient animal dyes of the world. The Chinese used lac dye, about 4000 years ago, for imparting attractive and lustrous red colour to silk. Lac is the resinous secretion of a kind of microscopic insect *Laccifer lacca* (syn. *Kerria lacca*), commonly known as the lac insect. The lac dye is obtained as a by-product while preparing shellac from the crude lac. It is acidic in nature and is generally present as its sodium/potassium salts, which are soluble in water. The pure lac dye is sparingly soluble in water and was found to be a mixture of five closely related anthraquinone derivatives. Their structures were elucidated and named as laccaic acid A, B, C, E, and laccaic acid D (Thomson, 1987, 1971).

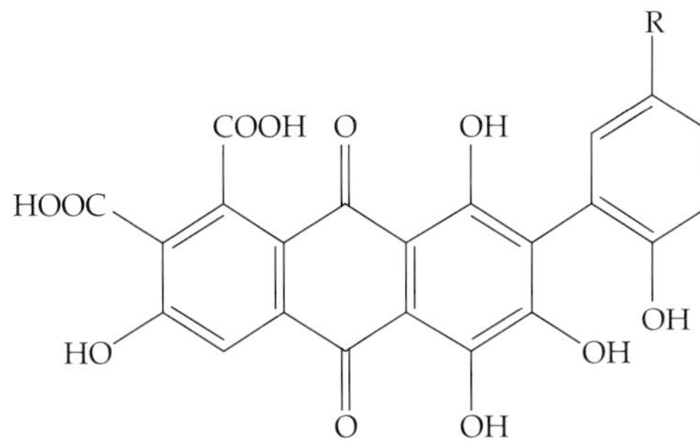

Laccaic acid A	$CH_2CH_2NH\ COCH_3$
Laccaic acid B	CH_2CH_2OH
Laccaic acid C	$CH_2CH\ (NH_2)\ COOH$
Laccaic acid E	$CH_2CH_2NH_2$

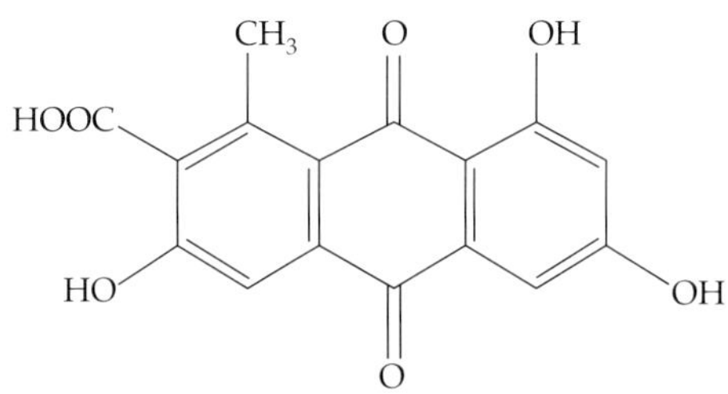

Laccaic acid D

(b) A red dye was isolated from various species of *Kermes* scale insects, especially *Kermes ilicis (syn. Coccus ilics* or *Kermococcus vermilis)* of the Coccoidea superfamily, which infest *Quercus ilex* (Holm Oak), and *Quercus coccifera* (Shrub Oak). *Kermes Vermilio* and some other *Kermes* species were also used for the isolation of dye. It is one of the oldest known dyestuffs (Leggett, 1964). The dye imparted a brilliant scarlet colour with alum as mordant. Although isolation of pure kermesic acid could be accomplished in end of the nineteenth century (Heise, 1895), its correct structure was elucidated in 1978 as 1,3,4,6-tetrahydroxy-7-carboxy-8-methylanthraquinone (Thomson, 1987, 1971; Cameron, 1981; Roberge and Brassard, 1978). The dye was highly priced but its cost was less in comparison to the famous Tyrian purple dye. It was used extensively until the sixteenth century, when it was replaced by another red dye isolated from the cochineal insect *Dactylopius coccus*.

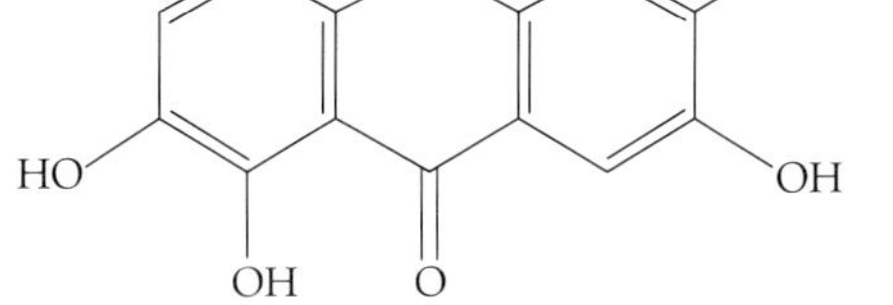

Kermesic acid

(c) A red dye was isolated from the body of the cochineal insect *Dactylopius coccus* occurring in Central and South America. The dye was isolated from the bugs either by drying them in sun (Baranyovits, 1978; Leggett, 1964) or by putting them in hot water or oven. The active principle (carminic acid) of the dye was isolated in the pure state (Schutzenberger, 1858) and its correct structure after synthetic confirmation was established as 1,3,4,6-tetrahydroxy-7-carboxy-8-methylanthraquinone-2-C-glucoside (Thomson, 1987, 1971). The dye was used for dyeing silk and wool using stannous chloride as the mordant.

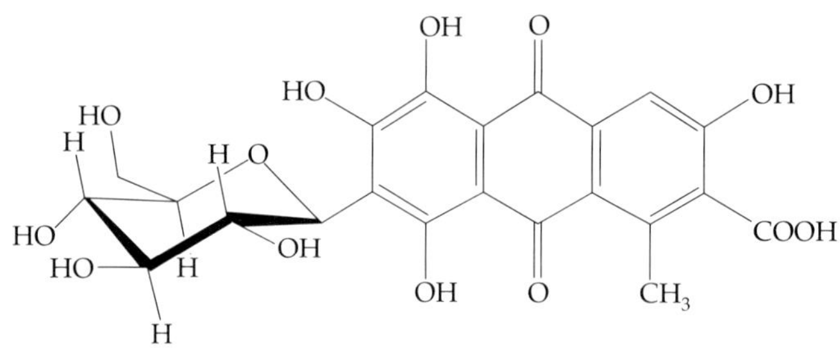

Carminic acid

(2) Naphthoquinone dyes

A large numbers of naphthoquinone compounds occur in nature but only a limited number, viz., lawsone, lapacol, juglone, alkanin, and shikonin occurring in different plants, could be used as dyes in the past.

(i) The powdered leaves of *Lawsonia inermis,* commonly known as *mehndi* or henna, have been used for colouring the hair and palm since ancient times in India. The isolation of the dye from the powdered leaves was carried out using aqueous sodium bicarbonate solution, and extracted dye imparted an orange shade to silk and wool (proteins fibres). Normally, a paste of *mehndi* in water is used by women folk. The active ingredient in the dye was found to be lawsone (2-hydroxy-1, 4-naphthoquinone) (Tommasi, 1920; La and Dutt, 1933; Thomson, 1971).

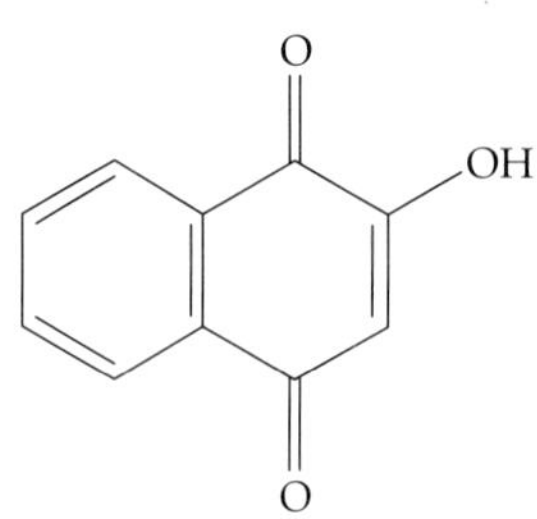

Lawsone

(ii) The husks of some *Juglans* species, namely, *Juglans cinerea* (butter nuts), *Juglans regia* (Persian walnuts), and *Juglans nigra* (black walnuts), were used for the isolation of the dye. The ancient Romans transported Persian walnuts from Asia Minor to Europe. The black walnut and butternut trees of eastern North America were used for the extraction of brown dye in 1664 (Buchanan, 1987; Adrosko,

1968). The structure of the active main constituent of the dye was elucidated as 5-hydroxy-1, 4-naphthoquinone, (Juglone) [Bernthsen, 1885; Thomson, 1971]. It is no longer used as a dye.

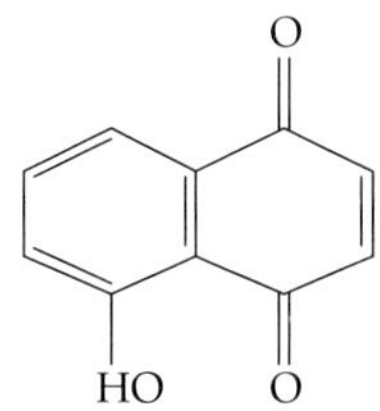

Juglone

FLAVONOID-BASED DYES

The origin of flavonoids can be traced to the latin word "flavus" meaning yellow. Flavonoid can be defined as the water-soluble polyphenolic compound consisting of 15 carbon atoms (C_6-C_3-C_6) and having 2-phenyl benzopyrone structure. The flavonoids are widely distributed in higher plants and account for the beautiful colours of flowers, fruits, vegetables, and herbs. The flavonoid-type dyes occur in different plant parts, viz. bark, fruits, flower petals, leaves, roots, and wood of higher plants, and do not appear to occur in fungi, lichens, and mosses.

Presently, the flavonoids can be mainly put under the following classes:

- Flavone (4H-1-Benzopyran-4-one, 2-phenyl)
  (i) Luteolin [4H-1-Benzopyran-4-one, 2-(3,4-dihydroxyphenyl)-5,7-dihydroxy]
- Flavonol (4H-1-Benzopyran-4-one, 3-hydroxy-2-phenyl)
  (i) Quercetin [4H-1-Benzopyran-4-one, 2-(3,4-dihydroxyphenyl)-3,5,7- trihydroxy]
  (ii) Kaempferol [4H-1-Benzopyran-4-one-2-(4-hydroxyphenyl)-3, 5,7 trihydroxy]

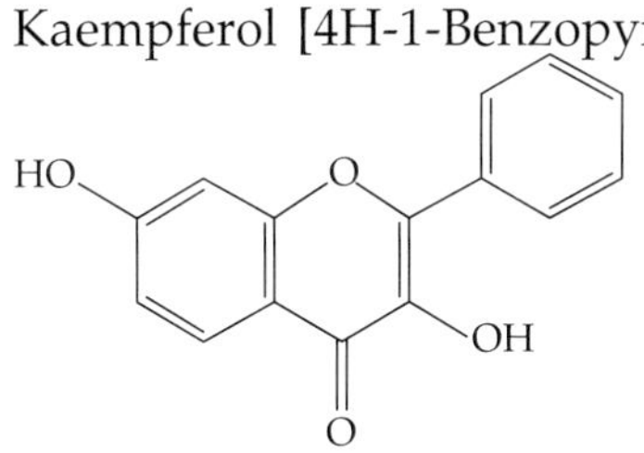

Flavone

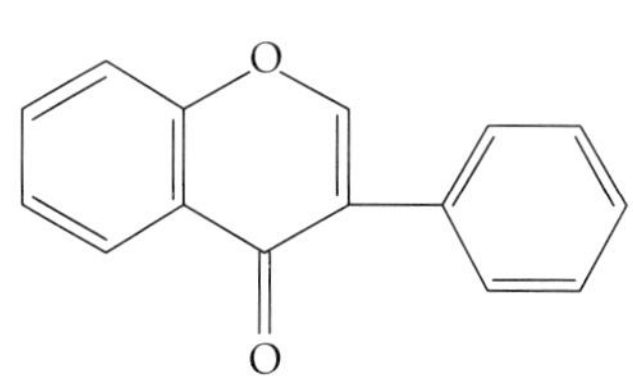

Isoflavone

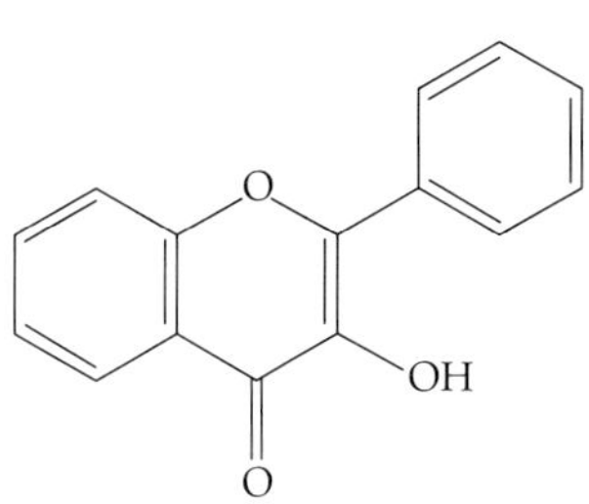

Flavonol

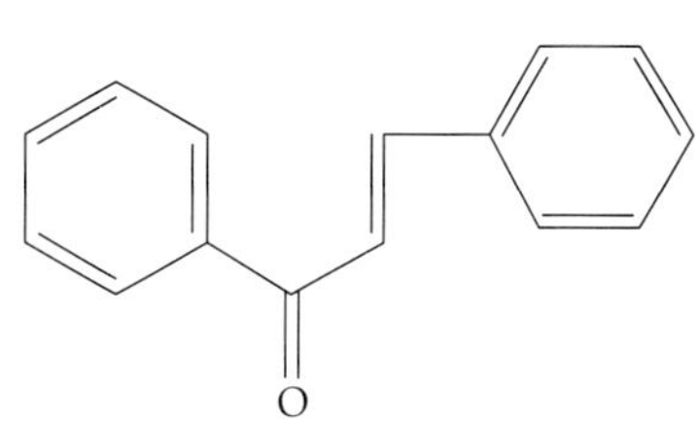

Chalcone

- Flavanone (4H-1-Benzopyran-4-one, 2,3-dihydro-2-phenyl)
- Isoflavone (4H-1-Benzopyran-4-one, 3-phenyl)
- Chalcone (2-Propene-1-one, 1-phenyl-3-phenyl)
  (i) 2,4,4-Trihydroxychalcone [2 Propen-1-one, 1-(2,4-dihydroxylphenyl)-3(4-hydroxyphenyl)]
- Flavan (2H-1-Benzopyran, 3, 4-didhydro-2-phenyl)
- Flavanols (2H-1-Benzopyran-3, 4 diol, 3, 4-dihydro-2-phenyl) and [2H-1-Benzopyran-3, 5, 7-triol, 2-(3,4-dihydroxyphenyl)-3, 4-dihydro]
- Anthocyanins

(1) Flavones and flavonols

Luteolin (4H-1-Benzopyran-4-one, 2-(3,4-dihydroxyphenyl)-5,-7-dihydroxy) isolated from *Reseda luteola* is the oldest known dye in Europe. The ancient Romans used luteolin for dyeing the garments and robes of Vestal Virgins (Kroschwitz and Howe-Grant, 1993; Meyers, 1990). The quercetin [4H-1-Benzopyran-4-one, 2-(3,4- dihydroxyphenyl)-3, 5, 7-trihydroxy], kaempferol [4H-1-Benzopyran-4-one-2 (4-hydroxyphenyl)-3,5,7-trihydroxy], and other flavonoid-containing dyes were also used for dyeing. The dyes generally occur as glycosides, which are soluble in water. The flavonoid dyes are mordant dyes and the presence of a carbonyl group at position 4 and a hydroxyl at position 5 make them capable to from chelates with different metallic mordants to give coloured insoluble complexes. The number of hydroxyls may vary from compound to compound depending on its structure.

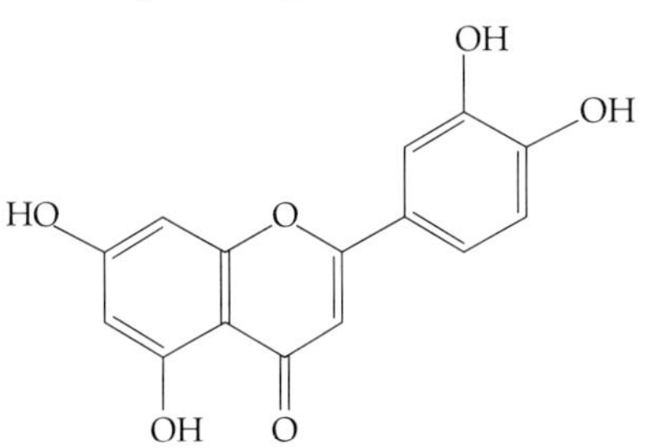

Luteolin

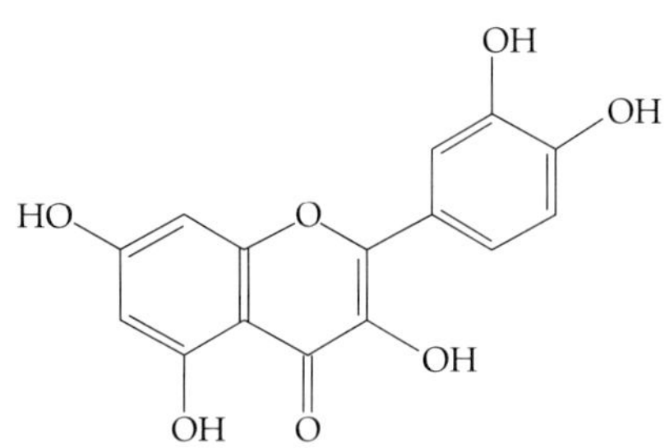

Quercetin

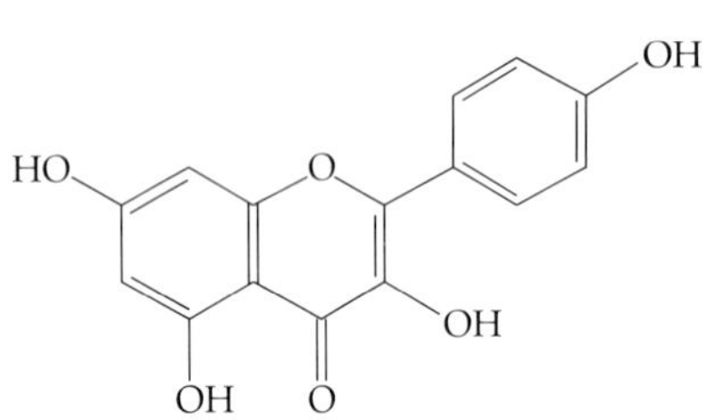

Kaempferol

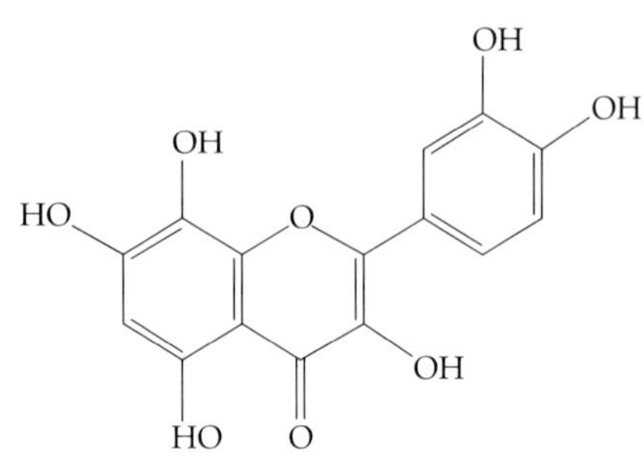

Gossypetin

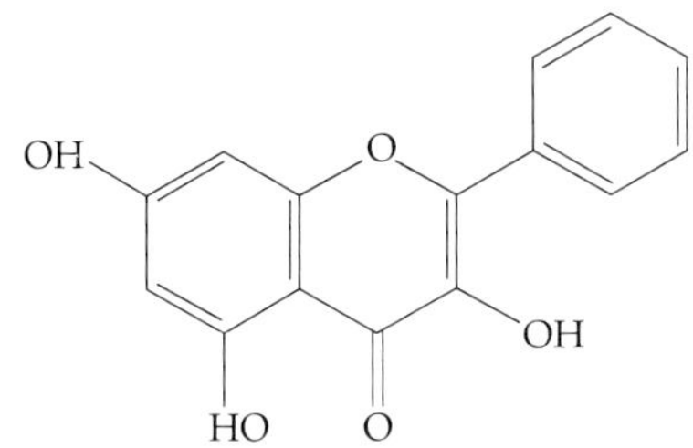

Chrysin

(2) Anthocyanins

The anthocyanins form the major group of water-soluble pigments in nature and are present in all higher plants. These can be found in all parts of plants mainly in fruits and flowers. The origin of the term "anthocyanin" can be traced to the Greek words *antho* for flowers and *kyanos* for blue and was introduced in 1835 by Willstatter and Everest. In the beginning, it was used for the blue colour of the cornflower (Marquart, 1835). Anthocyanins can be defined as the polyphenolic compounds responsible for the attractive blue, pink, red, and violet colours occurring in different plants having a C_6–C_3–C_6 carbon skeleton. These occur widely as glycosides in nature.

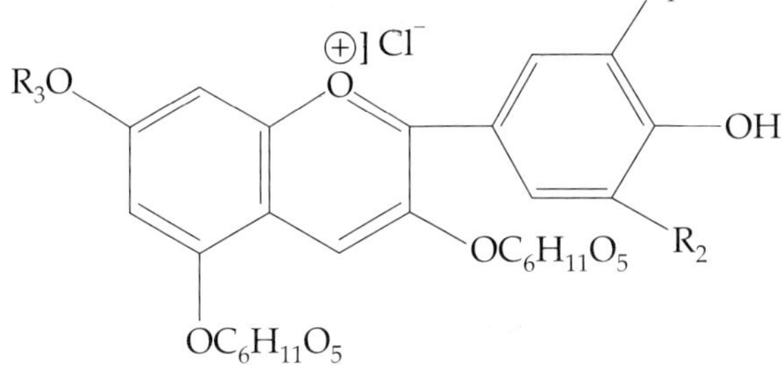

	R_1	R_2	R_3
Pelargonin chloride	H	H	H
Cyanin chloride	OH	H	H
Delphin chloride	OH	OH	H
Peonin chloride	OCH_3	H	H
Malvin chloride	OCH_3	OCH_3	H
Hirsutin chloride	OCH_3	OCH_3	OCH_3

The sugar-free anthocyanins are known as anthocyanidins and the basic compound is 2-phenylbenzopyrylium chloride or flavylium chloride. Anthocyanidins belong to three basic groups, namely, pelargonidin, cyanidin, and delphinidin.

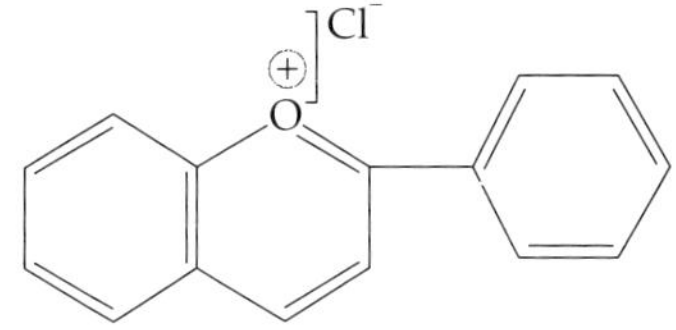

Flavylium chloride

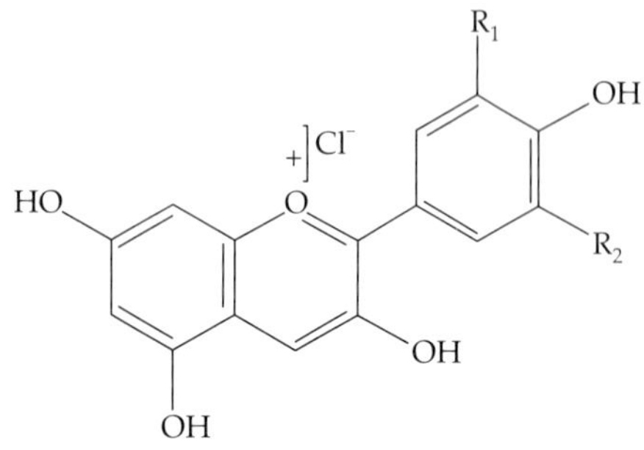

Pelargonidin	$R_1 = R_2 = H$
Cyanidin	$R_1 = OH$, $R_2 = H$
Delphinidin	$R_1 = R_2 = OH$

Seventeen anthocyanidins are known in the nature, but only six, namely, cyanidin, delphinidin, malvidin, pelargonidin, peonidin, and petunidin, occur widely (Finar, 1996; Nollet, 2004).

Characteristics of Anthocyanins

(i) Anthocyanins are pH-sensitive.

(ii) On boiling with hydrochloric acid, they yield anthocyanidin and sugar moiety. The commonly occurring sugars are glucose, rhamnose, galactose, and arabinose. Most prevalent sugar is glucose.

(iii) Commonly, sugar moiety occupies 3-position and occasionally 5-position.

(iv) The alkaline degradation of the anthocyanidins, belonging to the three basic groups, using concentrated potassium hydroxide at 140-150°C leads to formation of phloroglucinol and phenolic benzoic acid.

(v) Due to their pH sensitivity, anthocyanins could not to be used as commercial dyes.

(vi) Some anthocyanidins having quinoidal form, namely, carajurin and dracorhodin, were used as dyes in the past (Pelletier and Caventou, 1818).

(vii) The anthocyanins possess antioxidant, anti-inflammatory, anti-atherosclerotic, and anticarcinogenic activities (Piatkowska et al., 2011).

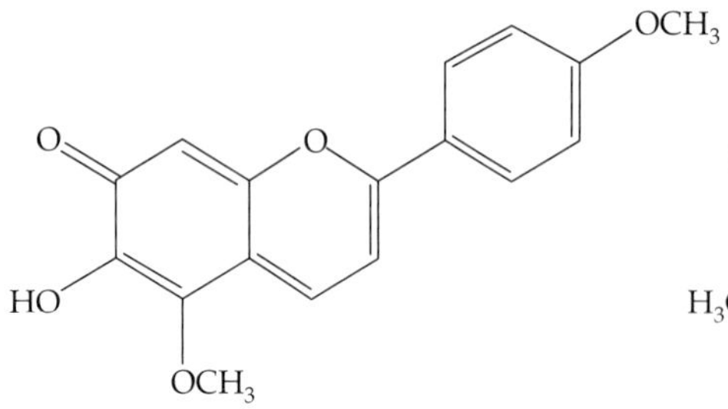

Carajurin

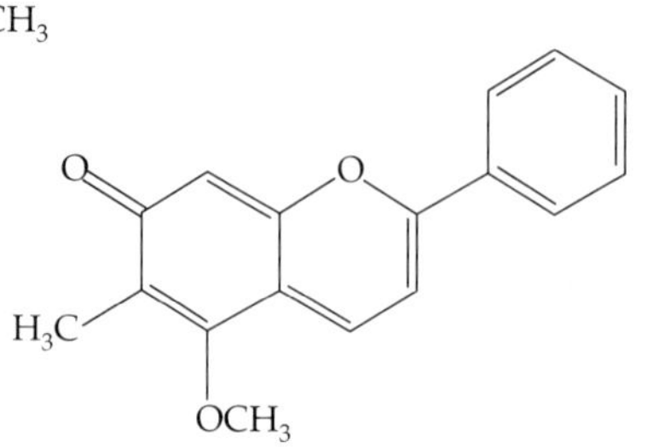

Dracorhodin

3. Pyran-based dyes(Chromenes)

(a) The wood of *Hematoxylon Campechianum* was used for the isolation of dye in the past. As the dye was obtained from the logs of wood, it was known as logwood dye. The wood chips were extracted with hot water under pressure or without pressure to obtain the dye. Removal of water yielded the dye in the powdered form. The dye solution containing leuco form of haematoxylin on concentration yielded the dye in the crystalline form (Haematein), and its structures could be elucidated after detailed studies (Perkin and Robinson, 1920; Mayer, 1943). It was used for blue dyeing and was the last natural dye to survive after the availability of synthetic dyes due to desirable bluish-black hue it imparted to the chrome-mordanted fibers. Its commercial use was discontinued due to long dyeing process and higher cost. Wood of *Caesalpinia sappan,* known as Brazilwood, and of other *Caesalpinia* species was used for the isolation of dye by extracting with hot water at normal pressure or under pressure. The dyed fabrics possessed poor fastness properties and therefore its use was discontinued. The aqueous dye solution contained Brazilin in leuco form, which got oxidized to brazilein during concentration (Kroschwitz and Howe-Grant, 1993., Ref. 49, 81, Ch. VA).

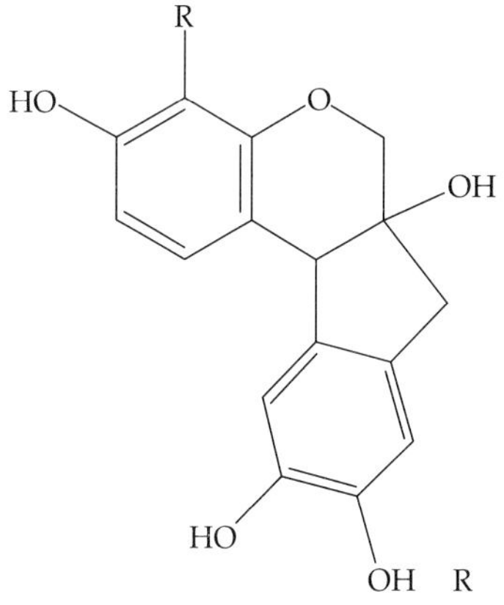

	R
Haematoxylin	OH
Brazilin	H

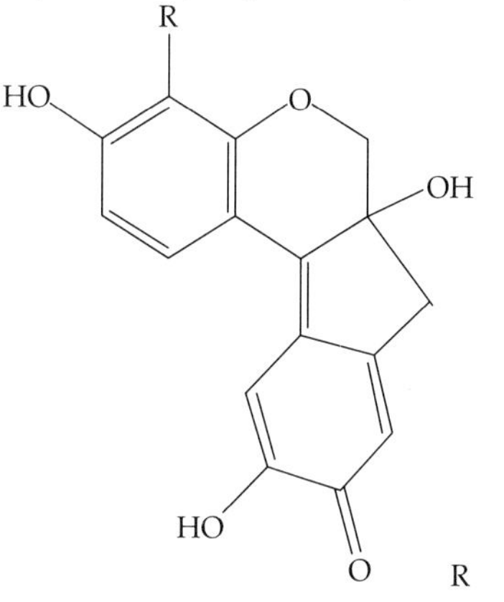

	R
Haematein	OH
Brazilein	H

(b) Kamala dye, a reddish-brown resinous powder, is isolated from the red pubescence of ripe fruits of *Mallotus philippensis.* The chief colouring constituents are the salmon-coloured rottlerin and its yellow isomer isorottlerin along with a minor constituent homorottlerin. The Kamala dye was used earlier for silk and wool dyeing and imparted a bright orange or flame colour (Ref. 94, Ch. VA).

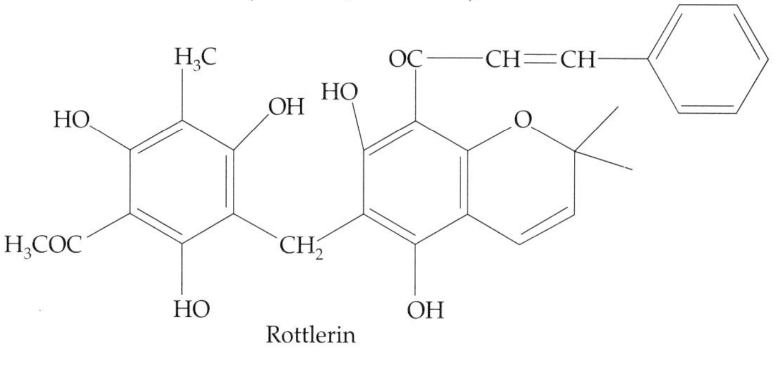

Rottlerin

CAROTENOID-BASED DYES

Isoprene is the basic building block of terpenoids. The combination of two isoprene units produces a monoterpenoid, three isoprene units produces sesquiterpenoids, four isoprene units produces diterpenoids, six isoprene units produces triterpenoid, and eight produces tetraterpenoid. Carotenoids consist of 40 carbon atoms (eight isoprene units) and thus come under the category of tetraterpenoids. Carotenoids can be defined as one of the most widely occurring groups of pigments found in nature. They occur in the chloroplasts and chromoplasts of plants and some other photosynthetic organisms like algae, some bacteria, and some types of fungus. They also occur in egg yolk, dairy products, and seafood. About 600 carotenoids have been isolated and characterized from the nature and may be broadly put under two classes, namely, carotenes, which are hydrocarbons without oxygen, and xanthophylls, which contain oxygen. Carotenoids generally absorb blue light and serve two roles in plants, viz., absorption of light energy for use in photosynthesis and protection of chlorophyll from photo-damage. The details about the chemistry of carotenoids have been given in a number of reviews (Zechmeister, 1960).

Bixa orellana occurs in India. Its seeds yield a red dye known as annatto dye. The principal constituent of the dye is bixin, and due to its non-toxic nature, this dye was and is still used in food industry for colouring dairy products, sugar confectionery, soft drinks, fish, meat, and other food items. It is also used in textile industry for dyeing cloth, yard threads, and sometimes in paint, varnish, and lacquer industry. In fact, recently, it has found use in cosmetics and soap industry (Yumnam and Tripathi, 2012). The use of washcloth dyed with bixin, to keep skin-related diseases at bay, is unique.

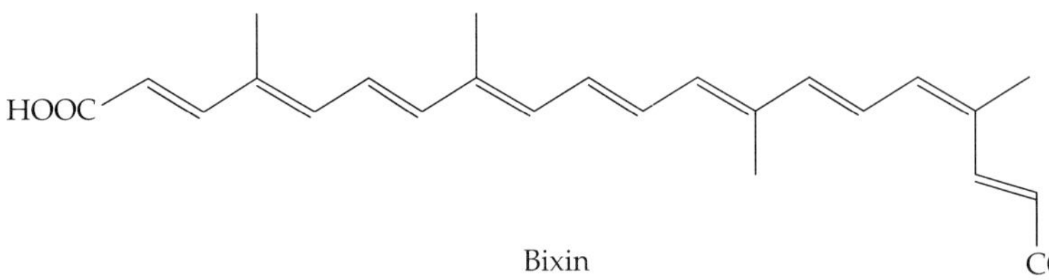

Bixin

Saffron is obtained from the flowers *Crocus sativus*. A gram of good quality saffron is obtained from about 170 flowers. The chief constituent of saffron is crocin (the digentiobiose ester of crocetin). On a number of occasions, saffron is confused with safflower, which is sometimes known as bastard saffron. Saffron is a popular food dye in India. It could not be used for textiles due its cost. During ancient period, it was not only used as a dye but also as a spice. Saffron is one of the most expensive dyes of ancient times due its non-availability in sufficient quantity (Kroschwitz and Howe-Grant, 1993, Ref. 60, Ch. VA). The plant grows mainly in limited areas of Jammu and Kashmir in India.

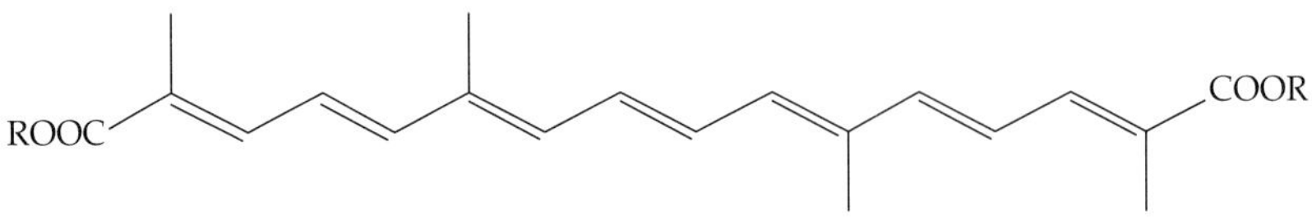

Crocetin R = H
Crocin R = gentiobiose

INDIGOID-BASED DYES

(1) Indigo is one of the oldest dyes (~5000 years old), with its origin in India. Its official name is indigotin and was obtained from plants of *Indigofera* species, mainly from *Indigofera tinctoria* (indigo) and *Isatis tinctoria* (woad). These plants do not contain indigotin as such but indican ($C_{14}H_{17}NO_{6,}$ a glycoside), which on hydrolysis gives D-glucose and indoxyl, and the latter on exposure to air is converted to indigotin ($C_{16}H_{10}N_2O_2$; 2,2′-bis [2,3-dihydro-3-oxoindolyliden]). The hydrolysis of indican by emulsion indicated it to be indoxyl-β-glucoside. Indigotin is now prepared synthetically on a large scale.

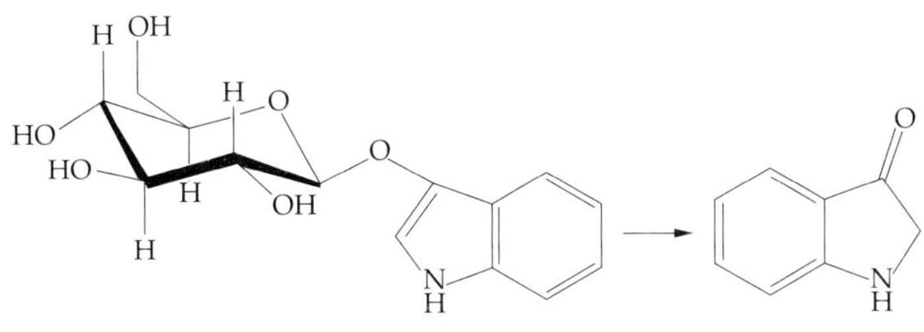

Indican Indoxyl

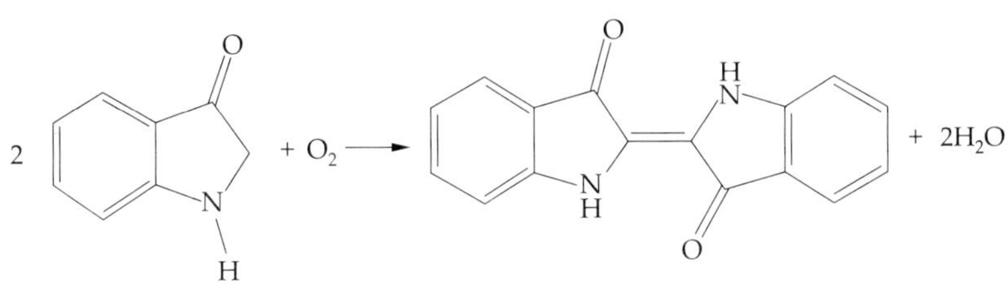

Indoxyl Indigotin

Indigotin is a dark-blue powder with a coppered gloss and is insoluble in water. Its paste, on agitation with alkaline sodium hyposulphite in large vats, is reduced to soluble leuco-compound, indigotin white. The fabric to be dyed is soaked in this alkaline solution and thereafter exposed to air, thereby regenerating the blue dye in the fabric (Finar, 1996, Ref. 82, Ch. VA).

(2) Tyrian purple: A purple-red natural dye, extracted from sea snails (shellfish) mentioned in texts dating about 1600 B.C., was known as Tyrian purple since ancient time. It was also called royal purple, imperial purple, or imperial dye and was probably first produced by ancient Phoenicians in the city of Tyre and hence the name Tyrian purple. As per the writing of Julius Pollux (Roman mythographer) in the second century B.C., the purple dye was first discovered by Heracles while walking with his dog along the coast of the Levant. It was observed that the dog's mouth turned purple on chewing the snails. The discovery of large number of Murex shells on Crete indicated that the Minoans might have perfected the extraction of Imperial purple centuries before the Tyrians. Examination of collocated pottery indicated that the dyes might have been produced during the middle Minoan period in the 20th to 18th century B.C.

(Reese, 1987; Stieglitz, 1994). The dye was very expensive (Theopompus) due to its low yield (~ 0.12 mg per snail) and thus became a status symbol for the kings. It also had the unique characteristics of becoming more intensive on exposure to sunlight and weathering. The purple colour was the colour of distinction and limited to its use by the Royal in the ancient Roman Empire. In the fifth century, only the emperors used the purple colour and violators were given severe penalty.

The dye is a mucous secretion from the hypobranchial gland of a medium-sized predatory snail *Bolinus brandaris* the spiny dye-murex (earlier known as *Murex brandaris*)—found in the eastern Mediterranean (Ziderman, 1986). The secretion can be obtained from snails either by poking, physically attacking them (renewable method), or crushing them completely (destructive method). The material obtained was salted for three days, followed by boiling the whole mass for 10 days. According to David Jacoby, 12,000 *Murex brandaris* snails are required to produce about 1.4 g of pure dye, which is sufficient to colour a small fragment of a garment (Jacoby, 2004). The dye was also extracted by crushing the smaller mollusk *Murex trunculus*. The production of purple dye from the shellfish ceased around 1453 A.D. due to the fall of Eastern Roman Empire by Turks, resulting in the destruction of Tyrian dye plants. It was substituted by other low-cost dyes like lichen purple and madder. The chief chemical constituent of the dye was found to be an organobromine compound (Friendlender, 1909), which was characterized as 6,6´- dibromoindigotin (Wouters and Verkecken, 1991). The same was synthesized earlier in 1903. Normally, the bromine compounds are found in algae and some other sea creatures, but rarely in the animals inhabiting land. However, it is strange to note that the dye was not synthesized commercially (Cooksey, 2001).

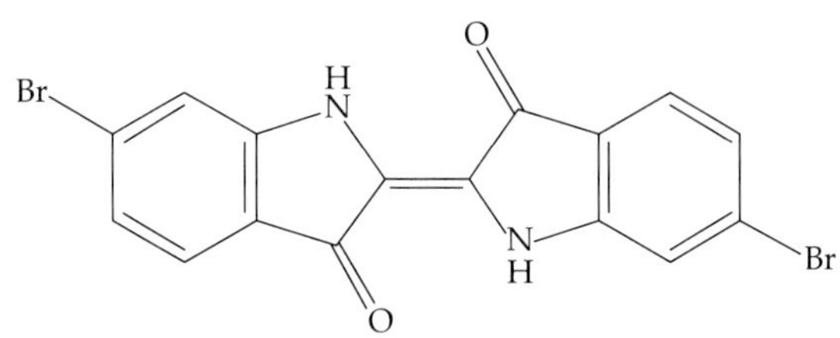

6,6′-dibromoindigotin

CHLOROPHYLLS

The green pigment of leaves was named as chlorophyll by Pelletier and Caventou in 1818. Chlorophylls are the commonly occurring pigments in nature. Every organism having the capability to undertake photosynthesis, including living plants, algae, and photosynthetic bacteria, has its green colour due to the presence of pigment chlorophyll. Chlorophyll pigments are non-covalently linked to proteins and escorted by carotenoids in higher plants. These consist of 10 conjugated double bonds and absorb strongly in the blue and red regions of the spectrum, thus producing the green colour. Chemically, chlorophylls are the compounds of natural porphyrins, which consist of tetrapyrrole ring, formed by four pyrrole residues joined together by methine groups, and four nitrogen atoms are coordinated with the central magnesium atom. Chlorophyll also contains a fifth isocyclic ring and a long hydrophobic side chain acquired from propionic acid at C_7, which in turn is esterified to phytol, a C_{20} terpenoid alcohol. More than 50 structures of chlorophyll are already known, of which chlorophyll-a and chlorophyll-b occur widely in nature. Chlorophyll-b has an aldehyde group at C_3 in contrast to methyl group in chlorophyll-a. The colour of chlorophyll-a is blue-green, while that of chlorophyll-b is yellow-green. The molecular formulae of chlorophyll-a and chlorophyll b are $C_{55}H_{72}N_5O_4$ Mg and $C_{55}H_{70}N_4O_6$ Mg, respectively, and the two compounds have different UV absorption spectra, which is useful in estimating the quantity of each chlorophyll in the mixture (Finar, 1996; Nollet, 2004).

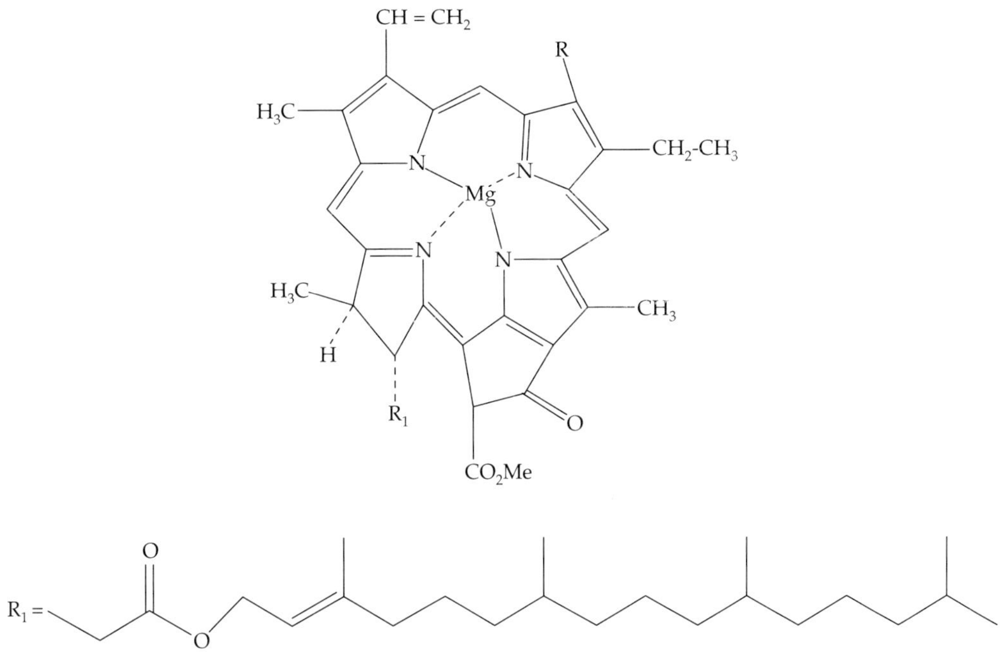

	R
Chlorophyll a	CH_3
Chlorophyll b	CHO

FOOD COLOURS FROM NATURE

Natural dyes have been used for colouring food items since ancient times. Some of the natural dyes like *haldi* (curcumin), saffron (crocin), annatto (bixin), and Cochineal (carminic acid) obtained from *Curcuma longa, Crocus sativus, Bixa orellana,* and *Dactylopius coccus,* respectively, have stood firm with time and are still in use for colouring food items. Saffron is a costly dye. Vegetable extracts were used for colouring wine to give it attractive colours in order to fetch more value for it. Food colours also have disadvantages as these can be used to hide the inferior products.

Haldi or turmeric, a yellow dye having curcumin as its chief chemical constituent, is obtained from the roots of *Curcuma longa*. It is a well-known dye in India and was also used for dyeing in other parts of the world during ancient period known to the ancient Romans and Greeks, who used it to dye cotton, silk, and wool. The dyed fabrics have poor colour-fastness properties. The dye is soluble in oil and is used in India for colouring vegetables in every house, as food colourant and at many other occasions. Curcumin possesses anticancer properties (Kroschwitz and Howe-Grant, 1993, Ref. 63 and 64 of Chapter VA).

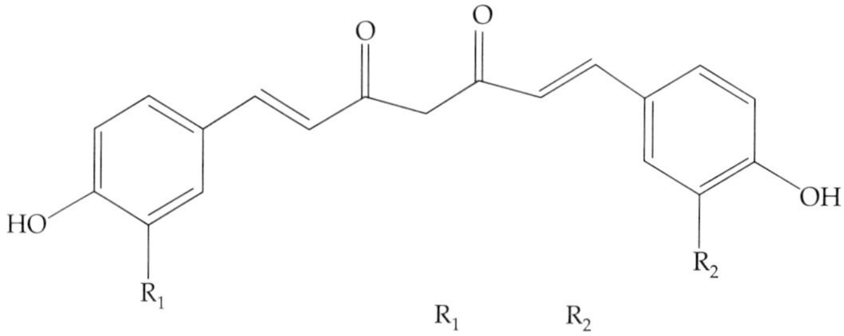

	R_1	R_2
Curcumin	OCH_3	OCH_3
Demethoxy Curcumin	H	OCH_3
Bis Demethoxy Curcumin	H	H

A red anthraquinone-based dye carminic acid is obtained by boiling the powder of cochineal bug with hot water. The dye (carmine) is precipitated from the extract with ethanol, after its treatment with aluminum salts. The dye obtained is soluble in water and is used for colouring cosmetics, pharmaceuticals, and food items.

The carotenoids account for orange, red, and yellow colour of fruits and vegetables and also exist in many dark-green vegetables. They are insoluble in fats. Carotenoids are not pH-sensitive in the normal food range of 2 to 7 and also do not change in the presence of vitamin C, which makes them useful for formal drinks. The findings of toxicity evaluation and controlled tolerance on pure carotenoids indicated them as safe for food colouring (Bagdon et al., 1960).The carotenoids occur in plants and can be isolated from them. But, it is economically viable to synthesize carotenoids on a commercial scale rather than isolate them from plants. β-Carotene, canthaxanthin, and β-apo-8′-carotenal are being manufactured for their commercial use (Isler et al., 1961). A microcrystalline powder of β-carotene is used for colouring margarine, butter, and cake mixtures. Canthaxanthin have the structure of β-carotene, with ketone groups at 4 and 4′ positions, possess tomato colour-enhancing properties, and thus can be used in peach ice cream, pink grapefruit beverages, pizzas, and sauces. β-Apo-8′-carotenal is used for colouring cheese (Kroschwitz and Howe-Grant, 1993).

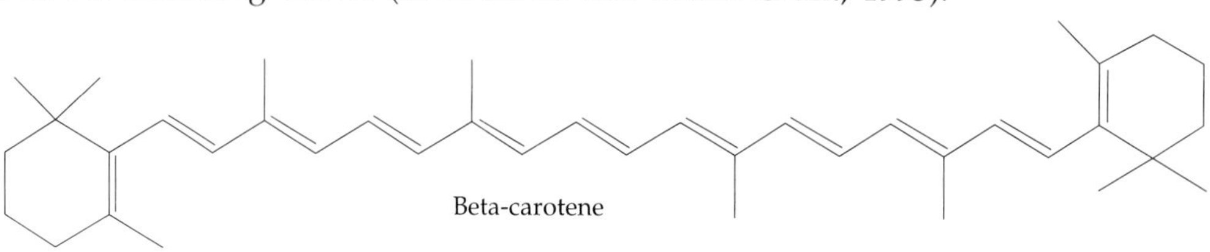

Beta-carotene

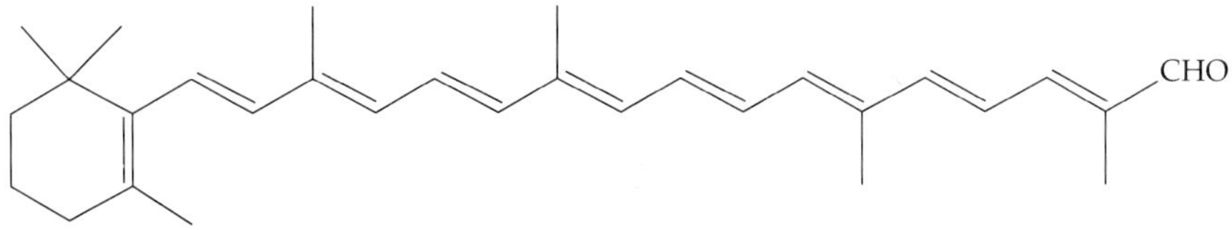

Beta-apo-8′-carotenal

References

1. Adrosko, R.J. (1968). *Natural Dyes in the United States.* Smithsonian Institution Press, Washington, D.C.; p. 39.
2. Anon (1972, reprinted in 2009). *The Wealth of India Raw Materials.* Council of Scientific and Industrial Research, New Delhi; pp. 82-85.
3. Bagdon, R.E., Zhinden, G., Studer, A. (1960). *Toxicol. Appl. Pharmacol.* 2, 225.
4. Baranyovits, F.L.C. (1978). *Endeavour* 2, 85.
5. Bernthsen, A. and Semper, A. (1885). *Ber. Deut. Chem. Ges.* 18, 203.
6. Buchanan, R. (1987). *A Weaver's Garden.* Interweave Press, Loveland, Colourado; p. 97.
7. Cameron, D.W, et al. (1981). *Aust. J. Chem.* 34, 2401
8. Cooksey C.J. (2001). Tyrian purple: 6,6′-Dibromoindigo and related compounds. *Molecules,* 6(9), 736-769.
9. Finar, I.L. (Reprinted in 1996). *Organic Chemistry,* Vol. 1, Vol. 2, ELBS with Longman; p.895-898.
10. Friedlaender, P. (1909). "Zur Kenntnis des Farbstoffes des anitiken Purpurs aus *Murex brandaris.*" (Contribution to our knowledge of the ancient purple dye from *Murex brandaris.*). *Monatshefte fur Chemie,* 30, 247.
11. Haller, H.L. (1940). *Helv, Chim. Acta.* 23, 466.
12. Heise, R. (1895). *Arbeit. Kaiserl. Gesund.* 11, 513.
13. Isler, O., Ruegg, R., Schudel, P. (1961). *Chima,* 15, 208.
14. Jacoby, D. (2004). "Silk Economics and cross-cultural artistic interaction: Byzantium, The Muslim World, and the Christian West." *Dumbarton Oaks Papers,* 58(197-240), p.210.
15. Kroschwitz, J.I. and Howe-Grant, M. (1993) [Editors]. *Encyclopedia of Chemical Technology,* vol. 8, 4th edition. A Wiley-Interscience publication, New York, Brisbane, Toronto, Chichester, Singapore; p. 784-809.
16. La, J.B. and Dutt. S. (1933). *J. Indian Chem. Soc.,* 10.
17. Leggett, W.F. (1964). *Ancient and Medieval Dyes.* Chemical Publishing Co; New York; p. 69, 82.
18. Marquart, L.C. (1835). *Eine Chemisch.Physiol-Abhandlung.* Bonn.

19. Mayer, F. (1943). *The Chemistry of Natural Colouring Matters*, ACS Monograph No. 89, translated and revised by A.H. Cook, Reinhold Publishing Corp, New York; p. 241.
20. Meyers, K. (1990). *Dyes from Nature*, Brooklyn Botanic Garden, New York; p. 11.
21. Nollet, Leo M.L. (2004). *Handbook of Food analysis* volume 1, Marcel Dekker, Inc. New York, U.S.A., pp. 850-860.

21a Pelletier, P.J., Caventou, J.B. (1818). Ann. Chim. Phys., 8, 250.

22. Perkin, W.H. Jr. and Robinson, R. 1920). *J. Chem. Soc.* 93, 489.
23. Piatkowska, E., Kopec, A., Leszczynska, T. (2011). Anthocyanins: Their profile, occurrence and impact on human body. *Zywnosc*, 18(4), 24-35.
24. Reese, David S. (1987). Palaikastro Shells and Bronze Age purple dye production in the Mediterranean Basin. *Annual of the British School of Archaeology at Athens*, 82, 201-206.
25. Richter, D. (1936). *J. Chem. Soc.*; 1701.
26. Roberge, G. and Brassard, P. (1978). *J. Chem. Soc.* Perkin Trans; 1041.
27. Schutzenberger, P. (1858). *Ann. Chim. Phys.* 54, 52.
28. Schweizer, H.R. (1964). *Kunstliche Organische Farbstoffe und ihre Zwischenprodukte*, Springer-Verlag, Berlin; p. 303.
29. Stieglitz, Robert R. (1994), "The Minoan Origin of Tyrian Purple,"*Biblical Archaeologist*, 57, 46-54.
30. Theopompus, cited by Athenaeous (12:526) around 200B.C. according to Gulick, Charles Barton 1941 *Athenaeous, The Deipnosophists.* Cambridge: Harvard University Press.
31. Thomson, R.H. (1987, 1971). *Naturally Occurring Quinones*, third and second editions, Academic Press, Inc.; New York, 1987, p. 469, 471., 1971, pp. 202-221, 373-419, 443-459, 462-472.
32. Tommasi, G. (1920). *Gazz. Chim. Ital.* 59, 263.
33. Venkataraman, K. (1952). *The Chemistry of Synthetic Dyes*, Vol. 1, Academic Press, Inc. New York; p. 69.
34. Wouters, J. and Verhecken, A. (1991). High performance liquid chromatography of blue and purple indigoid natural dyes. *Journal of the Society of Dyers and Colourists*, 107(7-8), 266-269.
35. Yumnam, J.Y. and Tripathi, D.P. (2012). A unique way of using washcloth dyed with bixin by the *Meiteis* of Manipur. *Current Science*, 102(12), 1621.

35a. Zechmeister, L. (1960). Fortschritte der Chemic der Organischer Naturstoffe, vol. 18, Verlogvon Julius Springer, Wein, p. 223.

36. Zemplen, G. and Bognar, R. (1939). *Ber. Deut. Chem. Ges*, 72, 913.
37. Ziderman, I.I. (1986). Purple dye made from shellfish in antiquity. *Review of Progress in Colouration*, 16, 46-52.

6 Mordants

INTRODUCTION

The dictionary meaning of mordant is caustic, biting, smarting, corrosive, or cleansing (Elliott et al., 2006). A chemical compound that fixes the dye on the fabric, by forming a coordination complex with the dye, which in turn attaches to the fabric, is known as mordant. The mordant has the affinity for both fibre and the dyestuff and forms the bond between them, thereby fixing the natural dye on the material to be dyed. Earlier, it was presumed that mordant promoted or encouraged or helped the dye bite onto the fibre in order to increase its colour-fastness properties. Mordant can also be defined as the polyvalent metal ion having the ability to form coordination complexes with certain dyes (Llewellyn, 2005). The metallic salts of some metals, such as aluminium, chromium, iron, and tin (alum, potassium dichromate, ferrous sulphate, and stannous chloride), are some of the examples of mordants (Gulrajani and Gupta; Tewari, 1994).

Natural dyes obtained from some plants and other natural sources can be used to dye the fabric, wood, and many other items directly without the use of any other chemical. These dyes are known as "substantive" dyes, direct dyes, or non-mordant dyes. Although a substantive dye imparts colour to the fabric directly, the use of mordants enhances the colour-fastness properties of the dyed fabric as compared to solo use of a natural dye. A range of colours can be obtained using different mordants. Natural dye, which needs a chemical compound (mordant) to fix it on the fabric, is known as mordant dye, adjective dye, or indirect dye. Most of the natural dyes are mordant dyes.

It appears that the use of mordants first started in the Middle Kingdom of Egypt (between 2200 B.C. and 1500 B.C.), as evident from the textiles found on preserved mummies in the Egyptian tombs. A number of salts, several forms of soda, wood ashes, and other chemicals have been used as mordants by earlier dyers. It has been observed that the vessel used for dyeing may itself serve as mordant. It has been reported that the use of copper or tin vessels brightens the colour, while iron kettle or iron filing makes them dull. The aluminum vessel or aluminum piece with little soda in a dye pot may be used as an alternative to alum mordant. Mordants are of three types, as described below:

1. Metal salts or metallic mordants
2. Natural mordants
3. Oil mordants

METALLIC MORDANTS

Some salts of metals such as aluminum, chromium, iron, tin, and copper are covered under this category. The metal in the metallic mordant should have at least two valencies. The dyes are attached to the mordant by means of a covalent and a coordinate bond. This process is known as chelation. The chelate (coordination complex) formed between a mordant dye and a polyvalent metal is called a lake, and owes its origin to the lac dye obtained from Indian lac insects *Laccifer lacca* (Llewellyn, 2005a). With the passage of time, the term "lac" has changed to lake and become the generic term for all dye-mordant complexes. The reaction between a dye and a mordant leads to the formation of two types of bonds, namely, a covalent bond, with a hydroxyl oxygen, and a coordinate bond, with another electron-donor oxygen. The common interaction between chromium mordant and a yellow natural dye containing luteolin is given below.

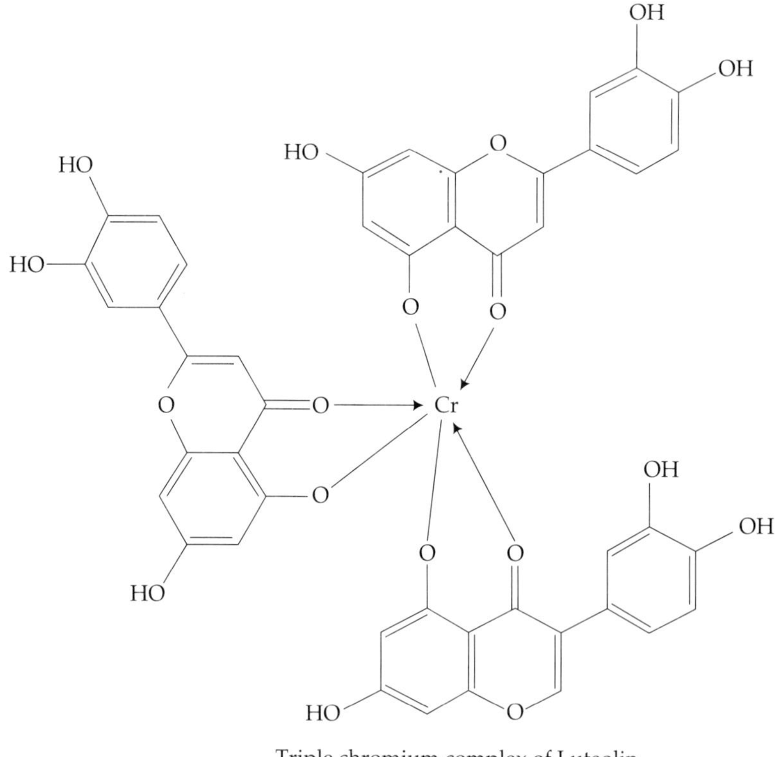

Triple chromium complex of Luteolin

The valencies of metal ions determine the number of dye molecules it can attach with itself. It is possible to attach three molecules of dye with aluminium due to its trivalent nature. However, the same is not observed in practice since the attachment of the metal mordant is also to take place with the substrate (fabric). Availability of more than one chelating point in a dye molecule leads to the formation of dye–mordant complexes involving several molecules of dye. The colour change in the dye solution on addition of mordant is due to the incorporation of the metal atom into the delocalized electronic configuration of the dye. The energy levels of metals are relatively low and that is why their involvement into a delocalized system results in the lowering of the overall energy. The colour of the lake is attributed to this process. The characteristics of some metals used as mordants are given in Table 6.1. A brief description of metals forming salts, which are used as mordants, is given below (Prakash, 1969).

Table 6.1 Characteristics of some mordants' metals

S.No.	*Mordant metal*	*Atomic weight*	*Atomic number*	*Oxidation states (numbers)*	*Ground state electron configuration*	*Common oxidized forms*
1.	Aluminum (Al)	26.981	13	3+	$1s^2, 2s^2, 2p^6, 3s^2, 3p^1$	Al^{3+}
2.	Chromium (Cr)	51.996	24	2+,3+,6+	$1s^2, 2s^2, 2p^6, 3s^2, 3p^6, 3d^5, 4s^1$	Cr^{2+}, Cr^{3+} Cr^{6+}
3.	Copper (Cu)	63.546	29	1+,2+	$1s^2, 2s^2, 2p^6, 3s^2, 3p^6, 3d^{10}, 4s^1$	Cu^+, Cu^{2+}
4.	Iron (Fe)	55.847	26	2+,3+	$1s^2, 2s^2, 2p^6, 3s^2, 3p^6, 3d^6, 4s^2$	Fe^{2+}, Fe^{3+}
5.	Tin (Sn)	118.710	50	2+,4+	$1s^2, 2s^2, 2p^6, 3s^2, 3p^6, 3d^{10}, 4s^2, 4p^6, 4d^{10}, 5s^2, 5p^2$	Sn^{2+}, Sn^{4+}
6.	Potassium (K)	39.098	19	1+	$1s^2, 2s^2, 2p^6, 3s^2, 3p^6, 4s^1$	K^+

(1) Chromium (Cr)

Chromium occurs in the form of chromite ore in nature and very little of it is processed for the production of ductile chromium, while major part is used in various intermediate forms. The chromium metal is produced on commercial scale by reduction of Cr_2O_3 by aluminium, silicon, or carbon, or by electrolysis of either CrO_3 or chrome alum solution. It belongs to the group VIB of the periodic table, along with molybdenum and tungsten, and has the atomic weight and atomic number of 51.996 and 24, respectively. Chromium has the ground electronic configuration as $1s^2, 2s^2, 2p^6, 3s^2, 3p^6, 3d^5, 4s^1$. This distribution of the outermost electrons is favoured over $3d^4$ and $4s^2$, because of the increased stability of the half-filled 3d shell, with one electron in each of the orbital. The orbitals project so close to the surfaces of the ions that the electrons in them interact with the chemical environment. Chromium is a typical transition element, which forms many coloured and paramagnetic compounds. The most common and stable oxidation states of chromium are 2+, 3+, and 6+. The lowest 2-, 1-, 0, and 1+ are formal oxidation states displayed by chromium in compounds such as carbonyls, nitrosyls, and organo-metallic complexes. The most stable and important oxidation state is 3+. The aqueous chemistry of chromium (III) is mainly the chemistry of coordinate and chelate compounds, of which thousands are known. The extensive chemistry of the chromium (III) state is dominated by the formation of stable, kinetically inert d^2sp^3

complexes; in which chromium is octahedrally six coordinate. Thousands of complexes are known and the ability of chromium (III) to coordinate with organic hydroxyl and amine groups finds commercial application in different fields, such as tanning, dyeing, and wood preservation (Anon, 1982).

In dyeing, especially of woollen clothes, sodium and potassium chromates, chrome alum, chromium chloride, chromium fluoride, and chromium acetate are employed. These salts act as mordants and produce the desired shades to woollen clothes. Large quantities of chromium salts are used in tanning industry. Chrome-tanned sole leather is stated to possess unique wearing qualities. Lead chromate is largely used in the pigment industry. Vermillion is produced by mixing lead chromate with potassium and sodium nitrates. The mixing of lead chromate with Prussian blue yields a green pigment. All the compounds of chromium (VI) are important and primary chromium compounds are used in textile dyeing, pigment manufacture, formulation of chrome tanning solutions, corrosion inhibition, metal finishing, wood preservation, and as an oxidant.

(2) Aluminium (Al)

Aluminium belongs to the group IIIA of the periodic table, along with boron (B), gallium (Ga), indium (In), and thallium (Tl), and has an atomic weight and atomic number of 26.981 and 13, respectively. Its occurrence in the earth's crust is maximum (7.5%), and it is the third most abundant element. Pure aluminium is manufactured from bauxite (Al_2O_3) or hydrated aluminium oxide ($Al_2O_3.XH_2O$). Aluminium ions can be reduced to Al by electrolysis in the absence of water only. Aluminium, m.p. 658°C, is a bluish-white metal with a brilliant lustre, which is soon destroyed by the formation of an oxide layer on it. The light metal is malleable, ductile, and a good conductor of heat and electricity. It is the most reactive of the post-transition metals and is cheap in comparison to other metals. Aluminium is also a strong reducing agent. The electron structure and oxidation state is given in Table 6.1. It forms a number of compounds, viz., oxides, halides, sulphates, nitrides, carbides, silicates, and many more. Potash alum ($KAl\ (SO_4)_2.12H_2O$) is used as mordant in dyeing and water purification, and is also put to a number of other uses.

(3) Copper (Cu)

Copper belongs to the group 1B of the periodic table, along with silver (Ag) and gold (Au). Mixed sulphides of copper and iron ($CuFeS_2$) and the basic carbonate [azurite, $Cu_3\ (CO_3)_2(OH)_2$; malachite, $Cu_2CO_3\ (OH)_2$] are the two main classes of copper ores, normally used for isolation of copper in the pure form. The atomic weight, atomic number, electron structure, and oxidation states of copper are described in Table 6.1. Copper exhibits a variable valency of one (Cu^+) or two (Cu^{2+}) by losing one or two electrons. The cuprous ion possesses tendency to revert to cupric by leaving metal copper. It forms a number of compounds, namely, oxides, hydroxides, halides, sulphates, sulphides, and many other compounds. Copper sulphate is used as mordant in dyeing, besides its other uses.

(4) Iron (Fe)

Iron is put in the group VIIIB of the Periodic Table, along with ruthenium (Ru), osmium (Os). Hematite (Fe_2O_3) and magnetite (Fe_3O_4) are two important ores of iron used for its manufacture on the commercial scale. Its atomic weight, atomic number, and electron structure are given in Table 6.1. Iron possesses valency of two (Fe^{++}) and three (Fe^{+++}) by losing two or three electrons from its outermost shells. Iron forms two series of salts, *-ous* and *–ic*, of which both are stable, the *–ic* salts being slightly more stable. Iron has numerous uses and forms a number of compounds, viz., oxides, sulphates, halides, and many others. Ferrous sulphate is used as a mordant in dyeing, besides other uses.

(5) Tin (Sn)

The metal tin belongs to the group IVA of the periodic table, alongwith carbon (C), silicon (Si), germanium (Ge), and lead (Pb). The main source of metal is its ore cassiterite (SnO_2), also known as tin-stone. Tin, m.p. 232°C, is a lustrous white metal and is malleable and ductile, but becomes brittle at 200°C. The atomic weight, atomic number, electron structure, and oxidation states are given in Table 6.1. The divalent tin ion Sn^{2+} (as in $SnCl_2$) is quite stable and is converted to stannic in redox reactions. Tin forms a number of compounds, and stannous chloride is used as a mordant in dyeing.

(6) Potassium (K)

Potassium belongs to the group of alkali metals (group 1A of the Periodic Table) and closely resembles the other elements of the group, namely, lithium, sodium, rubidium, cesium, and short-lived radioactive element francium. Potassium has a valence of +1, does not form complex ions, and joins to other elements or groups of elements in its compounds by ionic bonds. It is more reactive than sodium. Potash alum is used as mordant in dyeing with natural dyes (Anon, 1979).

Electronic configuration of mordant metals

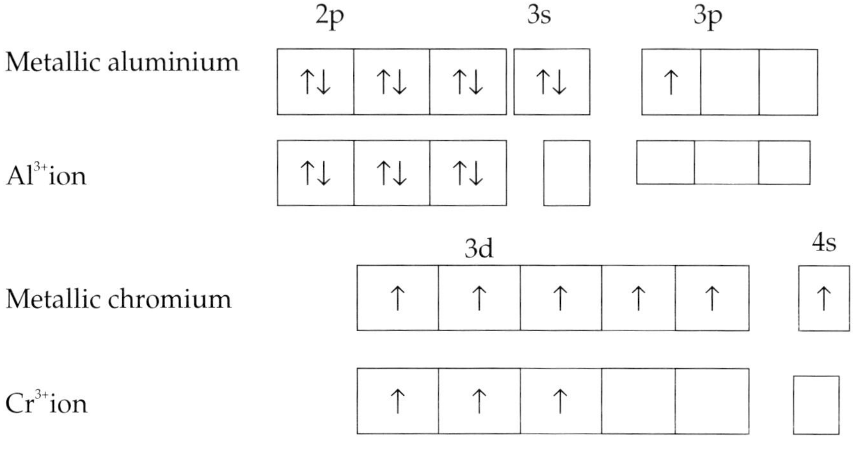

Table 6.2 Some metalic mordants

S. No.	*Mordant Name*	*Chemical formula*	*Colour*	*Uses*	*Permissible limit in the ultimate product*	*Nature*
1.	Alum (Potassium aluminium sulphate)	$KAl(SO_4)_2 . 12H_2O$	White	Mordant	No limit (To be used in optium quantity)	Non- Toxic
2.	Blue vitriol (Copper sulphate)	$CuSO_4 . 5 H_2O$	Blue	Mordant	Copper, Cu (50 ppm)	Toxic
3.	Chrome (Potassium dichromate)	$K_2Cr_2O_7$	Orange	Mordant	Chromium, Cr (2.0 ppm)	Toxic
4.	Iron (Ferrous sulphate)	$FeSO_4 .7H_2O$	Green	Mordant	No limit (To be used in optium quantity)	Non- Toxic
5.	Tin (Stannous chloride).	$SnCl_2$	White	Mordant	No limit (To be used in optium quantity)	Moderately toxic
	Chemical assistants	*Chemical Formula*	*Colour*	*Uses*	*Permissible limit*	
1.	Calcium carbonate	$CaCO_3$	White	Alkalinity	As per requirement	
2.	Baking soda (Sodium bicarbonate)	$NaHCO_3$	White	Alkalinity	As per requirement	
3.	Common salt (Sodium chloride)	NaCl	White	Dye absorption	As per requirement	
4.	Cream of tartar (Potassium bitartrate)	$KC_4H_5O_6$	White	Acidity	As per requirement	
5.	Glauber's salts (Sodium sulphate)	Na_2SO_4	White	Dye spreading	As per requirement	
6.	Lime (Calcium oxide)	CaO	White	Alkalinity	As per requirement	
7.	Oxalic acid	$(COOH)_2 .2H_2O$	White	Acidity	As per requirement	
8.	Vinegar (a weak solution of acetic acid)	CH_3COOH	Colourless	Acidity	As per requirement	
9.	Washing soda (Sodium carbonate)	$Na_2CO_3 .H_2O$	White	Alkalinity	As per requirement	

	3d					4s
Metallic copper	↑↓	↑↓	↑↓	↑↓	↑↓	↑
Cu^+ ion	↑↓	↑↓	↑↓	↑↓	↑↓	

	3d					4s
Metallic iron	↑↓	↑	↑	↑	↑	↑↓
Fe^{3+} ion	↑	↑	↑	↑	↑	

	4d					5s	5p		
Metallic tin	↑↓	↑↓	↑↓	↑↓	↑↓	↑↓	↑	↑	
Sn^{2+} ion	↑↓	↑↓	↑↓	↑↓	↑↓	↑↓			

	3p			4s
Metallic potassium	↑↓	↑↓	↑↓	↑
K^+ ion	↑↓	↑↓	↑↓	

CHEMICAL COMPOUNDS OF METALS

(1) Potassium dichromate

Potassium dichromate (*surkh kaih* or *lal potash*), a red-orange crystalline solid, with m.p.398°C, is a common reagent normally used as a powerful oxidizing agent in laboratory and industry. It is also known by other names, such as bichromate of potash, dipotassium dichromate, dipotassium salt, potassium bichromate, and lopezite (Potassium Dichromate Listing). Earlier it was directly manufactured by fusing chrome iron ore with potash carbonate and lime in free air. It is prepared from sodium dichromate by treating its solution with potassium chloride. Due to much less solubility, the potassium salt separates out from the solution.

$$Na_2Cr_2O_7 + 2KCl \longrightarrow K_2\,Cr_2\,O_7 + NaCl$$

It is soluble in water, and in the dissolution process, it ionizes as follows:

$$K_2Cr_2O_7 \longrightarrow 2K^+ + Cr_2O_7^{2-}$$

$$Cr_2O^{2-}_7 + H_2O \rightleftharpoons 2CrO^{2-}_4 + 2H^+$$

Potassium dichromate is extensively used as mordant in dyeing, chrome-tanning, and calico-printing. In chrome tanning, the hides are soaked in weakly acidified solution of potassium dichromate, followed by its immersion in a bath containing hypo or some other reducing agent. The leather gets tanned due to the precipitation of chromic oxide within its pores.

(2) Copper sulphate

Copper sulphate or cupric sulphate is a well-known chemical compound, with the molecular formula as $CuSO_4.5H_2O$. Its blue colour is due to the association of five molecules of water with it. It is also known as blue vitriol or blue stone (copper sulphate MSDC). It is generally obtained by roasting copper pyrites in air at a regulated temperature. The iron of the pyrites is converted to the insoluble oxide, and copper sulphate is then extracted with water. It is also manufactured by dissolving copper scrapping in hot dilute sulphuric acid in the presence of air.

$$2Cu + H_2SO_4 + O_2 \longrightarrow 2CuSO_4 + 2H_2O.$$

Copper sulphate loses all its water of crystallization at 230°C, while at 100°C, it partly remains as monohydrate $CuSO_4.H_2O$. The crystalline salt is blue, while the anhydrous form of salt is white. The different hydrates forms of copper sulphate are given below:

- $CuSO_{4.}5H_2O$ ------- Blue
- On exposure to air, $CuSO_4.3H_2O$ ----------- Pale blue
- On heating at 100°C, $CuSO_4.H_2O$ -------- Bluish white
- On heating at 230°C - 250°C, $CuSO_4$--------- White

It forms double salts with alkali sulphate as $K_2SO_4.CuSO_4.6H_2O$ and $(NH_4)_2SO_4.CuSO_4.6H_2O$. Copper sulphate is used as a mordant with vegetable dyes. It also finds application in calico printing, electroplating, electrotyping, and in the preparation of other copper salts. It is also used as fungicide and germicide in agriculture.

(3) Ferrous sulphate

Ferrous sulphate occurs as heptahydrate, with chemical formula as $FeSO_4.7H_2O$ (molar mass 278.05 g/mol) and m.p.64°C. It is one of the very useful ferrous salts and is also known as iron (II) sulphate, green vitriol or *hira kasis*. It is found along with copper sulphate, in the drained water of pyrites, which forms an excellent source of its manufacture. It also occurs native as *melanterite* or *coppears,* which is formed from the oxidation of pyrites with water in the presence of atmospheric air.

$$2FeS_2 + 7O_2 + 2H_2O \longrightarrow 2FeSO_4 + H_2$$

The chemically pure iron along with dilute sulphuric acid may be used for preparing the volumetric grade salt. Ferrous sulphate on heating first loses water of crystallization, followed by its decomposition to ferric oxide, sulphur dioxide, and sulphur trioxide.

$$2\ FeSO_4 \longrightarrow Fe_2O_3 + SO_2 + SO_3$$

Ferrous sulphate is a good reducing agent and in combination with ammonium sulphate forms a double salt $FeSO_4\ (NH_4)_2\ SO_4.6H_2O$, known as Mohr's salt, which is used in preparing the standard solutions. Ferrous sulphate is used as a reducing agent, nutritional supplement, colourant, and is put to a number of other uses. It is used as a mordant in dyeing of wool, silk, cotton with vegetables dyes, and also in the manufacture of ink, an example for this is the iron gall ink, which was used in the Middle Ages until the end of the eighteenth century. One of the methods known as china blue method, involving direct application of indigo by using ferrous sulphate, was developed in the U.K.

(4) Stannous chloride

Stannous chloride, also known as tin dichloride or tin protochloride, is a white crystalline solid, with m.p. 247°C (anhydrous) and 37.7°C (dihydrate). Its molecular formula is $SnCl_2$and has a molar mass of 189.60g/mol (anyhydrous) and 225.63g/mol (dihydrate). Stannous chloride is prepared commercially by the reaction of hydrochloric acid on tin and is crystallized out as monoclinic prisms with two molecules of water. It can be prepared in anhydrous form by the action of hydrochloric acid gas on heated tin or by heating the mixture of tin and mercuric chloride (Armarego and Chai, 2009).

$$Sn + 2HCl \longrightarrow SnCl_2 + HCl$$

It is soluble in water (less than of its own mass), alcohol, and ether and is a very good reducing agent. Dilution of the water solution leads to the formation of an insoluble basic salt due to its hydrolysis.

$$SnCl_2 + H_2O \rightleftharpoons Sn\ (OH)\ Cl + HCl$$

It is advisable to dissolve stannous chloride in hydrochloric acid to obtain its clear solution. Stannous chloride has a lone pair of the electrons and the molecule is bent in the gaseous phase. It also behaves as a Lewis acid, forming complexes with legends.

Stannous chloride is used as mordant in the textile dyeing with vegetable dyes to impart brighter shades, e.g., cochineal, lac, and other dyes.

(5) Alum

Any chemical compound having the general empirical formula $XY(SO_4).nH_2O$ normally comes under the category of alum, where X is usually a monovalent alkali metal or ammonium,Y is a trivalent metal (aluminium, chromium, or iron), and n is often 12 (Locke, 1902). The hydrated form of potassium aluminum sulphate $[KAl(SO_4)_2\ .12H_2O]$, i.e., potash alum is generally called alum. Alum can be prepared on commercial scale by boiling bauxite with sulphate acid and potassium sulphate in tanks of the size depending upon the quantity to be produced. The cleaned liquor reaction on evaporation and crystallization yielded the alum crystals, which were purified by washing, and followed by centrifuging and blocks formation. Alunite or alumstone, $K_2SO_4.Al_2\ (SO_4)_3.4Al\ (OH)_3$ on calcinations and treatment with sulphuric acid followed by mixing with extra potassium sulphate was also used for the preparation of alum on commercial scale.

$$K_2SO_4.Al_2\ (SO_4)_3.4Al\ (OH)_3 + 6H_2SO_4 \longrightarrow K_2SO_4 + 3Al_2\ (SO_4)_3 + 12H_2O$$

$$K_2SO_4 + Al_2\ (SO_4)_3 + 24H_2O \longrightarrow K_2SO_4Al_2\ (SO_4)_3.24H_2O$$

The soda alum, Na Al $(SO_4)_2.12H_2O$, is found in nature as mineral mendozite and is soluble in water. The different types of alums and their uses are given below:

S. No.	*Name*	*Chemical formula*	*Uses*
1.	Potash alum	$KAl(SO_4)_2.12H_2O$	As mordant in vegetable dyeing, leather tanning, as a flocculant in water purification, and many other uses.
2.	Soda alum	$NaAl\ (SO_4)_2.12H_2O$	In baking powder and as an acidulent in food.
3.	Ammonium alum	$NH_4Al(SO_4)_2.12H_2O$	Mordant in dyeing of fabric, tanning, water purification, and many other uses. Pure alumina is obtained on strong heating.
4.	Chrome alum	$KCr(SO_4)_2.12H_2O$	In tanning and dyeing.
5.	Selenate-containing	$KAl(SeO_4)_2.12H_2O$	As antiseptic and other uses.
6.	Ferric alum (Mohr's salt)	$NH_4Fe(SO_4)_2.12H_2O$	For making standard solution in volumetric analysis, and as mordant in dyeing.

The very common metallic mordants and their use in dyeing with natural dyes are given in Table 6.2.

The general methods of mordanting are given below:

(i) Pre-mordanting or onchrome method: In this method, the fabric to be dyed is treated first with mordant and the mordanted fabric is subsequently subjected to dyeing with the dye.

(ii) Simultaneous mordanting or meta-mordanting or metachrome method: In this method, separate hot mordant and dye solutions are mixed together in a dye bath container, followed by dyeing of the fabric in it.

(iii) Post-mordanting or afterchrome method: In this method, the cleaned fabric is dyed first, followed by its direct transfer to the mordant bath for mordanting, viz. fixing the dye on the fabric.

The method of mordanting is chosen as per the nature of the dye used for dyeing. For example, pre-or post-mordanting methods are suitable for an acidic mordant and acidic dye as these cause less damage to the fabric. Simultaneous or meta-mordanting is suitable when the dye lake formed is stable. Each dye reacts differently with each mordant. The dye molecule forms complex with the mordant, which in turn attaches to the fabric. Different colours are imparted on the fabric by using the same dye and different mordants.

The cotton fabrics have low affinity for mordants. However, due to amphoteric nature of wool and silk, these can absorb acids and bases quite effectively and thus are receptive towards mordants. The wool reacts with a metallic mordant salt and hydrolyzes into acidic and basic constituents. The basic component reacts with carboxylic acid group (-COOH) of wool and silk, while washing with water removes the acidic component. The fine precipitates of the solution also adhere to the wool surface, which in turn are attached to the dye molecules, causing low colourfastness of the dyed wool. One of the amino acid (cystine) present in the wool has this groups (-SH), which reduce hexavalent chromium of potassium dichromate to trivalent form, which in turn complexes with fibre and dye. Therefore, potassium dichromate is not so effective for wool. The dipole moment of a molecule indicates the degree of charge separation within a chemical compound, and accounts for its polarity. The dipole moments of a compound can be determined by multiplying the distance, *d*, which separates the charges of equal magnitude having opposite sign, and the magnitude of charge, *c*. The permanent dipole-dipole interactions exist between polar covalent compounds due to attraction of + atoms of a molecule to the – atoms of another molecule. The very strong dipole-dipole interaction is indicated by the hydrogen bonds. They are not really chemical bonds in the formal sense. Strong hydrogen bonding exists in polar covalent molecules containing H and one of the three small, highly electronegative elements F, O, or N.

Limitations of metallic mordants

Natural dyes form metal complex with different metallic mordants to impart various colours on the dyed fabric. The limits to the use of metal salts have been put by the famous "German Ban". Accordingly, the indicative maximum permissible quantities of different metals in the ultimate product are as follows: arsenic, As (1.0 ppm); lead, Pb (1.0 ppm); cadmium, Cd (2.0 ppm); chromium, Cr (2.0 ppm); cobalt, Co (4.0 ppm); copper, Cu (50 ppm); nickel, Ni (4.0 ppm); zinc, Zn (20 ppm) (Anon, 1999). Therefore, it is advisable to use metallic mordants in the prescribed limits. However, there is no upper limit for the presence of aluminium, iron, and tin metals in the finished products; still the mordanting conditions should be optimized to avoid the excess use of metal salts, which may cause water pollution.

NATURAL MORDANTS

(1) Tannins

Tannins are generally organic chemical compounds consisting of C, H, and O. Some tannins contain smaller amount of other elements, such as N or P. The extensively distributed plant polyphenols containing adequate numbers of hydroxyl and other suitable functional groups (i.e., carboxyl) of fairly high molecular weight are known as tannins, and these usually occur as glycosides. The name "tanin" comes from the French word tanin (tanning substance) and is used for a range of natural polyphenols (Anon, 2009; Römpp, 1997). Tannins have the ability to convert animal hides into leather by a process known as tanning. These are complex polyphenols and are not based on a class of compounds with a common basic structure. The molecular weight of tannins vary in the range from 500 to 3000 for hydrolysable tannins (Gallic and ellagic acid esters) and up to 20,000 for condensed tannins (proanthocyanidins) (Bate-Smith and Swain, 1962). Tannins consist of amorphous polyphenols, which make colloidal solution in water with an astringent taste; they also bind and precipitate proteins from solution (Falbe and Regitz, 1995). Vegetable tannins are of two types, namely, hydrolysable tannins and condensed tannins. The hydrolysable tannins have a polyester structure and can be easily hydrolyzed by acids or enzymes into a sugar or related polyhydric alcohol and a phenol carboxylic acid. They are further subdivided into gallotannins and ellagitannins, depending on the involvement of gallic acid or ellagic acid in esterification and their hydrolysis gives gallic acid and ellagic acid, respectively. The ellagic acid is normally isolated as its stable dilactone. Therefore, the term "hydrolysable tannin"includes both the gallotannins and the ellagitannins (Kashiwada et al., 1992). The sugar portion of the hydrolysable tannins has most often been found to be D-glucose.

Non-hydrolysable oligomeric and polymeric proanthocyanidins were classified as condensed tannins (Würdig and Woller, 1989). The condensed tannins (proanthocyanidins; molecular weight upto 20,000), in contrast, do not readily break down with acids; instead, they undergo progressive polymerization under the action of acids to yield amorphous phlobaphanes or tannin reds. These tannins are formed by condensation or polymerization of flavan-3, 4-diols and flavan-3-ol precursors. Therefore, tannins may be classified into four prominent groups, namely, gallotannins, ellagitannins, complex tannins, and condensed tannins depending on their chemical characteristics (Anon, 2009; Khanbabaee and Ree, 2001).

- Gallotannins consisting of galloyl units or their meta-depsidic derivatives in combination with varying chemical moieties, namely, catechin-, polyol-, or triterpenoids.
- Ellagitannins contain at least two C-C coupled galloyl units to each other and not having a glycosidically attached catechin moiety.
- Complex tannins consisting of catechin moiety attached by a glycosidic linkage to a gallotannin or an ellagitannin moiety.
- Condensed tannins are polymeric in nature and produced mainly by the attachment of C-4 of a catechin moiety to C-6 or C-8 of another catechin moiety with the repetition of the process resulting in the formation of the polymers of varying molecular weight.

Tannin occurs in large number of plants and are common in Gymnosperms and Angiosperms. Tannins are situated normally in the vacuoles or surface wax of the plants. Naturally occurring tannins obtained from different parts of the plants, viz., wood, bark, fruits, and leaves, are an important non-wood forest product and are

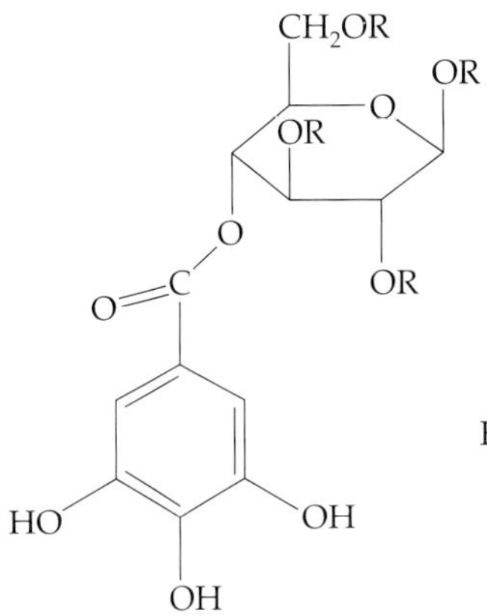

Gallotannins

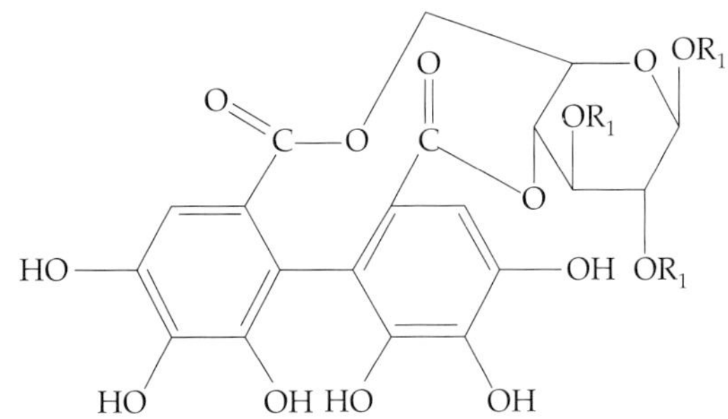

Ellagitannins

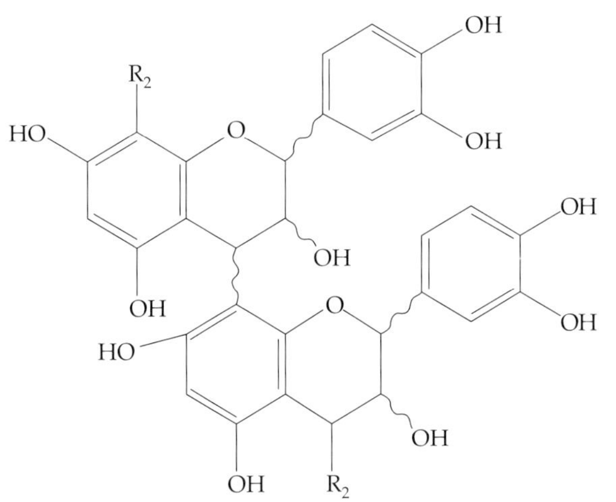

R_2 = (Catechin moiety)$_n$

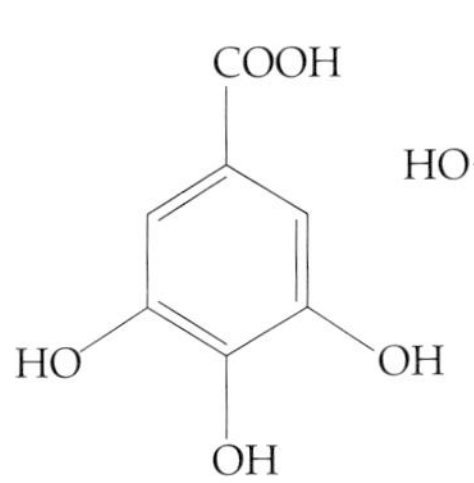

Gallic acid (R)

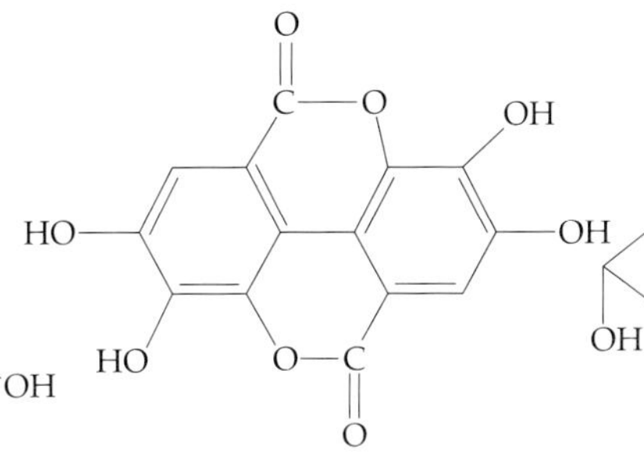

Ellagic acid (R_1)

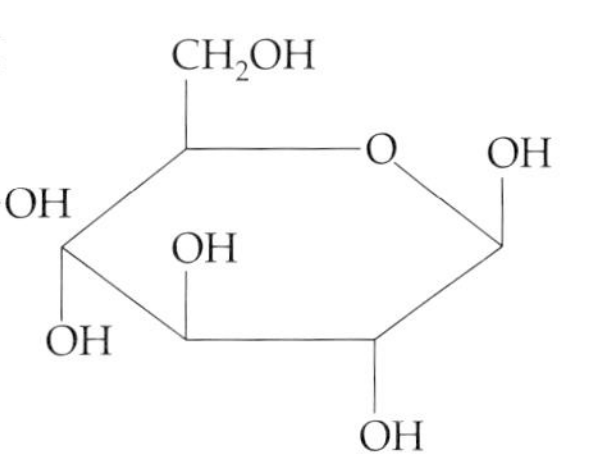

β-D-Glucose

used extensively in the leather industry for converting raw hides and skins to leather and also as a mordant in dyeing. Oak galls and wattle bark are an excellent source of tannins. Wooden vessels used for dyeing act as mordant into the dye bath. The presence of tannin in a reasonable concentration in a plant material is not the only criteria for judging its suitability as a tannin material. Tannins can be conveniently classified depending on the plant part from which these are obtained.

(i) Bark tannins
(ii) Fruit tannins
(iii) Leaf tannins
(iv) Wood tannins

(i) Bark tannins: Tannins are obtained from the barks of a number of trees growing in India. Bark of such trees are collected after they fall for timber. Some of the commercially important tannin-yielding barks are described below.

(a) *Acacia nilotica* ssp. *indica: Acacia nilotica* ssp. *indica,* popularly known as babul, is one of the most important sources of tannin in North India. It occurs throughout the drier parts of India and is an important social forestry species, which is planted extensively. The average tannin content of the bark supplied to the tanneries is 12%, which decreases on storage. The bark contains condensed tannins with (+) catechin, (–) epicatechin, (+) dicatechin, quercetin, gallic acid, (+) lecucocyanidin gallate, and other constituents responsible for the tanning potency of babul liquor (Tewari, 1994).

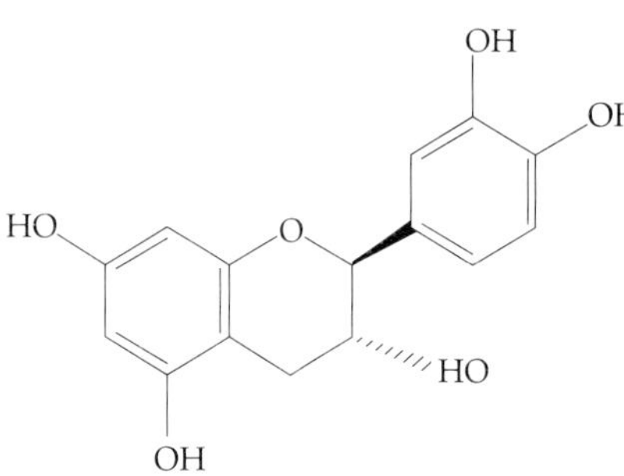
(–)-Catechin

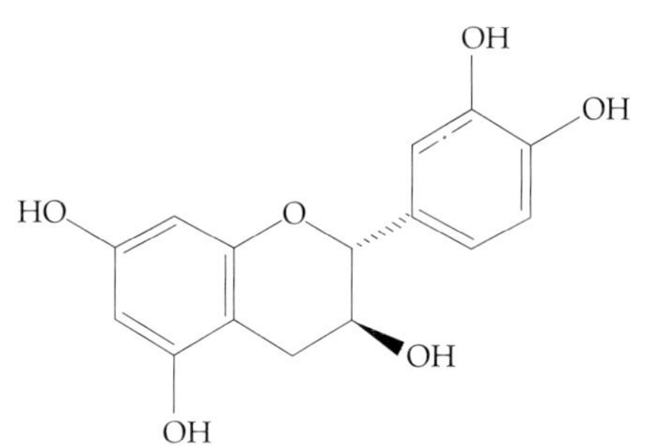
(+)-Catechin

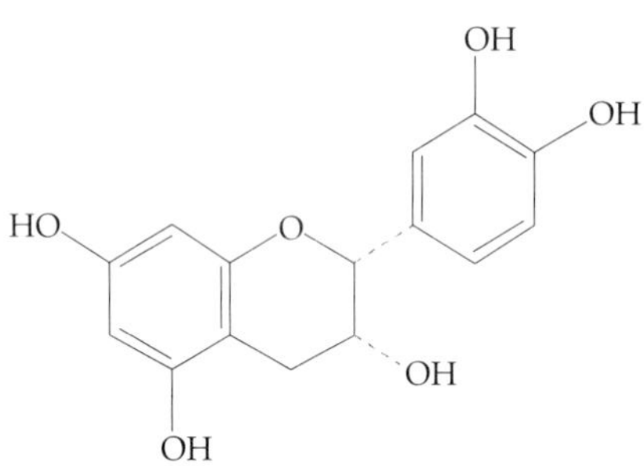
(–)-Epicatechin

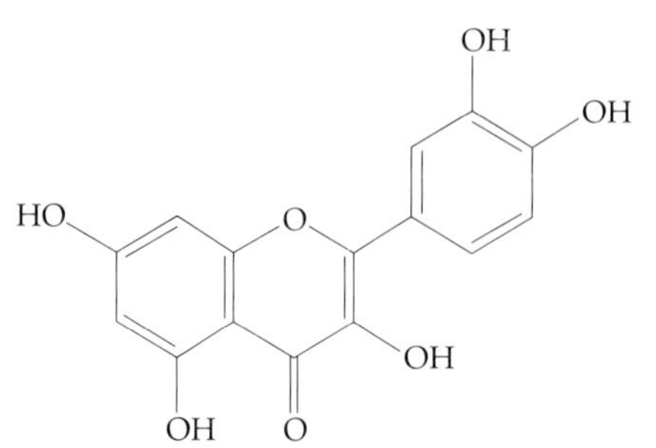
Quercetin

Acacia mearnsii

(b) *A. mearnsii: Acacia mearnsii*, popularly known as wattle, a native of Africa, was introduced in India in the middle of the nineteenth century near Kodaikanal. Presently, it is being cultivated in the hills of Nilgiris and Kerala in South India, Assam, and Nagaland. It thrives best at elevations above 1500-2000 m in areas where there is an annual rainfall of 150 cm, an average maximum shade of temperature of 23.9°C, and average minimum temperature of 1.6°C. The wattle bark is one of the richest vegetable-tanning materials and contains about 35% of condensed tannins and 10%-12% of non-tannins. Its constituents have already been identified.

(ii) Fruit tannins: The vegetable tannins are also obtained from the fruits of a number of trees, generally containing hydrolysable tannins. These tannins are usually being used in combination with bark and wood tannins.

(a) *Terminalia chebula: Terminalia chebula* occurs in sub-Himalayan tracts from Ravi eastwards to Bengal and Assam ascending up to an altitude of 1500 m and also in the moist forests of the West Coast and on the hills of Deccan and South India. The dried fruits of *T. chebula* are of commercial value and known as myrobalan in the trade. The ideal time of collection of fruits in India is January every year, and the myrobalans collected from the Salem district, Tamil Nadu (India), are regarded as one of the best in India for colour and tannin content. The myrobalan fruit contains 2%-4% tannins, which increases to 20% on the removal of stones of myrobalans, which constitute 20%-50% by weight of the fruits. Myrobalan contains hydrolysable tannins of pyrogallol type, with *chebulagic acid*, *chebulinic acid*, and *corilagin* as the major constituents.

(iii) Leaf tannins: The leaves of some plant species are rich in tannins and mainly contain hydrolysable tannins.

(a) *Anogeissus latifolia: Anogeissus latifolia* occurs in the sub-Himalayan tract from Ravi eastwards and in the hills throughout India, ascending to 1200 m. The leaves are used for tanning under the trade name "dhawa sumac." Tannin content in the mature leaves was found to be 32.50%, alongwith 10.50% non-tannins. The hydrolysable tannins containing 90%-95% of gallotannins were found to occur in the leaves.

(iv) Wood tannins: Cutch is the most important wood tanning material produced in India. It is obtained as a by-product of *Katha* industry, which utilizes the wood of *Acacia catechu,* popularly known as *khair*, as raw material. The *Acacia catechu* trees are widely distributed in India and occur mainly in Punjab, Himachal Pradesh, Uttar Pradesh, Madhya Pradesh, and Bihar. The heartwood of *khair* trees, containing more than 3% of *katha*, is an economically suitable raw material. Trees are felled for the extraction of *katha*, usually when they are 25 to 30 years old and are above 30 cm in diameter. Branch wood with heartwood of 2.5 cm and above in diameter can also be used for extraction. The main operations involved in the extraction of *katha* and *cutch* are: (i) conversion of the heartwood of *khair* into chips, (ii) extraction of chips with water, (iii) concentration of the extract, (iv) crystallization of *katha*, (v) filtration of *katha*, and (vi) preparation of *cutch*.

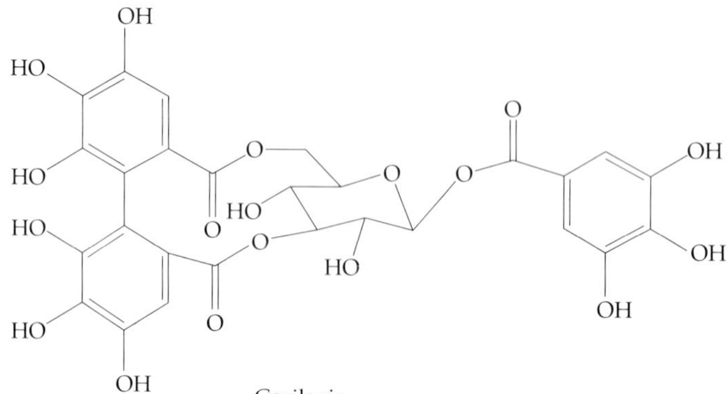

Corilagin

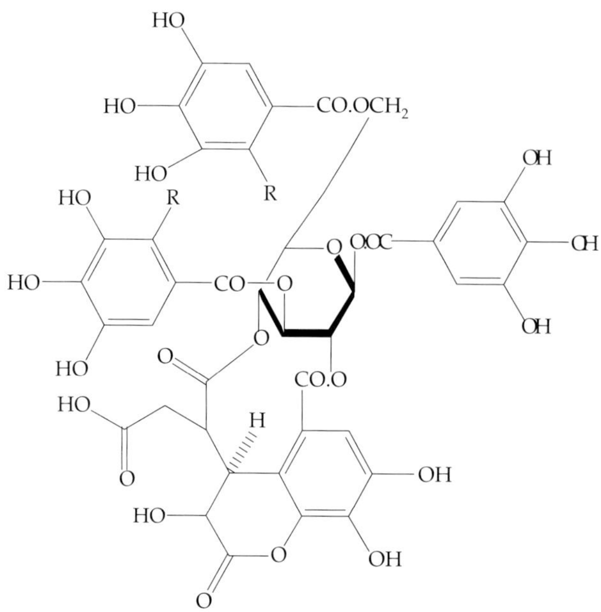

Chebulinic acid
R = H
Chebulagic acid
RR = Bond

In the factory, chips are extracted in a battery of steam-heated autoclaves at about 110°C by adjusting the steam pressure. The concentration is done at a lower temperature under reduced pressure until the extract attains a specific gravity of 1.05 to 1.07. It is then allowed to cool to the room temperature in big tanks and then to 0°C in aluminium or stainless steel containers kept inside cold room. *Katha* crystallizes out in about 10 to 15 days' time. *Katha* is filtered using a filter press or under vacuum and traces of cutch removed using a hydraulic press. It is then cut into tablets and dried in cold rooms maintained at about 10°C temperature and a relative humidity of 40 to 50. The filtrate consists of the *cutch* solution, which is further concentrated in triple effect evaporators under vacuum until it solidifies on cooling. The viscous mass while hot is poured into wooden frames of suitable size and dried in shade. Katha is mainly used as an ingredient of pan and pan masala (Karnik and Sharma, 1964).

The tannin content in *cutch* was found to be 57%-59%, but was rated as an inferior tannin due to its low solubility, high non-tannin content (27%-29%), low penetration into hide, and the colour it imparts to leather. *Cutch* is mainly used as a cheap dye for cotton and silk and in calico printing. It is also used in large quantities in oil-well drilling. Some important tannin-rich plant species are given in Table 6.3 (Tewari, 1994).

General Method of Tannin Extraction

(i) Collection of plant material
(ii) Air drying in shade
(iii) Chipping and coarse powdering of plant material of required size
(iv) Charging of plant material to the battery of autoclaves containing water in appropriate proportion
(v) Solid liquid extraction under standardized conditions
(vi) Settling and filtration of tannin extract
(vii) Concentration of tannin liquor under vacuum
(viii) Spray drying of concentrated tannin liquor to obtain powder

Functions of Tannins in Plants

(i) Biological role not fully understood.
(ii) Some contend it to be a by-product of metabolic process and has a definite function in plant.
(iii) Astringent property, prevents damage from browsing animals or creatures.
(iv) Protects trees from insects and fungal attack.
(v) Tannins or their precursors may serve as oxygen carriers in the oxidation system of plants.
(vi) Get transformed to produce coloured pigments.
(vii) Tannin-containing woods in general are dark and durable.

Uses of Tannins

(i) As natural mordants in dyeing of textiles with natural dyes.
(ii) In leather industry for conversation of animals hides into leather.
(iii) To reduce the viscosity of the drill mud in deep oil-well drilling in combination with caustic soda.
(iv) In the manufacture of tannin-formaldehyde and tannin-phenol-formaldehyde resins (wattle tannins).
(v) In boiler-water treatment for the prevention and removal of boiler scale.
(vi) In a preservative treatment of fishing net.
(vii) In flotation of certain ores.

Table 6.3 Some important sources of tannins

S. No.	*Name of plant species*	*Parts used for tanning*	*Family*	*Popular name*	*Tannin content (%)*	*Pattern*	*Distribution*
1.	*Acacia catechu* Willd.	Heartwood	Mimosaceae	Khair, Cutch Tree	57-59 (in cutch)	Tr	Assam, Gujarat, Maharashtra, Punjab, Uttar Pradesh, West Bengal
2.	*Acacia mearnsii* De Wild.	Bark	Mimosaceae	Common wattle, green wattle	30-35	Tr	Kodai Hills, Nilgiris, South India
3.	*Acacia nilotica* Delile subsp. *indica* (Benth.) Brenan	Bark, pods	Mimosaceae	Babul	12-18, 12 - 19	Tr	Throughout India
4.	*Caesalpinia coriaria* (Jacq.) Willd.	Pods	Fabaceae	Divi-Divi Plant	28-41	Tr	Tamil Nadu and Maharashtra
5.	*Cassia auriculata* Linn.	Bark	Fabaceae	Tanner's cassie	18	Sh	Gujarat, Maharashtra, South India
6.	*Cassia fistula* Linn.	Bark	Fabaceae	Amaltas, golden-shower	11-12	Tr	Throughout Tropical India
7.	*Ceriops decandra* Griff. Ding Hou	Bark	Rhizophoraceae	Goran	20-37	Sh or Small tree	Tidal forests of Sundarbans, East coast of India and Andamans
8.	*Ceriops tagal* (Perr.) C.B Robins.	Twig bark, bole bark	Rhizophoraceae	Goran, guttia	25.89, 41.22	Mangrove	Coastal forests of India

9.	*Cleistanthus collinus* Hook.f.ex Planch.	Bark	Phyllanthaceae	Oduvan	33	Sh	Distrbuted from the Deccan northwards up to Ganges
10.	*Emblica officinalis* Gaertn.	Stem bark, twig bark, fruits	Euphorbiaceae	Amla	8-9, 21, 28	Tr	Mixed deciduous forest of India
11.	*Rhizophora mucronata* Lam.	Bark	Rhizophoraceae	True mangrove	20-30	Tr	Tidal shores and creeks in India and the Andamans and Nicobar Islands
12.	*Terminalia alata* Heyne. ex.Roth.	Bark	Combretaceae	Asan, Laurel	18.7	Tr	Throughout the Himalayas from Kangra eastwards to the Goalpara Divisions of Assam and southwards throughout peninsula
13.	*Terminalia arjuna* (Roxb.) Wight & Arn.	Bark	Combretaceae	Arjuna	20-24	Tr	Throughout India
14.	*Terminalia bellerica* (Gaertn.) Roxb.	Fresh fruit pulp	Combretaceae	Bahera	25	Tr	Throughout India
15.	*Terminalia chebula* Retz.	Fruits (with stones of Myrobalan)	Combretaceae	Harra, Myrobalan	20	Tr	Throughout India

Tr = Tree., Sh = Shrub

(2) Alum root

The roots of *Heuchera americana* and *Heuchera micrantha* were used as mordant for vegetable dye in the past. The quantity of roots required was proportional to the weight of the yarn or fabric taken (Tewari, 1994).

(3) Mineral iron

Naturally occurring mineral iron (iron salts) was used as mordant in early times. It was found in many clays and iron-rich mud and in creeks and springs. The dyed fabrics when immersed in the mineral spring led to darkening of colour and made it fast to washing and light. The tannic acid in combination with iron salts produced black colour.

(4) Dunging

Mordanting with liquid animal manure has been carried out since time immemorial. Cows fed with dry fodder yielded the best dung. Cow dung was used for cotton mordanting.

(5) Milk

The use of the fatty substances have been practised since early times by the dyers. The primitive Hindus soaked cotton in milk followed by its drying in sun. The dried fabric was mordanted with alum and then dyed to red colour using a suitable vegetable dye.

OIL MORDANTS

The Turkey red colour from madder is obtained by using oil mordants. Due to the solubility of alum in water and less affinity toward cotton, it is easily washed out from the cotton fabric. The fatty oil consists of free long chain fatty acids, such as palmitic, stearic, oleic, and linoleic, as well as their glycerides. The carboxylic acid group (–COOH) of fatty acids react with metal salts and gets converted into –COOM, where M may be Al, Cr, or Sn. Treatment of fatty oils with sulphuric acid produces sulphonated oils, which possess better metal-binding capacity than the starting oil, due to introduction of sulphonic acid group, ($-SO_3H$). This can react with metal salt to produce $-SO_3M$. The bound metal forms a complex with the mordant dye, such as madder, to give Turkey red colour of superior fastness and hue. The sulphonated oils are called Turkey red oils (TRO). Presently, the term TRO is used for sulphonated castor oil (Gulrajani and Gupta).

APPLICATIONS OF MORDANTS

The vegetable dyes are chiefly the mordant dyes. The most common fabrics used by humans are silk, wool, cotton, and polyester. The interaction between a fabric, mordant, and natural dye can clearly be understood by knowing their structures. A brief description about silk, wool, cotton, and polyester fabric is given and the dyes and mordant for dyeing can be selected according to the nature of the fibre.

(1) Silk

Silk is a unique gift of nature to mankind. It was first discovered in China. Its natural sheen, light weight, vibrant colour, and other such unique qualities have made it the queen of textiles. Silk is mainly produced by the silkworm *Bombyx mori*, a species of commercial importance (Cook, 1984) and consists of fibroin protein. The process of rearing silkworms for the production of silk is known as sericulture. Silkworm larvae are fed on mulberry (*Morus alba*) leaves, and after 35 days, they spun a silk cocoon, which on immersion in hot water frees the silk filaments. The silkworm larva secretes the silk through its two salivary glands present in its head, along with a gum (sericin), which fixes the two filaments together and accounts for up to 25% of the mass of the fibre. It is ejected in the liquid form, through openings called spinnerets, and solidifies on contact with air. A silkworm spins of filament and completely encloses itself in a cocoon in about two or three days. The outside ends of the filament are found by brushing the cocoons, which are then wound on a reel. One cocoon comprises about 1000 yards of filament and the silk at this stage is known as raw silk. The raw silk fibre consists of two continuous filaments of a protein (fibroin), which are cemented together and surrounded by a gum known as sericin. A yarn can be made by combining several threads of silk. The lustre and softness normally associated with silk appear after removal of the gum. This may be carried out by putting yarn or fabric in a soap solution (with added alkali) at 95°C for at least two hours (Trotman, 1984; Asquith, 1977). Only the small amount of cocoon's silk is usable; thus, 5500 silkworms are needed to produce about 1 kg of silk. China and Japan are the two dominant producers, manufacturing together more than 50% of the world's production each year (Anon, 2007).

Silk fibroin, like wool keratin, is composed of polypeptide chains, produced by the condensation of α-aminoacids. The fibroin protein contains 17 amino acids, of which chemically simpler compounds, namely, glycine, alanine, and serine, are the major ones and tryptophan, histidine, and cystine are the minor ones. Due to virtual absence of sulphur in fibroin, the disulphide bridges between the long chain molecules are absent, in contrast to wool. The orientation of polypeptide chains in silk is parallel to the fibre axis, with neighbouring chain running in the opposite directions and held by hydrogen bonding to form a sheet (Dickerson, 1969). The fibre obtained is very strong, as the resistance to tension is borne directly by the covalent bonds of the polypeptide chains (Lewis, 1992). The fibre is quite flexible as the sheets are held together with sole van der Waals forces of attraction between them. The silk obtained from silkworm *Bombyx mori* contains about 60% of protein in crystalline form. The silk fibroin contains less number of amino acids, having strongly basic residues, such as arginine, histidine, and lysine (Asquith, 1977; Truter, 1973) as compared to wool, thus resulting in adsorption of less amount of acid than wool. The number of amino groups present in silk is about 20% in comparison to wool, yet sufficient number of dyeing sites are available for dye retention. Glutamic acid, cystine, serine, glycine, and leucine were found to major amino acids in the Merino wool (Truter, 1973., Lewis, 1992). The basicity of silk and wool, as measured in equivalent number of basic groups per kg, is 0.15 and 0.82, respectively (Vickrestaff, 1954).

(2) Wool

Humans use considerable amounts of fibres obtained from different animals, of which the wool from sheep is commercially the most valuable. The wools obtained from different breeds of sheep are classified according to the fibre diameter and length. The premium fine wool is obtained from Merino sheep breed of Spain. Different impurities present in the wool are sand and dirt, vegetable matter, grease and suint. Wool grease consists of esters of fatty acids, while its suint contains potassium salt of fatty acids plus some phosphate, sulphate and nitrogenous matter (Truter, 1956). Wool scouring is an essential step before its dyeing.

Keratins comprise a group of proteins and wool is a member of this group. Keratins have been the classified as hard or soft as per their tactile properties. The hard keratins of wool, hair horns, features, etc., contain higher concentration of sulphur (more than 3%) in comparison to soft keratins of skins (wool/cotton blends). Sulphur is present mainly in the form of residues of the amino acid cystine. The X-ray diffraction pattern of keratins classifies them into two types, namely, alpha or beta. Wool contains about 82% of keratinous protein with high cystine content and approximately 17% of the non-keratinous proteins, with relatively low cystine content. It also contains waxy lipids (1%) and a small amount of polysaccharide material. The non-keratinous proteins and lipids are concentrated in specific regions of the structure only. There are about equal numbers of basic amino and acidic carboxyl groups in the side chains of wool, accounting for the amphoteric nature of the fibre, and is thus capable of combining with large amounts of acid and bases (Lewis, 1992).

(3) Cotton

Cotton is the most important of the naturally occurring cellulosic fibres that has textile applications (Preston, 1986; Lewis, 1992). The molecular formula of cellulose is $(C_6\ H_{10}O_5)_x$, and its hydrolysis with concentrated hydrochloric acid yielded D-glucose (96%), indicating its structure to be based on D-glucose units. The inter-glucose units in the pyranose form have β-linkage involving C_1-C_4links.

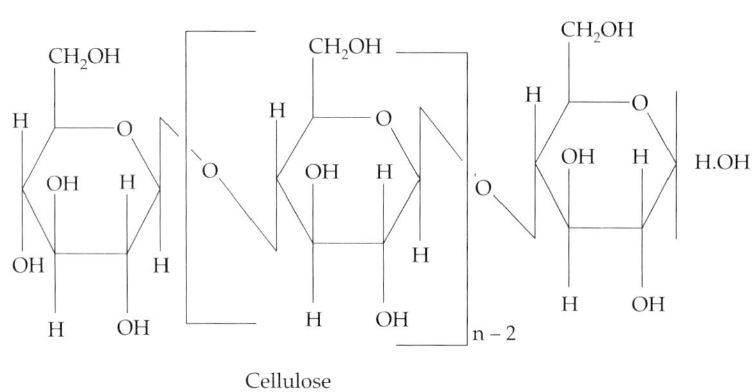

Cellulose

The molecular weight determination indicated the presence of about 10,000 glucose units in cellulose. The cellulose molecule is non-planar, but possesses a screw axis, with each glucose unit being at right angle to the previous one. It is a rigid chain molecule and the long chains are held together by hydrogen bonding, thus having a three-dimensional brickwork. The partly amorphous character of cellulose accounts for the fibre flexibility.

The cotton fibre contains three principal components:

(i) A network of cellulose fibrils constitute the primary wall situated on the outside portion of the fibre, which is covered with an outer layer of pectin, protein, mineral matter, and wax, thus comprising about 5% of the mass of the fibre. A cleaned cellulosic material is required for dyeing and printing.

(ii) The bulk of fibre is made of secondary wall, consisting of fibrils of cellulose arranged spirally around the fibre axis.

(iii) The cavity left by the dried cell sap is known as lumen and is lined with a thin film of residual protein matter.

There is a considerable swelling in the cotton fibres in water due to partial disruption of the hydrogen bonds by water. A textile fibre gets a negative surface potential (zeta potential) on immersion in the neutral aqueous solution and the degree of the same increases with the hydrophobicity of the fibre. The potential can be decreased or reversed in an acidic dye bath medium for wool, silk, nylon, and acrylic fibres due to their ability to take up hydrogen ions. The absence of sites in the cotton fibre, which can absorb hydrogen ions, and the presence of negative surface charge provide a barrier for the adsorption of anionic dyes. Addition of electrolyte (sodium chloride or sodium sulphate) to the dye bath decreases the extent of surface potential, thus easing the approach of the dye anion within the range at which van der Waals forces become effective (Johnson, 1989; Vickrestaff, 1954; Trotman, 1984). The electrolyte also increases the dye activity in solution and reduces the osmotic work that must be done in bringing sodium ions into the fibre during the absorption of the dye (Vickrestaff, 1954).

(4) Polyester

The polyester fibre is the most important of all synthetic fibres, due to its high tensile strength, excellent durability to wear, and resistance to attack by chemicals (Doughty, 1986). It is manufactured by condensation of ethylene glycol and dimethyl terephthalate (or terephthalic acid), followed by polymerization. The wool/polyester blends are used widely by humans according to their requirement (Lewis, 1992).

Linen or flax, hemp, jute, and ramie are the other fibres that are used for making industrial textiles. Regenerated fibres, such as viscose, are also of major importance.

References

1. Anon. (1979, 1982). *The story of the elements potassium and chromium* (part I, May-June, 1979., part I and part II, September-October, November- December, 1982). Sarabhai M Chemicals, Baroda, India.

1a. Anon. (1999). *Convention on Natural Dyes*, Department of Textiles, Indian Institute of Technology, New Delhi, p. 18.
2. Anon. (2007). Silk Route, *Down to Earth Supplement*, April, 2007, p. 74.
3. Anon. (2009). Training module on "*Capacity building of communities involved in sustainable forest management.*" Forest Research Institute, Dehra Dun, Uttarakhand, p.145-150.
4. Armarego, W.L.F., Chai, C.L.L. (2009). *Purification of laboratory chemicals* (6th edn.). The United States of America: Butterworth-Heinemann.
5. Asquith, R.S. (1977). *Chemistry of Natural Protein Fibres*, Ed. (New York: Plenum Press).
6. Bate-Smith and Swain. (1962). Flavonoid compounds. In: Florkin M., Mason H.S. *Comparative Biochemistry III*. New York: Academic Press, pp. 75-80.
7. Copper (II) Sulfate MSDC. (http://ptcl.chem.ox.ac.uk/MSDSCO/copper_II_sulfate.html). Oxford University. Retrieved 2007-12-31.
8. Cook, J.G. (1984). *Handbook of Textile Fibres*, Vol. 1, *Natural Fibres*. Shildon, Durham: Merrow.
9. Dickerson, R.E. and Geis, I. (1969). *The Structure and Action of Proteins*. Menlo Park, California: W A Benjamin.
10. Doughty, S.M. (1986). *Rev. Prog. Colouration*, 16, 25.
11. Elliot, J., Knight, A., Gowley, C. (2006). *Oxford Dictionary & Thesaurus III*. Published by Manzar Khan, Oxford University Press, YMCA Library Building, Jai Singh Road, New Delhi.
12. Falbe, J. and Regitz, M.(1995).*CD RÖMPP Chemie Lexikon, Version 1.0*, Georg ThiemeVerlag, Stuttgart/New York.
13. Gulrajani, M.L. and Gupta, D. *Natural Dyes and Their Application to Textiles*. Indian Institute of Technology, New Delhi.
14. Johnson, A.(1989). *The Theory of Colouration of Textiles*, 2nd edn. Bradford: SDC.
15. Karnik, M.G. and Sharma, O.P. (1964). An improved method for the estimation of Catechin in Kattha and Khair wood samples. *ISI Bulletin*, 16 (8), 325-327.
16. Kashiwada, Y., Nonaka, G.I., Nishioka, I., Chang, J.J., Lee, K.H. (1992). Anti tumor agents, 129. Tannins and related compounds as selective cytotoxic agents. *J. Nat. Prod.*, 55,1033-1043.
17. Khanbabaee, K., Ree, Teunis van. (2001). Tannins: Classification and definition. *Nat. Prod. Rep.*, 18, 641-649.
18. Lewis, David, M. (Editor, 1992). *Wool Dyeing*, published by Society of Dyers and Colourist, P.O Box 224, Perkin House, 82, Grattan Road, Brandford, West-Yorkshire, England on the Behalf of Dyer's company, Publications, pp. 3-29, 258-331.
19. Llewellyn, Bryan D. (May, 2005). "Stain Theroy-How mordant work"(http://web.archive.org/web/20070814015208/stainsfile. info/Stains File/theory/mordant.htm)

20. Llewellyn. Bryan D. (May, 2005a)."StainTheory-Lac" (Http://web.archive.org/web/200708132208461 stainsfile info/Stains File/dyes/75450.htm.)
21. Locke, J. (1902). On some double sulphates of thallic thallium and caesium. *American Chemical Journal*, 27, 281.
22. "POTASSIUM DICHROMATE LISTING". "http:/ www,epa.gov/osw/hazard/wasteypes/wasteid/inorchem/docs/pot-dich.pdf).USEPA.
23. Prakash Satya, Tuli, G.D., Basu, S.K. (1969). *Advanced Inorganic Chemistry*, S Chand and Company, New Delhi, India, pp. 211-253 (K), 269-280(Cu), 377-399(Al), 417-476 (Sn), 659-666(Cr), 735-758(Fe).
24. Preston, C. (1986). *The Dyeing of the Cellulosic Fibres*, Ed. Bradford: Dyer's Company Publications Trust.
25. *Römpp, Lexikon der Chemie.*, (1997). CD-version. Thieme, Stuttgart.
26. Tewari, D.N. (1994). *Tropical Forest Produce*. International Book Distributors, DehraDun, India, pp. 529-561
27. Ttrotman, E.R. (1984). Dyeing and *Chemical Technology of Textile Fibres*. High Wycombe: Charles Griffin.
28. Truter, E.V. (1956). *Wool Wax*, London: Cleaver-Hume Press.
29. Truter, E.V. (1973). *Introduction to Natural Protein Fibres: Basic Chemistry*. London: Paul Elek.
30. Vickrestaff, T. (1954). *The Physical Chemistry of Dyeing*, 2nd edn. London: Oliver and Boyd.
31. Würdig, G. and Woller, R. (1989). *Chemie des Weines*, Eugen Ulmer GmbH, Stuttgart, Germany.

7 Methods of Dyeing

INTRODUCTION

The use of natural dyes by humans started with their appearance on the earth. In the beginning, they started putting colour on the walls of the caves, utensils, and on themselves. The evidence of dyeing in India in the ancient times can be traced back to the Indus Valley Civilization. The recovery of a dyed piece of cotton cloth during excavations at the archaeological sites of Mohenjo-daro indicated the existence of natural dyeing during that period (Bhardwaj and Jain, 1982). The dye used at that time was characterized as madder. Dyeing with vegetable dyes and animal dyes existed in China about 5000 years back (Goodwin, 1982). Natural dyeing was also carried out during the Bronze Age in Europe. Some prominent ancient vegetable dyes used in the past include blue indigo from the leaves of *Indigofera tinctoria*, red dye (madder and manjith) from the roots of *Rubia tinctoria* and *Rubia cordifolia*, and yellow dye from the flowers of *Crocus sativus* and *Reseda luteola* (whole plant), while the animal dyes included Tyrian purple from the Mediterranean shellfish *Murex brandaris, Murex trunculus*, kermes from *Kermes ilics*, cochineal from *Dactylopius coccus*, and lac dye from the insects *Laccifer lacca*. Tyrian purple was the costliest of all dyes due to its low yield and unique colour, which was mainly used by the members of royal family. Until the middle of the nineteenth century, nature was the primary source of dyes and these were obtained from different parts of the plants, animals, and minerals. The first synthetic dye Mauveine was discovered by W.H. Perkin in 1856.

The process of adhering colour to the fibres, yarns, and fabrics by the application of dyes is known as dyeing. To start with, people might have started rubbing the dye-bearing parts or their juices containing pigments onto the cloth. Eventually, improved dyeing methods were developed to obtain charming colours on different fabrics. At present, dyeing is a complex, specialized field, and synthetic dyes are mainly used for colouring textiles and for other uses. However, synthetic dyes cause environmental pollution during their production and use and is thus a matter of concern. The azo dyes which released 22 aromatic amines cause harmful effects to humans. The safest option may be the use of natural dye, which are eco-friendly, renewable, aesthetically appealing, and can produce a number of shades that are difficult to obtain with synthetic dyes. It is not possible to substitute synthetic dyes with natural dyes for such a large population of the world due to the availability of limited plant sources, which can be used as dyes. A simple method involves the extraction of dye from the dye-bearing material by a suitable solvent, followed by its application on the textiles products either directly or by using a mordant to improve its colourfastness properties.

It is possible to dye different types of materials or fibres using natural dyes but with varying degree of colour-fastness properties. Natural dyes are mainly used for dyeing of natural fibres obtained from animal origin or vegetable origin. The former includes wool, silk, and some other less-known fibres, while the later includes cotton, flax or linen, jute, hemp, and many others. All animals' fibres are based on proteins, while cellulose is the basic ingredient of the plant fibres. The chemical nature of different fibres, yarns, or fabrics determines the type of the dye to be used for its colouring. Protein fibres consisting of amino acids, such as silk and wool, and nylon requires acid dyes for colouring them. The polyester yarn requires disperse dyes, while acrylic can be dyed with basic dyes. The cotton fibre consisting of cellobiose as the base moiety requires direct dyes, modern synthetic reactive dyes, and vat dyes for its dyeing. Different mordanting parameters are required for each category of fibre. The conditions can be varied according to requirement.

DYEING METHODS

The following methods are generally used for dyeing textiles with natural dyes (Tewari, 1994). It is essential to perform proper scouring and soaking of the material (fibre, yarn, fabric, garment, etc.) before its dyeing. Dyeing of yarn requires its pre-skeining to avoid tangling.

(1) Direct dyeing method for direct or substantive dyes
(2) One-pot dyeing and mordanting method for mordant dyes
(3) Separate mordanting dyeing method for mordant dyes

(1) Direct dyeing method

Natural dyes are of three types, namely, direct dye, mordant dye, and vat dye. The direct dye or vat dye fixes itself on the textile directly, while the second one needs a chemical compound known as mordant for fixing itself to the textile. The process of textile dyeing with a direct dye, such as turmeric, is known as direct dyeing. The process involves the extraction of dye-bearing plant part with water or alkaline or acidic aqueous solution for a predetermined time and temperature for the isolation of dye. The dye solution is separated by filtration and kept at an appropriate pH as per the requirement of the dyeing. The scoured and presoaked textile is introduced into the dye bath and dyeing continued for a predetermined time and temperature. The dyed material is taken out after cooling the dye bath to room temperature, washed with water, and dried in shade. Dyeing of the fabric can also be done using the isolated dye powder. Indigo, a vat dye, is another example of direct dye. Indigotin or indigo is obtained from *Indigofera tinctoria* by fermentation of indican present in the leaves. Indigo is insoluble in water and is made soluble by reduction, followed by dyeing and its regeneration on the textile by drying in the air. The synthetic vat and sulphur dyes are also chemically reduced before their application on the fabric. Direct dyes are preferred by the dyers.

(2) One-pot dyeing and mordanting method

Natural dyes that require a mordant (metallic salt or an organic mordant) for fixing it on the fabric are known as mordant dyes or adjective dyes. The majority of the natural dyes are mordant dyes. Dyeing of textile using this method requires no separate mordanting. In this method, the scoured textile and dye material are soaked overnight, followed by simmering the combined material for 15 minutes. The textile is removed temporarily for the addition of mordant, returned to the pot, and again simmered for about 45 minutes. The time varies according to the nature of the fibre and plant material taken for dyeing. The contents are cooled overnight, followed by rinsing the textiles with water and drying in shade. This method is very useful for strong and textile-receptive natural dyes. Some dye materials require long cooking for the extraction of dye, and thus, dye extract should be prepared before the addition of the textile and mordant to it. The large-scale dyeing requires isolated dye powder. The mordant and dye solution prepared separately in minimum quantity of water are added simultaneously to the dye bath at about 60-70°C, followed by dyeing of fabric in the bath at a temperature and time suitable for the fabric. The method is not suitable for plants fibres like cotton, linen, and jute, which require mordant for fixing. The conditions can be varied according to requirement.

(3) Separate mordanting dyeing method

The mordanting of the textile or fabric can be done before (pre-mordanting), after (post-mordanting), or before and after its dyeing. This method is applicable to the dyes that require a more intensively mordanted textile. The process involves dissolving the requisite amount of mordant in minimum quantity of hot water, followed by its mixing with water in a large pot, according to the material-to-liquor ratio. The mordanting is done at a particular temperature range for a specified time. The mordanted textiles can be kept for future dyeing after its rinsing and drying. In the post-mordanting method, the textile is dyed first, followed by its mordanting with an appropriate mordant, depending on the colour to be produced on the fabric. The pre- and post-mordanting of the undyed and dyed fabric can also be done in order to obtain darker shades.

DYEING PROCEDURES

Dyeing can be carried out during any stage of textile manufacturing. Textiles may be dyed as fibre, as yarn, as fabric, and as garments, depending on the type of fabric or garment being made. It is much easier to dye fibre or yarn rather than fabric, mainly because of the exposure of their large surface to the dye molecules, which leads to maximum retention of the dye, thus resulting in uniform colour and enhanced colourfastness. Some of the dyeing procedures include garment dyeing, piece dyeing, stock dyeing, top dyeing, yarn dyeing, resist dyeing, chain dyeing, cross-dyeing and random dyeing.

DYEING WITH NATURAL DYES

After the discovery of synthetic dyes in 1856, the use of natural dyes started decreasing and reduced to minimum by the beginning of the twentieth century. However, natural dyes have their own merits —these are aesthetically appealing, eco-friendly, renewable, and involve practically no or very mild chemical reactions in their use. The countries with rich forest flora and good research support on natural dyes can start exploiting natural dyes through sustainable harvesting of the dye-yielding plants. Wool and silk of animal origin made up of proteins can be dyed readily with natural dyes, while plant fibre, consisting of cellulose, can also be dyed, but not so readily. The synthetic fibres are difficult to dye with natural dyes. Natural dyes are mainly vegetable dyes, which are obtained from different parts of the plants, namely, bark, flowers, leaves, seeds, roots, and wood. Therefore, dyeing of a fabric with vegetable dye can be accomplished by pursuing the following steps:

(1) Selection of the plant material according to the colour to be imparted
(2) Collection of the plant material
(3) Processing of the plant material
(4) Extraction of dye from the plant material
(5) Dyeing of textile

(1) Selection of the Plant Material

Almost all plants contain colouring material, but it may not be suitable for dyeing the textile. A number of plants suitable for dyeing have been identified and given in Chapter IV A of the book. The plant selection can be done according to the colour to be obtained on the fabric. Proper identification of plant is essential by preparing a herbarium sheet and its comparison with authentic specimen.

(2) Collection of the Plant Material

Normally, it is advisable to collect the plant material preferably from the mature plants. Sustainable harvesting is required to avoid damage to plants. The fresh plant material can be directly used for dyeing after its collection and chopping into small pieces. However, one can collect the plant material at a time and store it after processing, which involves its cleaning, proper shade drying, and powdering in a pulverizer.

(3) Processing of Plant Material

The freshly collected plant material, after removal of sand and foreign particles and followed by chopping into small pieces, can be directly used for the extraction of dye for home dyeing. In case the plant material is required for future dyeing, it should be air-dried in shade, followed by cutting into small pieces. The plant material is to be coarsely powdered and stored in air-tight containers under cold conditions to avoid any fungal decay, but it may not be stored for long.

(4) Extraction of Dye from the Plant Material

Vegetable dyes can be extracted from various parts of the plants using the following methods:

(i) Aqueous extraction (extraction with water)
(ii) Solvent extraction (extraction with organic solvent)
(iii) Supercritical fluid extraction (extraction with carbon dioxide)

(i) Extraction with water: Vegetable dyes can be extracted from different parts of plants by using water as the solvent. Water is a polar solvent and dissolves a number of polar compounds of the dye material. Hot water dissolves phenolics, tannins, flavonoid glycosides, quinone glycosides, and other compounds present in the dye material. The appropriate material-to-liquor ratio is required for the efficient extraction of the dye. Time required for the isolation of dye is also to be determined. Generally, extraction may be carried out by gently heating the solution at a temperature of 85-90°C, with stirring for about 1 hour from the time the bath attained this temperature. The dye can also be extracted by using alkaline or acidic solution of water. The stainless steel vessels are generally used for the extraction of dye, as the dye changes its colour while extracting it in other metal (Fe, Al, or Sn) vessels. Glass apparatus may be used for laboratory-scale extraction.

- Determination of material-to-water ratio and extraction time: This can be done by extracting the varying amount of material in aqueous medium for a fixed time. This process can be illustrated by taking an example of the isolation of dye from a bark sample. The varying percentage of the bark in water (2%-20%) at an interval of 2% may be extracted for an hour by gentle heating (85-90°C), filtered, and made to 100 ml. One millilitre of aliquot can be diluted to 100 ml and the optimum amount of bark-to-water ratio for extraction of dye was determined by measuring the optical density. The maximum optical density of a solution observed by extracting a particular amount of bark, for example, 10 g in 100 ml of water for 1 hour, will indicate the optimum material-to-liquor ratio. Similarly, the time of bark extraction may be determined by extracting this particular amount of bark in 100 ml of water for varying intervals of time (15-120 minutes) at an interval of 15 minutes and measuring the optical density in the way cited above. The maximum optical density of a solution obtained by extracting the material for a particular interval of time, for example 60 minutes, will indicate the time of extraction. Thus, the dye can be extracted by taking the material (bark)-to-water ratio (1:10) and heating it for 1 hour at 85-90°C. Removal of the water will yield the dye, which can be used for dyeing textiles. These dye-extraction parameters are to be determined for each dye material to be used for the isolation of dye (Dayal and Dobhal, 2001a; Dayal et al., 2006). However, the varying percentage of bark in water, the temprature and time of extraction and dilution of the aliquot are not fixed and may be varied as per the requirement. The dye can be further fractionated into ether, ethyl acetate, and n-butanol soluble portions, from which pure active principles can be isolated by chromatography (TLC, column, and HPLC) and characterized by spectral data (IR, UV, NMR, and MS).

(ii) Solvent extraction of dye: The processed plant material can also be extracted with organic solvent at room temperature and under hot conditions using a Soxhlet apparatus. The Soxhlet extractor was invented by Franz von Soxhlet in 1879, to be used for the separation of lipid from a solid material. Normally, the Soxhlet extraction is needed when a particular compound present in a matrix (plant material) has limited solubility in the solvent to be used for the extraction and the undesired material remains insoluble. The coarsely powdered plant material containing the dye may be either placed in a paper thimble or directly into the main chamber of Soxhlet extractor, which is fitted onto the mouth of a round-bottom flask containing the solvent kept in a heating mantle of appropriate size. The flask is then equipped with a double-surface or bulb-reflux condenser. The solvent is heated to boiling and the solvent vapours thus formed travel up and condense in the thimble or solid in the chamber. The condenser does not allow the solvent vapours to escape. When the chamber is filled with the solvent, it siphons out from the side tube, taking the solute along with it. The process is repeated till the extraction is complete. Removal of the solvent on a water bath yields the extracted dye (Soxhlet, 1879). The solvent for the extraction can be chosen depending on the polarity of the dye molecules. Petroleum ether is used for extracting the non-polar or less-polar dye molecules, acetone for flavonoids, and other phenolic compounds. The flavonoid glycosides and other higher-molecular-weight compounds are extracted with methanol or ethanol or a blend with water containing either of them. The extraction can also be done by putting the coarsely powdered plant material directly into 5 or 10 litres round-bottom flask fitted with a double-surface or bulb condenser. The proportionate quantity of the solvent is added to the flask and put under reflux by keeping it on a heating mantle. The extraction has to be repeated three to four times by taking out the solvent and recharging with recovered or fresh solvent each time. It is possible to extract considerable amount of plant material by using this process. The dye extracted with organic solvents may not be readily soluble in hot water and thus may face difficultly while dyeing the textiles. However, it is comparatively easy to isolate and characterize the active principles from the organic extracts of the dye material.

(iii) Supercritical fluid extraction of dye: Supercritical fluid extraction is the latest method of extraction of the chemical compounds from their source of occurrence by using supercritical fluid. A substance above its critical temperature and critical pressure is known as supercritical fluid. In the region above the critical temperature, however, the substance cannot be liquefied by increasing the pressure, and there is no such phase transition from gas to liquid, or vice versa. The substance in this region is generally referred to as supercritical fluid, irrespective of the fluid density. The critical point represents the highest temperature and pressure, at which the substance can exist as a vapour and liquid in equilibrium. At critical point, the distinction between gas and liquid disappears, resulting to the same densities of the two phases. The process of extracting a component (extractant) normally consisting of a number of chemical compounds from the base material by using a supercritical fluid as the solvent is known as supercritical extraction (SFE). Generally, the extraction is carried from solid matrix, but is also possible from liquids. Carbon dioxide is a suitable fluid for supercritical fluid extraction because it is inexpensive, non-toxic, non-flammable, and has comparatively low critical temperature, (31.05 ± 0.05°C), and pressure, (73.858 ± 0.005 bar). Water has a critical temperature of 647 K and a critical pressure of 220 bar due to its high polarity. The polar and high-molecular-weight compounds are not soluble in carbon dioxide and thus are unable to extract them from the material. The drawback of carbon dioxide can be overcome by adding small amounts of polar cosolvents, such as methanol or ethanol, known as modifiers.

The cosolvents increase the solvating power of carbon dioxide and thus help in extracting the polar compounds. The selective extraction can be carried by using a supercritical fluid by altering its properties by varying pressure and temperature. For example, the essential oil can be extracted from the plant material by carbon dioxide at a low pressure (100 bar), while lipids are isolated at high pressure, using ethanol as a modifier. The extraction by a solvent is based on its diffusion into the matrix in order to diffuse out the material into the solvent. The supercritical fluids are used for faster extraction due to their higher diffusivity and lower viscosity. The time of extraction is reduced considerably.

A supercritical extraction equipment requires the following four essential components:

(a) A pump for carbon dioxide

(b) The sample container pressure cell

(c) Pressure-maintaining device for the system

(d) The extracted material collection cell

The extraction using the equipment can be carried by fast pumping the liquid to a heating zone, followed by its heating to critical conditions. Thereafter, it is taken into the extraction vessel for its diffusion into the material, to dissolve out a mixture of chemical constituents (extractants). The pressure requirements of the vessel vary according to the nature of the compounds to be extracted. Normally, the supercritical fluid extractions are carried out below 350 bar. Higher pressure (800 bar) is required for the extraction of vegetable oils. The supercritical fluid containing the extracted material is removed from the extraction cell into a separator at a lower pressure for the removal of extracted material. Thereafter, carbon dioxide can be recycled after cooling and recompression or made to escape into the atmosphere. The supercritical extraction method can be used for the isolation of natural dye from plant material without any chemical degradation. The equipment is quite costly and therefore may be used for thermally labile or costly natural dyes (Saito and Yamauchi).

(5) Dyeing of Textile

The following preparations are required to be done before proceeding for dyeing of the textiles.

(i) Selection of the dye vessel: Stainless steel or glass vessels are always preferred for dyeing with natural dyes. The vessels of other metals, such as iron, copper, aluminium, tin, are avoided, as these may act as mordants during dyeing of textile, thus resulting in the change of shade and colour of the dyed textile.

(ii) Quality of water: Water with minimum hardness and neutral pH may be used for dyeing of textiles with natural dyes to obtain charming colours. It may not be possible to reproduce the shade with varied hardness and pH.

(iii) Preparation of textile material for dyeing: The fibre, yarn, or fabric of animal origin or vegetable origin should be made free from foreign materials, dirt, fats, oils, waxes, and any other unwanted material before dyeing. The presence of impurities may affect dyeing of textiles. Therefore, different fibre or yarn or fabric or garments must be scoured before dyeing. The process of cleaning the garments with hot water and cleansing agents is known as scouring. The cleansing agent may vary for different textiles:

(a) Wool: Raw wool contains dirt, lanolin, dead skin, sweat residue, pesticides, and vegetable matter. The yellow-coloured lanolin, also known as wool grease or wool wax, is secreted by the sebaceous glands of wool-bearing animals. Lanolin is a complex mixture of different types of lanolin esters. The vegetable matter is removed by hand picking in smaller samples, while chemical carbonization is useful for commercial samples. The wool scouring is done to remove these substances by boiling in soap solution, soda, and ammonia. The scoured wool is yellow in colour. Pure white wool can be obtained by bleaching the wool with hydrogen peroxide, followed by washing with water. The bleached wool is suitable for dyeing.

(b) Silk: Raw silk, emitted from two salivary glands in the head of a silkworm *(Bombyx mori)* larva, consists of two main proteins, fibroin and sericin. The first one (fibroin) is the structural centre of silk, while the latter (sericin, $C_{30}H_{40}N_{10}O_{16}$) a sticky material, surrounds it. Fibroin is made up of the amino acids and forms beta pleated sheets. Sericin makes about 25% of the weight of raw silk. The scouring of silk involves the removal of sericin by heating in hot water containing small quantities of wetting agent and soda ash as per the requirement.

(c) Plant fibres: Cellulose, a polymer of glucose, is the major component of all the plant fibres, namely, cotton, jute, and linen. These fibres have comparatively low affinity for natural dyes than animal fibres (wool and silk). Therefore, thorough scouring of these fibres is to be carried out before their dyeing with natural dyes. This may be carried out by gently boiling the skeins of the fibres in soap solution containing small amount of washing soda for two hours. The material is left overnight in the scouring bath, followed by its rinsing with hot water. Jute or linen skeins may also be washed in washing machine. Insufficient scouring leads to poor dyeing.

MORDANT

The majority of the natural dyes may fade severely unless the yarn is treated first with a chemical to help fixing the colour to the yarn. This chemical is known as mordant and the process of using such a chemical is called mordanting. It can be carried out before/after or during the process of dyeing. Some metallic salts, such as potassium aluminium sulphate, ferrous sulphate, copper sulphate, and potassium dichromate, are examples of mordants. Naturally occurring tannins also act as mordants. Some of the ancient mordants, such as wood ashes, clays, rock salts, vinegar, and urine, were used throughout the world. When the yarn is simmered in a solution of mordant, certain amount of mordant will adhere to the yarn and the dye then combines with the mordanted fibre to form a coloured yarn. Different mordants allow different chemical reactions (complex formation) between the dye and the mordanted yarn, each resulting in a different colour.

(1) Preparation of dye and mordant solution

The required amount of dye powder and mordant are dissolved in soft, warm water separately in smaller vessels and thereafter diluted to the desirable volume with water. The agitation of dye powder solution is necessary to avoid lump formation and complete dissolution of the dye.

(2) Quantity requirement

The quantities of dye, mordant, and chemical assistant are determined on the basis of weight of the material (fibre/yarn/fabric) to be dyed. Lesser quantities of dye and mordant are needed for imparting lighter colour to the material. Normally, the material should float freely in the dye and mordant baths, respectively. The material-to-liquor ratio is to be calculated accordingly with the predetermined concentration of dye and mordant.

(3) Mordanting methods

(i) Pre-mordanting

In this method, the yarn is treated first with the mordant solution. The time and the temperature of mordanting is not the same for all mordants and textiles and has to be determined by experimentation. A general method of wool mordanting is being described here. Dissolve the mordant, for example, alum, in the minimum quantity of hot water and thereafter add this solution to the remaining amount of water and mix thoroughly. Enter the cleaned and thoroughly wet sample to the warm solution and bring bath to simmer in about an hour with occasional stirring. The bath is further simmered for about half an hour after lowering the heat by stirring it gently from time to time. The sample is removed with a stainless steel rod kept over the pot for some time. Excess of water is squeezed out after cooling it. The mordanted sample is transferred to the dye bath for dyeing. It can also be kept damped in bath for next-day dyeing or can be stored for future use after dyeing. The mordanting time and temperature may vary with the mordant. Similarly, the mordanting of the fabric with other mordants can done. The percentage of mordant and chemical assistant may be varied as per the convenience of the dyer; however, the excess quantities should be avoided to check the environmental pollution.

(ii) Simultaneous mordanting

In this method, the scoured fabric is introduced into the predetermined optimum dye concentration solution with appropriate material-to-liquor ratio. The dyeing vessel may be simmered for 15 minutes, followed by temporarily removing the textile for the addition of mordant. The textile is re-entered into the dye bath and can be simmered for about 45 minutes or for more time, depending on the nature of the textile to be dyed. The dyeing of fabric can also be done by introducing the fabric into the dye bath containing the calculated amount of dye and mordant.

(iii) Post-mordanting

The dye bath is cooled to room temperature after dyeing the textile, which is subsequently removed from the bath and kept over for some time for the removal of excess dye and is transferred to the mordant bath. The excess of adhering dye solution also may be removed by squeezing the textile. The mordanting of the dyed textile is completed as per the optimized conditions. The textile is then removed from the bath, washed thoroughly with water, and dried in shade.

DYEING PROCESS

The scoured yarn is ready for dyeing. Different fabrics may take dye differently, depending on their basic building units. For example, a dye solution with same mordant will impart varying colour on silk and wool. Rayon, nylon, and other synthetic fibres will dye differently than natural fibres. It is essential to soak the cleaned and dried fabric for about 30 minutes in water before dyeing. It helps to wet the fabric thoroughly for better penetration of the dye. Fabric such as silk, which does not saturate readily, require about 12 hours of soaking in water. The wet fabric fibre should be free from air.

The dye solution is prepared in water as per the weight of fabric/fibre to be dyed. The natural dye powder contains a number of other constituents besides the active dyeing compounds. Therefore, the quantity of natural dye required for dyeing a particular amount of fibre is more in comparison to synthetic dye. The material-to-liquor ratio is to be adjusted so that the fabric freely floats in the dye solution. The less quantity of liquor may affect the uniformity of the shade. The optimum dye concentration may be determined as given below.

(1) Determination of optimum dye concentration

The optimum dye concentration can be determined by using varying percentage of dye solutions for a fixed weight of silk. Dye solutions of different concentrations (0.4%-

2.0%) at an interval of 0.2% may be prepared and the optical density of each solution measured by diluting 1.0 ml of each solution to 100 ml. Thereafter, silk samples of equal weight, for example, of 1.0 g each, can be dyed in the solution of different concentrations at 70-80°C for 30 minutes. The optical density of each solution may be measured by making it to 100 ml and thereafter diluting 1.0 ml of the each solution to 100 ml. The absorption percentage of the dye by silk for each dye solution will be determined. The dye solution indicating the maximum percentage of dye absorption is the optimum dye concentration and generally it can be used for other fabrics also. The maximum quantity of silk that can be dyed using this solution without colour variation can be determined by dyeing trials of the varying amounts of silk. The dilution effect of the solution of the optimum dye concentration can also be determined by diluting the solution with simultaneous dyeing trials till the colour variation in the dyed fabric is observed. (Dayal and Dobhal, 2001a; Dayal et al., 2006). However, the interval of dye concentration, dilution of the dye solution, weight of the silk, dyeing temprature are not fixed and may be changed as per the requirement.

(2) Dyeing

The dyeing of the fabric may be accomplished using the following method. The damp fabric was introduced into the dyestuff solution of optimum concentration, kept in the dye bath, and its temperature may generally be raised to 80-85°C. Dyeing was continued for specified time as per the nature of the yarn to be dyed. The dye bath was cooled to room temperature and dyed material taken out and washed with water to remove the sticking dye. The soap solution of appropriate concentration in water can also be used for washing as per the need. This process is applicable for direct dyes. While using an adjective dye or mordant dye, the dyed fabric is directly transferred to mordant bath usually maintained at 70-85°C for fixing the dye into the fabric for a specified period. The mordant bath was cooled to room temperature and the mordanted yarn was taken out and washed with soft water to remove the sticking mordant. The time and temperature of dyeing are higher for plant fibres and may be varied according to nature of fibre and requirment. In case of pre-mordanting dyeing method, the yarn is mordanted before dyeing. Glauber's salt (sodium sulphate) is useful for the complete removal of dye from the dye bath, while the cream of tartar and tin salt brighten the colour and iron salt make it dull. It is always advisable to lift out the fibre from the dye while adding a chemical during the course dyeing and return it to the bath after thoroughly mixing the chemical (Dayal and Dobhal, 2001a; Dayal et al., 2006). However, temprature is not fixed and may be changed as per the need.

(3) Fixing

The colour of the dyed fabric may change on exposure to the alkaline solution. The treatment of the dyed fabric with a dilute solution of washing soda will remove this deficiency; this process is known as fixing. Sodium carbonate solution (0.25%) with appropriate material-to-liquor ratio usually at 75-85°C may be used as fixing bath, and the process normally requires about 20 minutes for its completion. The samples were taken out, washed with water, and dried in shade. The alkali concentration, time and temprature of fixing may be changed according to the requirement.

(4) Determination of colour-fastness properties

The colour-fastness properties of the dyed fabrics to washing, light, crocking, and perspiration were determined and evaluated using standard methods.

EXPLORING THE NEW SOURCES OF NATURAL DYES

There are about 400 plants species, different parts of which may be used as sources of vegetables dyes. However, to augment the use of natural dyes, the plant material should be readily available to the users and the dye obtained should impart fascinating shades to the textiles with very good colour-fastness properties. Dayal et al. undertook the studies on some abundantly occurring plant species for their possible use as the additional sources of natural dyes. The following nine plant species were selected for the study.

(i) *Ageratum conyzoides* Linn. (Whole plant)
(ii) *Cassia tora* Linn. (seeds)
(iii) *Eucalyptus* hybrid mainly *E. tereticornis* Sm. (leaves and bark)
(iv) *Grewia optiva* Drumm (seeds)
(v) *Lantana camara* Linn. (leaves)
(vi) *Parthenium hysterophorus* Linn. (whole plant)
(vii) *Pinus roxburghii* Sarg (needles, bark)
(viii) *Populus deltoides* Marsh (bark)
(ix) *Shorea robusta* Roxb.ex Gaertn.f. (bark)

The systematic studies involved the following steps:

(i) Extraction of dye: Extraction of dye in aqueous solution from the above-cited plant species involves the following steps:
   (a) Determination of optimum material-to-liquor ratio for the extraction of dye.
   (b) Determination of optimum time of extraction by using the optimized material-to-liquor ratio.
   (c) Temperature range for the extraction of dye with water.
(ii) Dyeing
   (a) Determination of optimum dye concentration for dyeing of fabric
   (b) Dyeing of fabric using optimized conditions
   (c) Mordanting of the dyed fabric
(iii) Determination of the colourfastness to washing, light, perspiration, and crocking of the dyed fabric.

A brief description about the plants, extraction of dye, dyeing trials, mordanting, and colourfastness determination is given below.

(1) About the plants

(i) Ageratum conyzoides: *Ageratum conyzoides* Linn. (Family: Asteraceae), an erect annual herbaceous about 1 m tall plant, is native to tropical America and occurs as a weed throughout India. It grows upto an altitude of 1800 m and also in the middle

Andaman (Anon, 1985). The plant is of medicinal and insecticidal value and is found to contain the different class of secondary metabolites, namely, flavonoids, terpenoids, alkaloids, coumarins, essential oils, and tannins (Solanki, 2010). *A conyzoides* is widely utilized in traditional medicine worldwide. Isolation of dye from the aerial parts of *Ageratum conyzoides*, its dyeing characteristics on different fabrics, and determination of its fastness properties were undertaken (Dayal et al., 2007).

(ii) Cassia tora: *Cassia tora* Linn. (Family: Caesalpiniaceae), an annual herb, up to 1-2 m in height, occurs throughout India up to an altitude of 1500 metres in the Himalayas. It is known as *chakunda* in Hindi. The seeds find use in the preparation of sweets and can also be consumed in time of scarcity. The seeds are of medicinal value and contain 7.65% of gum in its kernels. A number of anthraquinone compounds, namely, emodin, aloe-emodin, physcion, chrysophanic acid, rubrofusarin, and its 6-β-gentiobioside, nor-rubrofusarin, and others have been characterized in the seeds. The seeds are available in plenty and have been reported to contain yellow dye (Anon, 1992). The studies were undertaken for the isolation of dye from the seeds (Dayal and Dobhal, 2001).

***(iii) Eucalyptus* hybrid:** *Eucalyptus* was first introduced in India in 1790, when a number of species were planted in the palace garden at Nandi Hills near Mysore. Regular trials were started in 1843 when Captain Cotton of the Madras Engineers successfully introduced *Eucalyptus globulus* at Wellington in the Nilgiri Hills. Since then, a number of *Eucalyptus* species have been introduced in India. *Eucalyptus* hybrid (Mysore gum, mainly *Eucalyptus tereticornis*) is extensively grown in India under social forestry programme. *E. grandis, E. camaldulensis,* and *E. globulus* are the other major species grown in India. The success of this fast-growing species is mainly due to its eco-adjustment under varied agro-climatic conditions and its capacity to endure adverse biotic and climatic conditions. It accounts for nearly 90% of the *Eucalyptus* plantations raised in the country, as a result of which, *Eucalyptus* hybrid wood and bark are now available in plenty. Wood has numerous uses and is also used in the manufacture of pulp and paper. The bark contains tannins and other polyphenols and is not of commercial use, except fuel. The leaves have been found to contain essential oil consisting of the varying amount of cineole and α- and β-pinenes as the major constituents (Dayal et al., 1997) and medicinally important ursolic acid (Dayal, 1987; Varshney et al., 2012). As the leaves and bark are available in plenty, the studies were undertaken to explore their use for the isolation of dye (Dayal and Dobhal, 2001).

(iv) Grewia optiva: *Grewia optiva* Drummond syn *Grewia oppositifolia* Buch. Ham. (Family: Tiliaceae) is a tree, up to 14 m high and 1.4 m wide, which occurs from Punjab to Bengal, ascending to an altitude of about 2100 m in the Himalayas. It is commonly called *Bhimal* in Hindi and often cultivated in the Himalayas. The leaves are used as fodder and trees are profusely chopped in the winter season due to scarcity of green fodder. The ripe fruits are edible and have a pleasant acidic taste. The fruits are dark brown in colour and contain a dye that can be extracted with water. The tree is often planted in hedges and field boundaries (Anon, 1956, 2009). The studies were undertaken for the isolation of the dye from the seeds (Dayal and Dobhal, 2001).

(v) Lantana camara: *Lantana camara* (Family: Verbenaceae), a genus of herbs, undershrubs or shrubs, has been reported from tropical and sub-tropical America. A limited number of species occur in India, and *Lantana camara* Linn. is one of them. It is a hairy, sometimes prickly, shrub that attains a height of about 1.8 m and is cultivated as a hedge plant. *Lantana camara* is an invasive species and has spread in large forest areas in India and other countries. Phytochemical examination of its different parts indicated the presence of triterpenoids, flavonoids, and essential oils (Anon, 1962). Studies were undertaken for the isolation of dye from the leaves (Dayal et al., 2006).

(vi) Parthenium hysterophorus: *Parthenium hysterophorus* Linn. (Family: Asteraceae) commonly known as congress grass, an annual herb, which aggressively colonizes disturbed sites, was accidentally introduced into India. It is light green, with branching stems, finely lobed leaves, and grows up to 1.5 m occasionally reaching 2 m in deep rich soils (Anon, 1966). Infestations of *Parthenium* weed can degrade natural ecosystems. It is regarded as one of the worst weeds because of its invasiveness, potential for spread, and environmental impacts. The whole plant is bitter and strong-scented, reckoned tonic, stimulating, and anti-hysteric. It was once a popular remedy in ague. Its odour is said to be peculiarly disagreeable to bees, and insects may be easily kept at a distance by carrying a handful of the flower heads. The study was undertaken to find its potential as a source of dye (Dayal et al., 2008).

(vii) Pinus roxburghii: *Pinus roxburghii* Sarg. syn. *Pinus longifolia* Roxb. (Family: Pinaceae) commonly known as Chir pine, is a tall tree (30-50 m) that occurs in the Himalayas from Kashmir to Bhutan and in the Shiwalik Hills at an altitude of 450-2400 m. In India, it occurs in Jammu and Kashmir, Punjab, Himachal Pradesh, Uttarakhand, and Sikkim. Chir pine is normally evergreen but becomes deciduous or partly so in arid situation or during dry season. It yields an oleoresin on tapping, which produces turpentine and rosin of commercial value. Huge quantity of pine needles are available in the pine forests during summer season and causes forest fire (Anon, 1969). The needles are highly inflammable due to the presence of a special type of wax called estolide. As the pine needles and bark are available in plenty, studies were carried out in order to explore their utilization as a source of natural dyes (Dayal and Dobhal, 1999; Dayal et al., 2001).

(viii) Populus deltoides: *Populus deltoides* (Family: Salicaceae), a genus of deciduous, dioecious trees, commonly known as Poplar, Aspens, and Cottonwoods occur mainly in the North temperate zone, while some species also occur in the sub-tropical regions. Poplar trees are of multipurpose use and occupy a unique and important position in the rural economy of India. They are known for their fast growth; easy, vegetative propagation; enrichment of soil with litter; and high production (10-30 m^3/ha/y) on a short rotation of 8 to 12 years. About 10 species which occur in India are wild. A number of exotic species have also been introduced in hills and plains in India. *Populus deltoides* is one such species and is being grown in the states of Uttar Pradesh, Himachal Pradesh, Punjab, Haryana, and North-Eastern states. The wood is used for match and other veneers, plywood boxes, and packing cases, pulp, and wood-wool. Thus, a huge quantity of bark is available for the development of value-added products from it. The studies were carried out for the isolation of dye from the bark (Dayal and Dobhal,1999, 2000).

(ix) Shorea robusta: The genus *Shorea* belongs to family Dipterocarpaceae. It is a large genus of resiniferous trees, extensively distributed in India, Burma, Ceylon, and other South-East Asian countries up to the Philippine Islands. The four *Shorea* species, namely, *Shorea assamica*, *Shorea robusta*, *Shorea roxburghii*, and *Shorea tumbuggaia* occur in India. *Shorea robusta*, commonly known as Sal, is a large sub-deciduous tree, widely distributed in parts of North-East and Central India and comprises about 13.3% of the total forest area in the country. It is a multipurpose tree as its leaves, resin, and seeds, available in plenty, are exploited for their varied applications (Anon, 1972). Terpenoids, polyphenols, flavonoids, and essential oils have been reported from the different parts of *Shorea robusta* (Varshney and Dayal, 2009). It is also the main source of timber in North and East India. Therefore, the Sal bark is available in considerable quantity in India and studies were carried out to find its use for the isolation of dye (Dayal and Dobhal, 2001a).

The details about the extraction of dye from the above cited plant species, dyeing of silk, wool and cotton samples using the extracted dye, and colourfastness evaluation of the dyed samples have been described (Dayal et al., 2007; Dayal and Dobhal, 2001; Dayal et al., 2006; Dayal et al., 2008; Dayal et al., 2001; Dayal and Dobhal, 1999, 2000; Dayal and Dobhal, 2001a).

(2) Extraction of dye

The extraction of dye from the different parts of above plant species was accomplished in aqueous medium using appropriate material to liquor ratio, time and temperature as determined for each plant species. Removal of the solvent yielded the respective dye. The details have been given in the above cited references.

(3) Dyeing

Dyeing of silk, wool and cotton samples was carried out using the isolated dyes with optimum dye concentration determined for each dye, appropriate material-to-liquor ratio, temperature and time. The above cited references contain the complete details.

(4) Mordanting

Post-mordanting of the above dyed samples of silk, wool and cotton are carried out using alum, copper sulphate, ferrous sulphate, potassium dichromate and stannous chloride as mordants in optimum concentration with appropriate material-to-liquor ratio, temperature and time. The full description is given in the above cited references.

(5) Colourfastness determination

The colourfastness properties of the dyed samples to light, washing, perspiration and crocking were determined using standard methods. The colourfastness data as described in the above cited references indicated good to very good colourfastness for the silk, wool and cotton dyed samples. The colourfastness of the cotton samples was found to be on the lower side in comparison to silk and wool samples.

The results of the dye yield, dyeing trials and colourfastness determination as described in the above cited references indicated that these plant species may be used as sources of dye for imparting various colours on silk, wool and cotton using different mordants.

RECENT DYEING METHODS

(1) Dyeing using ultrasound technique

The sound wave having higher frequency in comparison to the upper limit of the human hearing range of approximately 20 kHz is known as ultrasound. Its frequency range varies from 20 kHz to a number of gigahertz. Ultrasound finds use in detection and ranging, acoustic microscopy, imaging, and other fields. Efforts have also been made for its use in textile dyeing due its harmless nature. The conventional dyeing requires large quantity of water, electricity, and thermal energy. The high temperature (90-95°C) is required for cotton dyeing, while synthetic fibre dyeing requires higher temperature (~130°C) and pressure, with longer time leading to increase the cost of dyeing of fabric. As a result of this, a number of non-traditional techniques, such as low-temperature dyeing in the presence of carriers, infrared dyeing, and low-temperature dyeing in the presence of ultraviolet rays, were carried out on laboratory scale but could not be commercialized due to various reasons. The latest dyeing technique using supercritical carbon dioxide is costly, and moreover, it is not feasible to include it in the existing dyeing machines. Attempts have been made on laboratory scale to use ultrasonic energy for several wet processes in textiles, namely, dyeing, desizing, scouring, and washing of natural and synthetic fibres, based on the principle of cavitation. The longitudinal vibrations of the molecules in liquid causes compression (high pressure) and rarefaction (low pressure). The later leads to the formation of cavities or bubbles, which enlarge, and during compression, the phase breakdown briskly, producing shock waves. The occurrence of bubble formation and collapse is known as cavitation and is caused by ultrasonic waves in solid/liquid or liquid/liquid systems. The cavitations in alcohol or methanol are reasonably high due to increased vapour pressure of the solvent. Besides the review of reports about the use of ultrasound in the textile processing (Mathur, 1999) and dyeing, the up-scaling of dyeing process was not undertaken using the ultrasonic waves. The Synthetic and Art Silk Mills' Research Association (SASMIRA) took the initiative to modify the already-operational industrial open width dyeing machine by combining ultrasonic transducers, which operated in the frequency range of 20-40 kHz, along with other suitable heating mechanism modification in the machine. The dyeing trials were carried out using conventional and ultrasonic techniques and comparison of data indicated that the later technique has the potential of its use in the dyeing industry due to the conservation of energy and time, with better dyeing properties (Zhang et al., 2010). Ultrasonic technique requires low dyeing temperature (50-55°C), 30% less dyeing time, with uniform dyeing, enhanced exhaustion and fixation about 25%, lesser selvedge to selvedge variation, and lesser effluent (Mathur et al.).

(2) Microwave dyeing

Microwaves are electromagnetic radiation of longer to shorter wavelength, varying from 1 m to 1 mm and corresponding to variation in frequencies from 300 MHz (0.3 GHz) to 300 GHz. The microwaves are generated by using high-power microwaves sources using special vacuum tubes and find use mainly in communication, radar, navigation, spectroscopy, heating and power application, and in other fields. The microwave radiation of about 2.45 GHz, when passes through the food items in a microwave oven, causes dielectric heating principally by the absorption of energy in water. In industry, the microwave heating is used for drying and curing of end products. Efforts have also been made to use microwave in dyeing (Tiwari and Vankar, 2001; Elshemy, 2011).

The use of microwave frequency of 2450 MHz for dyeing in aqueous dye solution affects the vibrational energy of the water and dye molecules, leading to frictional heating caused by ionic conduction. The dyeing takes place due to the interaction of dye molecules with the constituting chemical moieties of the material to be dyed. The mordant helps in fixing the dye molecules onto the fabric. Microwave dyeing is eco-friendly, uses less power and time, and has enhanced dye retention.

(3) Supercritical carbon dioxide dyeing

The colouration of different fabrics using supercritical fluid carbon dioxide (SC-CO_2) is one of the recently developed methods of dyeing. A substance above its critical temperature and critical pressure is in the supercritical phase, wherein liquid and gaseous phases become indistinguishable and the phase is known as supercritical fluid, possessing the properties of both gas and liquid. Its gas-like viscosity and liquid-like density make it an excellent solvent, which dissolves non-polar solids, making them useful in various applications. The supercritical state of a substance enjoys many advantages and replaces water in the dyeing process. The research work has been carried out over the last decade for using SCF technology for different textile applications, including dyeing, bleaching, and scouring. Carbon dioxide is a supercritical fluid of choice due to its easily achievable critical pressure (73.858 ± 0.005 bar) and temperature (31.05 ± 0.05°C) (IUPAC, 1976), as well as non-toxic, non-flammable, and inert nature. The advantages of using supercritical carbon dioxide are given below:

(i) Readily available in plenty and never-ending resource
(ii) No release of volatile organic compounds by its use
(iii) Biodegradable due its utilization by plants
(iv) No disposal problem due its recovery and reuse from the dyeing process
(v) Non-corrosive and cheap

The use of supercritical carbon dioxide (SC-CO_2) in place of water for dyeing textile fibres was patented by Schollmeyer. The initially developed laboratory scale process using 0.4 litre autoclave has been up scaled (Bach et al., 1999) for the dyeing of cotton, wool, and cellulose fibres (Mishima and Matsuyama, 2004).

The polyester and other synthetic fibres can be dyed with modified disperse dyes, using supercritical fluid carbon dioxide, which causes polymer fibres to swell, making way for the disperse dye to easily spread widely within polymer, piercing the pore and capillary structure of the fibres. The intense penetration of the dye leads to effective colouration of the hydrophobic fibres. The dyeing and removal of the excess dye are performed in the same vessel. The residual dye can be reused. The equipment used for supercritical fluid dyeing conforms to package dyeing system, in which a roll of fabric is inserted into a high-pressure vessel filled with CO_2, and is able to attain pressure up to 250 bar.

Polypropylene fibres containing no sites for binding the dye molecules were also dyed using supercritical carbon dioxide (SC-CO_2) (Li-qui et al). The dyeing was carried out by putting the fabric and dye in the dyeing vessel, which was sealed and heated to a pre-determined dyeing temperature, with the simultaneous pumping of CO_2 to the settled pressure. Dyeing was continued for a specific time, followed by a gradual reduction of pressure to the atmospheric pressure, with constant dyeing temperature.

Dyeing of the fabric using SCF carbon dioxide can be done by wrapping it around a perforated stainless steel tube and mount inside the autoclave around the stirrer and placing the dye alone at bottom of the vessel (Sabat and Magdum, 2012). The dyeing trials of the yarn, synthetic fibres, and cellulosic fibres have been described in the chapter. The dyeing with supercritical carbon dioxide is beneficial over the conventional dyeing method due to complete elimination of water and auxiliary (surfactants and salts) chemicals, reduction in energy cost, rapid diffusion, and potential for high degree of dye exhaustions and recovery of the surplus dye from the system. The process is environment friendly but requires new costly machines, leading to higher initial investment (Prince, 2011).

The advantages of the supercritical dyeing are given below:

(i) Waterless dyeing process leads to water saving and no waste water disposal problem.
(ii) No drying of dyed textile.
(iii) Less energy consumption.
(iv) Less air emissions.
(v) Significant reduction in dyeing time.
(vi) Elimination of surfactants and auxiliary chemicals.
(vii) Efficient utilization of dye with very less dye residue, which can be reused.
(viii) About 95% of the carbon dioxide can be recycled.
(ix) Supercritical fluids possess higher diffusion coefficients and lower viscosities than liquids. Therefore, they allow better penetration into materials due to absence of surface tension.

PROCESS FOR THE ISOLATION OF NATURAL DYES

Natural dyes are mainly obtained from the different parts of the plants, viz., bark, flowers, fruits, leaves, roots, seeds, and wood. These dyes contain different classes of compounds, viz., alkaloids, carotenoids, flavonoids, indigoids, quinones, and other class of compounds, such as aglycones and glycosides, depending on the plant source. The dye containing the compounds can be extracted with pure water, water (alkaline or acidic), organic solvents, and latest supercritical fluid extraction (SFE) extraction technique. The water-extracted dye can be purified by fractionation between organic solvents of varying polarity, viz., ether, ethyl acetate, and n-butanol. Water or alkaline aqueous solution is the most suitable solvent for the extraction of dye due to high polarity, low cost, and non-toxic nature. The extracting power of the alkaline water solution is more than that of water. The main steps usually involved in the extraction of vegetable dye from the plant material on the large scale are given below (Dayal et al., 2009; Gupta and Gulrajani, 2001).

1. Plant material
   (i) Selection of plant material
   (ii) Collection of plant material
   (iii) Drying of plant material
2. Size contraction of the plant material
3. Extraction of plant material
4. Filtration to isolate dye solution
5. Concentration of dye solution
6. Spray drying of the concentrated dye solution
7. Packing of the dye powder

1. Selection and collection of plant material

The different plant parts are the main sources of natural dyes. The selection of the plant material is important for the isolation of dye. The plant material should be cheap, preferably readily available throughout the year, should have good dye yield, and should be able to impart charming colours on different fabrics with good colour-fastness properties. Collection of the material from the plants of same age, region, and processing under identical conditions, moisture, and ash content of dye and UV spectra will be helpful in obtaining the reproducible results of dyeing. Some important plant species whose different parts have been used for dyeing of textiles are given in Table 7.1.

2. Size contraction of plant material

The plant material, after its collection, is air-dried in shade, chopped into small pieces, and powdered in a grinder. The size reduction of the plant material is essential for the efficient extraction of the dye. It should not be a fine powder, as big lumps will be formed, resulting in bumping of the solution during extraction and retention of the excess quantity of the dye solution after the completion of the process. The material of mesh size may vary depending on the part to be used for the extraction of dye.

Table 7.1 List of dye yielding plant species

S.No.	Name of plant species	Family	Popular name	Dye bearing parts	Pattern	Vital colour
1.	*Acacia catechu* Wild.	Mimosaceae	Khair, Cutch Tree	Heartwood	Tr	Red, brown
2.	*Artocarpus heterophyllus* Lam.	Moraceae	Kathal, Jackfruit	Wood, roots, fruits	Tr	Yellow
3.	*Berberis aristata* DC.	Berberidaceae	Kilmor	Bark, roots	Sh	Yellow
4.	*Butea monosperma*(Lam.)Taub.	Fabaceae	Palas, Flame of the Forest	Flowers	Tr	Yellow-orange
5.	*Caesalpinia sappan* Linn.	Fabaceae	Sappan Wood	Wood	Tr	Red
6.	*Carthamus tinctorius* Linn.	Asteraceae	The Safflower, False Saffron	Flower	H	Red
7.	*Crocus sativus* Linn.	Iridaceae	Kesar, saffron	Flowers		Yellow
8.	*Curcuma longa* Linn.	Zingiberaceae	Haldi, Turmeric	Rhizome	H	Yellow
9.	*Haematoxylon campechianum* Linn.	Fabaceae	Logwood, Patang	Heartwood	Tr	Black
10.	*Indigofera tinctoria* Linn.	Fabaceae	Nil	Leaf, whole plant		Blue
11.	*Lawsonia inermis* Linn.	Lythraceae	Mehndi, Henna	Leaf	Sh	Yellow-red, Orange
12.	*Mallotus philippensis* Muell.Arg.	Euphorbiaceae	Kamala Tree, Sindur	Fruits	Tr	Orange
13.	*Morinda citrifolia* Linn.	Rubiaceae	Al, Ach	Root bark	Tr	Red, yellow
14.	*Oldenlandia umbellata* Linn.	Rubiaceae	Indian madder, Chay-Root	Roots	Woody Annual	Red
15.	*Pterocarpus santalinus* Linn. f.	Fabaceae	Red Sandal Wood, Raktachandan	Wood	Tr	Red
16.	*Punica granatum* Linn.	Punicaceae	Anar,Pomegranate	Rind, flowers	Tr	Yellowish brown, light red

Contd...

S.No.	Name of plant species	Family	Popular name	Dye bearing parts	Pattern	Vital colour
17.	*Quercus infectoria* Olivier	Fagaceae	Majuphal, Gall Oak	Galls	Tr	Yellow, dark grey
18.	*Reseda luteola* Linn.	Resedaceae	Weld,Dyer's Rocket	Whole plant	H	Deep yellow
19.	*Rheum emodi* Wall.ex Meissn.	Polygonaceae	Dolu, Himalayan Rhubarb	Rhizome	H	Reddish, yellow
20.	*Rubia cordifolia* Linn.	Rubiaceae	Indian madder, Manjit	Roots	Cl	Scarlet-red, orange
21.	*Rubia tinctorum* Linn.	Rubiaceae	European madder, Alizari	Roots	H	Red
22.	*Semecarpus anacardium* Linn.f.	Anacardiaceae	Bhilawa, The Marking Nut Tree,	Bark, fruits	Tr	Black
23.	*Terminalia chebula* Retz.	Combretaceae	Harra, Myrobalan	Fruits, barks	Tr	Yellow, brown

(Tr = Tree, Sh = Shrub, H = Herb, Cl = Climber)

3. Extraction of plant material

The extractor, for the isolation of dye from the plant material, should be made of stainless steel only. In case the extractor is made of other metals, for example, iron, aluminum, or tin, the colouring compounds of the dye may react with the extractor metal, resulting in the change of dye colour during extraction. The extractor should be a jacketed vessel to avoid heat loss, and its size depends on the quantity of the material to be extracted in an operation. It can be fitted with a reflux condenser if needed, which will not allow the solvent vapours to escape. The time, temperature, and pH of the water used for extraction are the important parameters responsible for the repeat dyeing results.

4. Separation of dye solution

The dye solution can be separated from the extractor by simple or vacuum filtration and the liquid collected in the liquor storage tank made of stainless steel. Fine filtration is needed for the removal of very small particles present in the dye solution to avoid contamination of the dye. Soft water devoid of metallic impurities should be used for extraction.

5. Concentration/precipitation

The filtered dye solution may be concentrated in an open container or in vacuum to the solution of suitable solid content fit for spray drying. The vacuum concentration is preferred due to sensitive nature of the dye. The dye solution obtained on extraction with alkaline solution can also be precipitated by changing the pH from alkaline to acidic. The precipitated dye is filtered, thoroughly washed with water, and dried in vacuum.

6. Spray drying

The concentrated dye solution with the suitable solid content may be spray-dried to obtain the dye in the powder form. The solid powdered dye may be packed in the air-tight non-reactive containers as the phenolic-compounds-rich dyes are hygroscopic in nature. In order to obtain reproducible results, the different lots can be mixed thoroughly. The dye characteristics with reference to melting point, colour with ferric chloride, solubility in cold water and methanol, ash content, pH, UV, and IR spectra and other parameters may be determined.

ITEMS FOR HOME DYEING

The simple types of utensils are required for home dyeing with natural dyes. The stainless steel vessels are most suitable for dyeing and mordanting. These are available in all the homes. The galvanized iron vessels and glass vessels can also be used. The copper, iron, aluminium, or brass vessels are to be avoided as these vessels act as mordant and modify the colour of the dye. Generally, a container of about 5 litres capacity is required for dyeing of about 100 g of fibre.

The following items are normally required for home dyeing:

(i) Stainless steel or glass vessels for dyeing
(ii) Measuring cylinders to measure the volume of dye solution
(iii) Plastic tubs for washing
(iv) Dyestuff to be used for dyeing the yarn
(v) Mordant for fixing the dye on the yarn
(vi) Gloves for hand protection
(vii) Safety glasses for eyes protection
(viii) Thermometer for measuring the temperature
(ix) Stainless steel or glass rod for stirring
(x) Weighing machine

The dye solution may be prepared from the processed plant material by soaking it in water for 24 hours, followed by its gentle boiling for 2 hours by using suitable material-to-liquor ratio. The dye liquor was cooled to room temperature and separated from the plant material by filtration. The appropriate dye concentration solution can be prepared by dissolving the dye powder in warm water with stirring, followed by dilution with water to the desired volume.

SAFEGUARDS FOR THE USE OF NATURAL DYES

(1) Natural dyes are commonly isolated using water as solvent and thus contain polar organic compounds (phenolics, flavonoids, quinones as aglycones, and their glycosides and other compounds), which are hygroscopic in nature. Therefore, proper storage and handling during the dyeing process are essential for natural dyes.
(2) The optimum quantities of dyes and mordants are to be used to avoid environmental pollution.
(3) Natural dyes are heat and pH-sensitive. Therefore, standardized temperature and pH are needed for dyeing.
(4) The mordants of same brand may be used to avoid the colour variation.

References

1. Anon (1956). *The Wealth of India, Raw Materials*, Vol. IV, 261, Reprinted 2009, Council of Scientific and Industrial Research, New Delhi, India.
2. Anon (1962). *The Wealth of India, Raw Materials*, Vol. VI, 31, Reprinted 2009, Council of Scientific and Industrial Research, New Delhi, India.
3. Anon (1966). *The Wealth of India, Raw Materials*, Vol. VII, 268, Reprinted 2010, Council of Scientific and Industrial Research, New Delhi, India.
4. Anon (1969). *The Wealth of India, Raw Materials*, Vol. VIII, 69, 211, Reprinted 2009, Council of Scientific and Industrial Research, New Delhi, India.

5. Anon (1972). *The Wealth of India, Raw Materials*, Vol. IX, 313, Reprinted 2009, Council of Scientific and Industrial Research, New Delhi, India.
6. Anon (1985). *The Wealth of India, Raw Materials,* Vol. 1: A, 108, Reprinted 2010, Council of Scientific and Industrial Research, New Delhi, India.
7. Anon (1992). *The Wealth of India, Raw Materials,* Revised, Vol. 3, 368, Council of Scientific and Industrial Research, New Delhi, India.
8. Bach, E., Cleve, E., Schollmeyer, E., Vardag, T., Korner, P. (1999). Experience with UHDE CO_2: Dyeing Plant on a Technical Scale, Part II: Concepts for the Development of the Pilot Plant in Respect of a Scaling Up of the Machine. *Melliand Int*, 2, 165-168.
9. Bhardwaj, H.C. and Jain, K.K. (1982). Indian Dyes and Industry during 18th-19th Century. *Indian Journal of History of Science* 17 (1), 70-81, Indian Science Academy, New Delhi.
10. Dayal, R. (1987). Occurrence of Ursolic acid and related compounds in *Eucalyptus* hybrid leaves. *Curr. Sci*, 56, 670.
11. Dayal, R., Parcha, V., Bartley, J.P. (1997). GC-MS Studies of Oil from *Eucalyptus* hybrid leaves. *Indian Perfumer*, 41 (4), 60.
12. Dayal, R. and Dobhal P.C. (1999). Some useful dye plants: Prospects and possibilities. *Book of Papers Convention on Natural Dyes,* held at Indian Institute of Technology, Delhi, from December 9-11, 83-94.
13. Dayal, R. and Dobhal P.C. (2000). Natural dye from *Populus deltoides* bark. *The Indian Textile Journal*, CX (7), 60-63.
14. Dayal, R., Dobhal, P.C., Parcha, V. (2001). Natural dye from *Pinus roxburghii* bark and needles. *Colourage*, Vol. XLVIII (3), 27-32.
15. Dayal, R. and Dobhal, P.C. (2001). Natural dye from Indian plants. *Colourage,* Vol. XLVIII(8), 33-38.
16. Dayal, R.and Dobhal, P.C. (2001a). Natural dyes from *Shorea robusta* bark. *Convention Proceedings Natural Dyes,* edited by Deepti Gupta and M.L. Gulrajani, Indian Institute of Technology, Delhi, December 17-18, 47-52.
17. Dayal, R., Dobhal, P.C., Kumar, R., Onial, P., Rawat, R.D. (2006). Natural dye form *Lantana camara* leaves. *Colourage*, Vol. LIII, No. 12, 53-56.
18. Dayal, R., Dobhal, P.C., Kumar, R., Onial, P., Rawat, R.D. (2007). Natural dye form *Ageratum conyzoides. Colourage*, Vol. LIV, No. 12, 42-45.
19. Dayal, R., Dobhal, P.C., Kumar; R., Onial, P., Rawat, R.D. (2008).Natural dye from *Parthenium hysterophorus. Colourage,* Vol. LV, No. 8, 75-78.
20. Dayal et al. (2009). Isolation of dye from *Eucalyptus* hybrid – A pilot plant experience. *Proceedings of Natural Products and Biodiversity Conference and utilisation*, Chemistry Division, Forest Research Institute (ICFRE), Dehradun, pp. 54-62.
21. Elshemy, N.S. (2011). Unconventional natural dyeing using microwave heating with cochineal as natural dyes. *Research Journal of Textile and Apparel*, 15(4), 26-36.

22. Goodwin, J. (1982). *A Dyer's Manual, Pelham*, ISBNO-7207–1327-7, pp. 11.
23. Gupta, D. and Gulrajani, M.L., (2001). Convention Proceedings of Natural Dyes, Department of Textile Technology, Indian Institute of Technology, Delhi. pp. 3-9.
24. IUPAC. (1976). *International Thermodynamic Tables of the Fluid State Carbon Dioxide* (Oxford: Pergamon Press).
25. Li-qiu Y.U., Shun-fen, Z., Liang H.E., Wei, M.A., Jin-zong, Y. The Dyeing of Polypropylene Fibres in Supercritical Fluid. *The Proceedings of the 3rd International Conference of Functional Molecules*, State Key Laboratory of Fine Chemicals, Dalian University of Technology, Dalian, China.
26. Mathur, M.R., Sankhe, M.D., Bardhan, M.K. Energy Conservation in Wet Processing: Development of Low Energy Dyeing Machine. The Synthetic and Art Silk Mills' Research Association, Worli, Mumbai.
27. Mathur, M.R. (1999). *Man-made Textin India*, March, 97.
28. Mishima, K. and Matsuyama, K. (2004). Dyeing technique using supercritical carbon dioxide. *Chorinkai Saishin Gijutsu*, 8, 35-38.
29. Prince, P.A. (2011). Dyeing of textiles in supercritical form. *Man-Made Textiles in India*, 5, 167-174.
30. Sabat, P.B. and Magdum, P.K. (2012). Supercritical carbon dioxide dyeing. *Colourage*, June, 41-47.
31. Saito, M. and Yamauchi, Y. Extract from *Handbook on Supercritical Fluid Extraction and Supercritical Fluid Chromatography*, Jasco Corporation, Japan, by Anatek Services, 8, Valmiki Apartment, C.S.T, Kalina, Santacurz, Bombay.
32. Solanki, J.D., Raiyani, K.D., Shah, B.N., Modi, D.C., Shahu, D. (2010). Phyto-pharmacological profile of *Ageratum conyzoides*. *Pharmacologyonline* (2, Newsletter), 987-1004.
33. Soxhlet, F.(1879). Die gewichtsanalytische Bestimmung des Milchfettes, Polytechnisches J (Dingler's) 232, 461.
34. Tewari, D.N. (1994). *Tropical Forest Produce.* International Book Distributors, Rajpur Road, Dehra Dun (India), pp. 317-358.
35. Varshney, V.K. and Dayal, R. (2009). Phytochemicals from *Shorea robusta*: A review, *Proceedings of Natural Products and Biodiversity Conference: Chemistry and Utilization*, Chemistry Division. Forest Research Institute, Dehradun, 68-75.
36. Varshney, V.K., Pandey, A., Onial, P.K., Dayal, R. (2012).Antifungal activity of phytochemicals from *Eucalyptus* hybrid leaves against some plant pathogenic and wood decay fungi. *Archives of Phytopathology and Plant Protection*, 1-8.
37. Zhang, H., Fang, G., Zheng, H. (2010) Ultrasonic-assisted extraction of lac dye. *Shipin kexue*, 31(20), 116-120.

8 Determination of Colourfastness

INTRODUCTION

The colourfastness of dyed textiles may be defined as the resistance offered by the colour to different interventions to which these dyed textiles may be subjected during manufacture and subsequent use. The change of colour and staining of undyed adjacent fabrics are evaluated as fastness ratings. The colourfastness is used to evaluate the two different characteristics of a dyed textile, namely, the degree to which it changes colour when subjected to a particular treatment or environment and the tendency of the textile to cause staining of the adjoining fabrics in its contact. The word "colourfast" was used for the first time in 1916. The colourfastness depends on the nature of dye, fibre, method of dyeing, construction of material, and shade depth imparted by the dye. The varying results are obtained by using the same dye on different textiles and method of dyeing. The best results are obtained by dyeing of fibre, followed by yarn and fabric dyeing. The maximum dye retention takes place in the fibre due to increased surface area and less thickness.

COLOURFASTNESS PROPERTIES

- Washing fastness
- Light fastness
- Rubbing fastness
- Perspiration fastness

The above colourfastness properties are determined by following the International Organization for Standardization (ISO)-described methods, using launderometer (washing fastness), xenoster (light fastness), crockmeter (rubbing fastness), and prespirometer (perspiration fastness). After the completion of testing of the dyed textiles using different instruments, the samples are evaluated on grey scales by determing the grades for colour change and staining which are given below. The lightfastness evaluation invloves the use of "Blue Wool Scale Cards" also (Anon, 1988).

Grades for colour change	**Grades of staining**
5 —No change in colour	5 — No stain
4 — Slight change	4 — Slight stain
3 — Noticeable change	3 — Noticeable stain

2 — Considerable change 2 —Considerable stain
1— Maximum change in colour 1 —Heavly stain

(i) Grey Scale for determination of colour change

The dresses made from different dyed fabrics are subjected to light exposure, washing, rubbing and perspiration at regular interval and suffer colour change, which can be measured by grey scale, made up of five pairs of standards. One half of each standard is kept of exactly the same chroma to the starting sample. The variation in the second half is maintained from starting chroma (no loss of colour) to white (entire loss of colour). The grey scale rating is determined by comparing the difference in colour between exposed and unexposed fabric, to one of the standard pairs. The nil colour fading and approximately all colours fading is indicated by 5 and 1 on this scale respectively. Determination of colour change, using IS 768:1982 method is based on ISO standard, ISO 105/A-1984 Textiles- Tests for colour fastness A02-Grey scale for assessing change in colour, issued by International Organization for Standardization (ISO). The ISO method "ISO 105-A 02:1993, Textiles-Tests for colourfastness-Part A02: Grey scale for assessing change in colour" formed the basis for implementation of grey change in Easy Match Textiles QC. This can act as an alternative to visual assessment as described in AATCC evaluation Procedure 1, "Grey scale for colour change" and may be used for assessing all samples except the one treated with fluorescent whitening agents (Anon, 1988., ASTM, 1994., CIE NO. 15.2, 1986., AATCC Technical Manual, 1994).

(ii) Grey scale for determination of staining

The transfer of colour from a test sample to adjacent sample can be determined by following a similar method to that of grey change evaluation by using five standard pairs, having one half of each standard as white, while the second half range varies from white (zero staining) to a grey, with the chroma value of the test sample (excessive staining). No staining and excessive staining is indicated by value of 5 and 1 respectively. Evaluation of staining, using IS769:1982 method, is based on ISO standard, namely, ISO 105/A-1984 Textiles-Tests for colourfastness A03-grey scale for assessing staining issued by the International Organization Standardization (ISO). An ISO method "ISO 105-A04: 1989, Textiles-Tests for colourfastness-Part A04: Method for the instrumental assessment of the degree of staining of adjacent fabrics" laid the foundation for the implementation of Grey Stain in the Easy Match Textiles QC. This can be used in place of visual assessment as described in AATCC Evaluation Procedure 2, "Grey Scale for Staining" for determination of staining on the adjacent fabric samples (Anon, 1988., ASTM, 1994., CIE NO. 15.2, 1986., AATCC, 1994).

(iii) Blue wool scale

The measurement and calculation of the colouring dyes fastness are being carried out by using the Blue Wool Scale. In the beginning, the test was devised for use in the textiles industry, but presently, it is also used in the printing and polymer industries for measuring light fastness of the colourants (Textile Fading Card, example 1 and 2). A blue wool textiles fading card contains eight strips of wool mounted side by side on a small card and each strip has been dyed with a blue dye that fades on exposure to a fixed amount of light. The desirable rating between 0 and 8 is assigned to the sample by recognizing a strip on the standard blue wool, which has faded equally to the sample. A rating of 0 indicates extensive fading (very poor light fastness) and the colourfastness increases with the increasing in rating. The 8 rating indicates no colour fading (very fast colour). The Blue Wool Scale cards are usually used in connection with Grey Scale cards for assessing the degree of change.

COLOURFASTNESS TO LIGHT

The coloured clothes are exposed to sunlight whenever we go out for work and the colour change may take place depending on their colourfastness to light. Therefore, the colourfastness of textile materials to daylight is of immense importance to consumers. Determination of colourfastness to light in sunlight is a time-consuming process and therefore is being evaluated by quicker standard methods using artificial light source, viz. Xenon-Arc lamp or Carbon-Arc lamp in the laboratory. However, in a few cases, there may be variation of results between artificial and daylight exposure. In such cases, method based on daylight exposure should be used.

Light-fastness testing involves exposing samples to intense artificial light to assess the impact on the material. The samples of the textiles materials to be tested are exposed to artificial light under recommended conditions along with eight dyed wool standard patterns for a certain time. The fastness is evaluated by comparing the change in colour of the sample with that of standard samples of the blue wool cloth having very low light fastness (No. 1) to very high fastness (No. 8). Determination of light fastness using IS 2454: 1985 method based on ISO 105/B-1984 Textiles-Tests for colourfastness BO2-colourfastness to light: Xenon arc, issued by International Organization for Standardization (ISO), can be performed as given below (Anon, 1988). The tests are being revised at regular intervals to include modifications in the method of testing. The light-fastness can be determined by using a Xenoster or Microsol light-fastness tester.

A test specimen of a fabric of not less than 1 cm × 4.5 cm dimensions is required for testing. In case of yarn testing, the specimen of the same dimension is to be prepared by winding the yarn closely on the card, while the loose fibre testing requires combining and compressing of the fibre so as to make a sheet that is to fasten on a card of not less than 1 cm × 4.5 cm dimensions. A complete test specimen was prepared by cutting the standard patterns of equal size with that of test sample and mounting them on a card along with the specimen in a such a way that each exposed and unexposed area is not less than 1 cm × 1 cm. An opaque cover AB was used for covering the samples. The samples were exposed to artificial light under the conditions as given in test. The effect of light on the standard patterns was observed periodically by lifting the opaque cover. The examination of specimens at a stage when the exposed portion of standard pattern 3 attains a change equal to grey scale grade 4-5, and comparing the occurrence of any change in them with changes of standard pattern 1,2 and 3 will give an idea about their tentative light fastness.

The exposure was continued until a change in colour of the exposed portion of the standard pattern 4 can just be recognized, equal to grey scale grade 4-5. At that time, an additional cover CD was placed in such a manner as to cover about half of the AB cover and the exposure was continued until a change in colour of the exposed portion of standard pattern 6 could just be recognized equal to the grey scale grade 4-5. At this time, an additional opaque cover EF was placed, covering about one-fifth of CD, while keeping the already-existing covers in position. The exposure was further continued for obtaining a contrast on standard pattern 7 equal to contrast illustrated by grade 4 on the grey scale or a contrast equal to grade 3 on the grey scale has been produced on the most resistant specimen, which ever happens first. The light fastness was evaluated by removing the three covers and comparing the changes in the colour of the specimens with those of the standard specimen, whichever happens first. The evaluation in numerical ratings is based on contrast equal to the grey scale grade 4 and or grade 3 between exposed and unexposed portions of the specimens. The light fastness of a specimen is the number of standard pattern that shows similar changes in colour by visual observation. A halfway rating will be assigned if the change in colour corresponds between the two standard patterns (Anon, 1988).

The chemical structure of the dye and its concentration on the dyed fabrics are responsible for the colourfastness of dyed textiles. The presence of a group along with multiple bonds in a dye molecule contributes to nature. Such type of groups are called "chromophores," e.g., N=N, NO, NO_2, and quinonoid structure. It was also observed by Witt that the introduction of certain groups in dye molecule deepens the colour of the dye molecule. These groups are called auxochromes, e.g., Cl, COOH, NH_2, and OH. The presence of certain groups in the molecules also caused lightening of the colour. The groups that caused deepening and lightening of the colour are known as bathochromic or hypsochromic groups, respectively. Therefore, the presence of a photostable chromophore maintains the light fastness of the dyed textiles.

The relatively higher fastness of anthraquinone dyes in comparison to triarylmethane acid dyes on wool may be directly attributed to photosensivity of the chromophore group. The same reasoning applies for the high fastness of anthraquinone-based dyes and low fastness of curcumin (*Curcuma longa*) and bixin (*Bixa orellana*) dyes. The light colourfastness of the luteolin-based dye is higher as compared to quercetin-containing dye, besides the presence of an additional hydroxyl group in quercetin. The reason being that the presence of a hydroxyl group at C-3 position makes it more susceptible to photochemical degradation. Examination of light fastness of hydroxy anthraquinones indicated that the fastness decreases with the increase of number of hydroxyl groups (Gebhard, 1909). As the concentration of dye increases in the textile, the colourfastness to light also increases as the specified number of dye molecules will be affected during a particular time.

CAUSES OF SUNLIGHT FADING

Electromagnetic radiation is a form of energy consisting of repetitively varying electric and magnetic fields. Sunlight reaching the earth constitutes a portion of the electromagnetic radiation given off by the sun and consists of all wavelength of visible light (44%), ultraviolet rays (3%) of shorter wavelength, and infrared rays (53%) of longer wavelengths (solar constant of ground level).

The visible light is electromagnetic radiation having wavelengths ranging from about 4×10^{-7} m (7.5×10^{14} Hz) [violet] to about 7×10^{-7} m (4.3×10^{14} Hz) [red] and is a very small segment of electromagnetic spectrum. The ultraviolet range (ultraviolet C to ultraviolet A) is from 100 nm to 400 nm, while the infrared range stretches from 700 nm to 10^6 nm (1 mm) and visible range from 380 nm to 780 nm, respectively. The visible range of light, as the name indicates, is visible to the naked eyes, while the ultraviolet and infrared rays are not visible to the eyes. Infrared rays cause heat sensation, whereas ultraviolet rays produce fluorescence in certain substances. The sunlight consisting of ultraviolet, visible, and infrared radiations causes fading of dyed textiles on their exposure to it. The ultraviolet radiation (3%) of the sun's energy results in 60% of the colour fading. It has been found that the fugitive dyes and those exhibiting excellent light fastness are degraded by visible light and ultraviolet rays, respectively.

NATURE OF FIBRE

The light colourfastness of a dyed textile depends on a number of factors, including chemical nature of the fibre, dye used for its dyeing, and the light source (Gupta, 1999). The silk and wool fibres consist of a number of amino acids, thereby possessing a number of sites for the dye retention. The amphoteric nature of wool fibre creates extra sites for dye absorption. Evaluation of lightfastness of the cotton and silk samples dyed using anthraquinone-based dyes indicated that the extent of colour fading on cotton was directly related to oxidation index of each dye, whereas different mechanisms operated for silk and wool (Saito et al., 1988). This was in agreement with the earlier findings, which reported that fading of cotton consisting of cellulose is an oxidation process, whereas fading of silk and wool made of proteins is a reductive one (Cummings et al., 1956). Dyeing trials followed by determination of lightfastness indicated that indigo imparted fast colour on wool in comparison to cotton, while the reverse happened with madder dye. The area-volume ratio of fibre along with its moisture regain characteristics are also important for light colourfastness. It has also been observed that the colour of fabrics may also contribute to their lightfastness, viz., blue and red colour occurring as the end colour of the visible spectrum are quickly affected by the ultraviolet and infrared rays than the central colours, such as green, yellow, or orange, but the physical and the chemical nature of fibre in the fabric composition and the dye used are the main factors accounting for lightfastness.

CLIMATE CONDITIONS

Humidity and temperature under normal conditions of sunlight affect the lightfastness of the dyed fabrics. Reduction in the relative humidity to 25% caused significant protection of the dyed textile to colour change. The rise in temperature causes increased rate of fading for the fugitive dyes. For example, the temperature increase of 10°C from 20°C caused an increase in fading rate by a factor of 1.5. It is also known that atmospheric gases, such as sulphur dioxide, oxides of nitrogen, and ozone, even react with the dyes without the presence of light. The creation of an inert atmospheric condition leads to reduction in fading of some but not of all the natural dyes. The dyed textile may also contain other substances besides the colourant. The presence starch, gums and turkey red oil (TRO) increase fading.

COLOURFASTNESS OF TEXTILE MATERIALS TO WASHING

The measure of dye transfer during washing is known as colourfastness to washing. This will help to find out the possibility of fabric bleeding and causing staining to other clothes during washing. Therefore, it is essential to know the washing fastness of dyed fabric. The colourfastness of a textile material to washing is determined by mechanically agitating a composite sample of textile specimen with adjacent undyed fabrics in the soap solution as per test conditions followed by rinsing and drying. Five standard tests are in use for the evaluation of wash fastness in the order of increasing harsh test conditions. The standard grey scale is used for assessing the change in colour of the specimen and the staining caused on the adjacent fabrics. Determination of washing fastness using IS 687: 1979 method based on ISO 105/C-1982 Textiles-Test for colour fastness COl – colour fastness to washing: Test 1, issued by the International Organization for Standardization (ISO) is one of the methods for determing the washing fastness. The stainless steel of good quality is used in the fabrication of the apparatus used for wash fastness determination. It is a suitable mechanical device consisting of water bath having a rotor for rotating the containers (500 ml) at a speed of 40 ± 2 revolution per minute. The water bath is fitted with a device for automatic regulation of its temperature. Launderometer developed by the Silk and Art silk Mills Industry Research Association or any other suitable mechanical equipment showing consistent results may be used for the wash fastness determination. The test specimen is surrounded by two adjacent fabrics of the size 10 cm × 4 cm, of which one is made of identical fibre to that of the textile to be tested. The second piece is made of the fabric, as indicated below (Anon, 1988).

First piece	Cotton	Wool	Silk	Linen	Viscose	Acetate	Polyamide	Polyester	Acrylic
Second piece	Wool	Cotton	Cotton	Cotton or Viscose	Wool	Viscose	Wool or Viscose	Wool or Cotton	Wool or Cotton

The soap solution (0.5%) used for testing should be as per specifications of the test. A composite specimen of fabric is prepared by placing a specimen of 10 cm × 4 cm between two adjacent fabrics and sewing along all the four sides. A yarn can be tested either by knitting in into a fabric or by making its parallel lengths in between two pieces of adjacent fabrics equal to about half the amount of their combined weight and sewing along all the four sides to hold it in place. The loose fibre can be tested by combining and compressing it into a sheet (10 cm × 4 cm) of about half of the mass of adjacent fabrics and enclosing it between two adjacent fabrics by sewing from all the four sides to hold the fibre in place and thereby forming a composite test specimen. Diagonals sewing can be done as per the requirement.

The wash fastness can be determined by placing the composite sample in the previously heated soap solution at a temperature of 40 + 2°C (IS 687: 1979) by keeping the material-to-liquor ratio of 1:50. The composite specimen was treated for 30 minutes at 40 + 2°C in the mechanical washing device (launderometer) followed by its rinsing twice in cold water and then in cold running water for ten minutes. The sample was dried in air at a temperature of not more than 60°C after removing the stitching from three sides. The colour change was evaluated in the treated test specimen and the extent of staining in the adjacent fabrics with the help of grey scales for giving the ratings (Anon, 1988).

COLOURFASTNESS TO PERSPIRATION

The change in colour of the dyed garments can be caused by the reaction between the dye present on them and the components of human perspiration such as skin waste. It varies for different persons and conditions. The colourfastness determining methods against perspiration have been established by ISO, AATCC, and various other standards. The climatic conditions of garments wearing along with concentration of perspiration in the presence of sunlight may cause varying changes in the dye molecules. These changes can be covered by determining the colourfastness to perspiration under acidic and alkaline conditions. Determination of colourfastness of textile materials to perspiration by following the standard IS 971:1983 based on ISO 105/E-1978 Textiles-Tests for colourfastness E04-Colourfastness to perspiration, issued by the International Organization for Standardization (ISO) is described below. The colourfastness tests are subjected to minor modifications as per requirements. The method involves the treatment of the textiles in contact with adjacent fabrics in two different solutions (alkaline and acidic) containing histidine for a specified time. The solution was drained out and the specimen placed between two plates under a specified pressure in a testing device (perspirometer). The change in colour of each specimen and staining on adjacent fabrics were assessed using grey scales after drying them separately.

The perspirometer is a suitable testing instrument for determination colourfastness to perspiration. It consists of a frame of stainless steel fitted with a piece weight capable of exerting a pressure of 12.5 kpa on the specimen, a base with glass or acrylic resin plates (0.15 cm thickness) of the same size. The test specimen is placed between two adjacent fabrics of the size 10 cm × 4 cm, each of which is made of identical fibre to that of textile to be tested or the predominant in case of blend as given in washing fastness. The alkaline and acid solutions for testing were prepared using histidine and other chemicals. A composite specimen of a textile is prepared by placing a specimen of 10 cm × 4 cm between two adjacent fabrics and sewing along one of the shorter side. A yarn can be tested either by knitting it into a fabrics or by forming a layer of parallel lengths of the yarn between two adjacent fabrics equal to half the amount of their combined weight and sewing along two sides to hold the yarn in place. The loose fibre can be tested by combining and compressing it into a sheet (10 cm × 4 cm) of about half of the weight of adjacent fabrics and enclosing it between adjacent fabrics by sewing from all the four sides to hold the fibre in place and thereby forming a composite sample test specimen.

A composite sample may be put in the above prepared alkaline solution by keeping the material-to-liquor ratio of 1:50 and allowed to remain in the solution for 30 minutes at room temperature. Thorough penetration of the solution should be ensured by pressing and moving the sample. The solution was discarded and excess of liquor from the specimen removed by pressing between two rods. Thereafter, the composite sample was kept between two glass or acrylic resin plates (11.5 × 6 cm) under a pressure of 12.5 kpa. The testing device along with composite sample was kept in oven for 4 hours at 37 ± 2°C. The composite specimen with only one shorter stitching was dried in the air at the temperature of not more than 60°C taking care that the three parts remain in contact only at the line of stitching. The second composite sample was treated in the acid solution instead of alkaline by using the same method. The change in colour of the treated test specimen and the degree of the staining on two pieces of adjacent fabrics was evaluated with help of grey scales to assign the fastness ratings (Anon, 1988).

COLOURFASTNESS OF TEXTILE MATERIAL TO RUBBING

The quality of a coloured fabric is related to colourfastness to rubbing, and the test for its determination was devised to determine the amount of colour transferred from the surface of coloured textile material to other surfaces by rubbing (dry or wet). The evaluation of colourfastness to rubbing requires rubbing a standard fabric (crocking cloth or cotton lawn) on the surface of the dyed fabric specimen. This is normally carried out using a crockmeter, which rubs a finger across the fabric specimen. Crockmeter is an instrument that is being used for testing the transference of colour from the surface of one material to another by either dry or wet rubbing. The rubbing fastness indicates the degree of dye fixation on the fabric. Therefore, very good rubbing fastness implies the improvement of washing fastness, durability, and other properties of the dyed fabrics. The rubbing fastness depends on the nature of colour and shade depth.

The standard methods usually used for the assessment of rubbing fastness are given below.

ISO 105/X-1984 (IS 766:1988): Colourfastness to rubbing

ISO 105-X 12 2002: Colourfastness to rubbing

AATCC 8:2005: Colourfastness to crocking (AATCC Crockmeter Method)

AATCC 116-2005: Colourfastness to crocking (Rotary Vertical Crockmeter Method)

AATCC 165-1999: Textile flour coverings (AATCC Crockmeter Method)

The dertermination of colourfastness to rubbing by using IS 766:1988 based on ISO 105/X-1984 Textiles-Tests for colourfastness X 12-colourfastness to rubbing issued by the International Organization for Standardization (ISO) is given below.

The testing is based on the rubbing of the textile specimens with dry and wet rubbing cloth. Two alternative size of rubbing finger are particularized, one for pile fabrics including textile floor coverings and the other for all other textiles. The staining of the rubbing cloth was assessed with grey scale. Crockmeter is used for determining the colourfastness to rubbing by using a rubbing finger of 3.2 cm diameter and flat area of 2.5 cm diameter with a circular transition of 0.32 cm radius. A downward force of 22 N is exerted by the rubbing finger by moving to and fro in a straight line along a 10.0 cm track, and is suitable for pile fabrics including textile floor covering. A rubbing finger consisting of a cylinder of 1.6 cm diameter with to and fro movement in a straight line, along a 10.0 cm path on the sample with downward force of 9 N, is used for all other textiles.

The test specimen for a fabric or textiles floor covering should be of at least of 14 cm × 5 cm size. Two specimens each are needed for dry and wet rubbing, of which one should have a direction parallel to the wrap yarns and the other parallel to weft. The testing of the yarn or fibre can be done either by knitting them into fabric to make four specimens (14 cm × 5 cm) or forming a layer of parallel strands by wrapping it lengthways on cardboard rectangle of appropriate measurements.

The testing may be carried out by fixing the test specimen in the crockmeter by clamps in such a way as the long direction of the specimens follow the track of the device. Dry rubbing test was carried out by putting the dry rubbing cloth at the end of the finger of the testing device and rubbing it to and fro in a straight line along a 10 cm long track on dry specimen, 10 times to and fro in 10 seconds with a downward force of either 22 N or 9 N on the finger for pile fabrics and other textiles, respectively. Wet rubbing testing was carried by putting wet cloth (100%) and the procedure repeated as in the dry rubbing. The staining on the rubbing cotton cloths was evaluated using grey scale for giving the rating of staining (Anon, 1988).

COLOUR MEASUREMENT

(A) Colour Space

The properties of the colour, namely, chroma (saturation or brilliance), hue (colour such as yellow, red, orange, etc) and value (lightness or darkness of colour) are naturally distinguishable by the human eye. The eye can recognize a limited number of colours but it is as good as any reference when building a colour space. The colour space can be defined as a mathematical model for mapping the colours that can be produced again by a device to standard colour model, normally the RGB model. The mathematically defined colour spaces, namely, CIE RGB and CIE 1931 XYZ colour spaces were created by the International Commission on Illumination (French Commission International del'Eclariage, CIE) in 1931 [CIE, 1932; Smith and Guild, 1931-1932]. The combination of a number of experiments performed by Wright and Guild resulted into the specification of the CIE RGB colour space, from which CIEXYZ colour space was originated, which uses non-real or imaginary stimuli (Wright, 1928; Guild, 1932). The invention allows all colours within its range to be expressed as positive numbers, unlike the RGB system, where some colours fall outside its range and are expressed as negatives. The CIEXYZ colour space contains all colour sensations to be experienced by an average individual. It provides a standard reference against which many other colour spaces are described. Three types of receptors (cones) are present on the retina of the eye with varying spectral sensitivities and of these one type of cone is specifically sensitive to red-orange (X), while the second and third types are more sensitive for green (Y) and blue (Z), respectively. This formed the basis for defining a standardized "normal human eye" with delineated spectral sensitivities for three receptors for a defined observation angle. This normally accepted definition of colour stimuli affords the basis for several other colour coordinate systems.

The tristimulus method determines the reflected light from the object using three sensors filtered to have the equivalent sensitivity as the human eye and thus directly calculates the tristimulus values X, Y, and Z. In case of spectrophotometric method, multiple sensors are used for measuring the spectral reflectance of the object at different wavelengths. The tristimulus values are calculated by the instruments microcomputer from spectral reflectance data by performing integration. The tristimulus values can be changed into derived colour coordinates (e.g., CIELAB, Hunter Lab, and CIELCh) by using suitable transformations, which indicate colour difference, that can be recognized by human eye, better than is possible with XYZ coordinates. The X, Y, and Z chromaticity co-ordinates are defined as x, y, and z.

(B) The CIELAB System

The international Commission on Illumination (CIE) accomplished the first stage of CIELAB in the year 1931. In the beginning it involved the development of an objectives colour assessment method which involves a light source, a sample and an observer. The first assessment was called the two-degree observer due to its dependency on a two degree field of vision. This method of observing a specimen is comparable to what the human eye would see from a distance. Two significant changes were made to the CIELAB system by the commission in 1965 by evolving the ten degree observer and describing artificial daylight (D65) as a standard illuminate. The use of ten-degree field of vision implies the viewing of sample at arm's length. The CIELAB system was later revised by the commission in the year 1976 and called CIE 1976.

The tristimulus values used by the CIELAB system are the artificial interpretations of human eye observing. These values, namely, XYZ represents red, green and blue stimulus of the eye, respectively. The lightness of the sample is also represented by the Y tristimulus. Basically, the CIELAB system compares a sample to a standard and makes a numerical determination based on perceived colour difference. It can also be interpreted that the colour difference is assigned a value and represented on a chart. The area recognized by CIELAB system is called the colour space. This notional space can be conceptualized as a three-dimensional cylinder with axis running through the centre from top to bottom. However, a square plane is used by CIELAB to represent the red/green and yellow/blue axes through the cylinder (Rite systems, colourants for polymers).

At present, the dimensions of the Hunter 1948 L, a, b colour space (L, a, and b) are used as familiar abbreviation for the CIE1976 (L^*, a^*, b^*) colour space (or CIELAB) (Hunter, 1948; Hunter 1948 a). The Hunter and CIE colour coordinates are based on the square root, and cube root modification of colour data respectively. A colour scale CIE $L^*a^*b^*$ or CIELAB was recommended the International Commission on Illumination in 1976 for everyone use for the comparison of colour value. It is a simplified cube root version of the Adams-Nickerson space produced by plotting the quantities of $L^*a^*b^*$in rectangular coordinates. It is approximately a uniform colour scale and due to which the variations between points plotted on a colour scale conform to the visual differences between the colours plotted. The CIELAB colour space is presented in cube form.CIE L Ch is a modification to the CIELAB scale, which plots in polar coordinates rather than rectangular ones.

The three coordinates $L^*a^*b^*$ of CIELAB describe the colour as given below.

Value	**Depiction**
L^*	Lightness or darkness of the colour
	L = 0, Black
	L = 100, White
a^*	Positive value – Red shade
	Negative value – Green shade
b^*	Positive value – Yellow shade
	Negative value – Blue shade
DL^* (L* sample- L* standard)	Positive value – Lighter than standard
	Negative value – Darker than standard
Da^* (a* sample- a* standard)	Positive value – More red than standard
	Negative value – More green than standard
Db^* (b* sample- b* standard)	Positive value – More yellow than standard
	Negative value – More blue than standard
DC^* ($C^*_{sample} - C^*_{standard}$)	Positive value – More saturated than standard
	Negative value – Less saturated than standard

$$C^* = \sqrt{a^{*2} + b^{*2}}$$

The delta values, viz., DL^*, Da^*, Db^* indicate the degree of variation of a standard and sample from one another in L*, a*, b* values. These delta values are useful for quality control or formula adjustment.

$$DE^* = \sqrt{DL^{*2} + Da^{*2} + Db^{*2}}$$

$$DH^* = \sqrt{DE^{*2} - DL^{*2} - DC^{*2}}$$

DE* represent the total colour difference and accounts for the variation between the L*, a*, b* of the sample and standards. It is not possible to know about the parameter(s) (L*,a*, b*) out of tolerances if DE* is out of tolerance. Also, in some cases when DL*, Da* or Db* is out of tolerance DE* is still within tolerance. Two more delta values, namely, DC* and DH*, are also related to this scale, which indicate the differences in chroma and hue angle between the sample and standard as described in polar coordinate system, respectively. It is possible to use CIELAB colour scale on any object for measuring its colour. The relationship between the CIE L* a* b* scale, and the CIEXYZ scale for any illuminant as per ASTM E308 references has been described(ASTM, 1994; CIE, 1978; CIE, 1986; AATCC, 1994).

(i) **L* Values:** The L* values represents the "lightness" of specimen and is based on the percent of light reflectance. It is denoted by the vertical axis running through the centre in the colour space diagram. The sample is black if L* value is zero and white for L*=100 as depicted on the bottom and top of the chart. A sample without hue and having reflectance between 0% and 100% will be a variation of grey, i.e., darker grey for a value closer to 0% and lighter grey for closer to 100%.

(ii) **DL* Values:** The Delta L* or "DL*" indicates the difference between the lightness/darkness of a specimen and a standard, and is obtained by deducting the L* value of a standard from the L* value of the sample. The positive or negative DL* value indicates the sample to be lighter or darker than standard, respectively.

(iii) **a* Values:** The red shade/green shade value is indicated by "a*" and it implies that sample have red or green shading. It is indicated by a horizontal plane (east to west) in the colour space chart. The sample having positive or negative a* value will appear in red or green region of colour space, respectively. A grey sample have the zero a* value and occupies the centre of red/green plane. As this plane also bisects the lightness/darkness axis therefore the intensity of the colour with an a* zero value is determined by its L* value.

(iv) **Da* Values:** The Delta a* or Da* value indicates the red shade/green shade difference between the standard and sample and is calculated by subtracting the a* value of the standard from a* value of the sample. A positive or negative Da* value indicates the colour variation to be red or green to standard, respectively.

(v) **b* Values:** The b* value represents the yellow shade/blue shade and not the yellow or blue colour of the sample. The horizontal plane from north to south in the colour space chart depicts the yellow shade/blue shade variable. A positive or negative b* value of the sample indicates its appearance in the yellow or blue part of the colour space, respectively.

(vi) **Db* Values:** The delta b* or Db* value indicates the yellow shade/blue shade variation between a standard and a sample and can be obtained by deducting the b* value of a standard from the b* value of a sample. In case of positive or negative Db* value the perceived colour difference will be yellow or blue to standard, respectively. The axis representing b* runs right angle to axis a* on the same horizontal plane implying that every colour sample being assessed possess red/green shade (a*) and a yellow/blue shade (b*). It is not possible to have one in the absence of other.

(vii) DE* Values: The Delta E* or DE* can be calculated as given below.

$$\text{DE}^* = \sqrt{DL^{*2} + Da^{*2} + Db^{*2}}$$

It has been observed that the two samples each having a DL* value of – 1.00 and + 1.00 with Da* and Db* value of zero will have the same DE* value. Therefore there may be deviation in colour results in some cases. Thus, DE* results should not be lone sources for setting pass/fail tolerances.

(viii) C* Values: The human eye may recognize a colour sample as having more colour than the standard and the sample may be expressed as being saturated with colour. This attribute of the colour is called chromaticity and the CIELAB system, designate this value as 'C*'. The C* value is identified on a plane starting at the centre of the colour space and is always a positive number. A grey sample without any hue would emerge in the centre of the colour space with a "C*" value of zero and having no saturation or chromaticity.

(ix) DC* Value: The Delta C* or DC* value obtained by deducting the standard C* value from the sample C* value and can be negative, which indicates lower saturation and chromaticity than the standard. A positive DC* value indicates higher saturation than standard.

(x) h Values: The real colour perceived and characterized by the human eye is known as hue denoted by *h* and measured by degrees of the angle and never a negative number. Due to dependency of chromaticity on hue (*h*) it is not possible to plot the C* value without locating the *h* value on the circumference of the circle. The C* value always appears on the same plane as the *h* value.

(xi) DH* Values: The Delta H* or DH* is obtained by subtracting the standard *h* value from the sample *h* value using the CIELAB system. If the *h* value of a sample indicates the basic colour to be blue than the negative or positive DH* would indicate the sample to be greener than standard or redder than standard, respectively. Also, it is not possible for a sample to be too blue or too yellow at the same time. Therefore, the DH* difference between the blue standard and blue sample would either appear as red or green.

Proper use of the colour computer is required to be done for the correct colour evaluation of the sample. The improper use may cause unnecessary selection or rejection of the sample. Thus, each and every sample requires visual assessment along with computer colour evaluation. After completion of dyeing, the idea about the dye uptake by the fabric can be estimated by studying colour strength and fastness of the dyed fabric. The CIELAB system is used for colour specification. The colour strength (K/S value) can be determined by using the Kubelka-Munk equation.

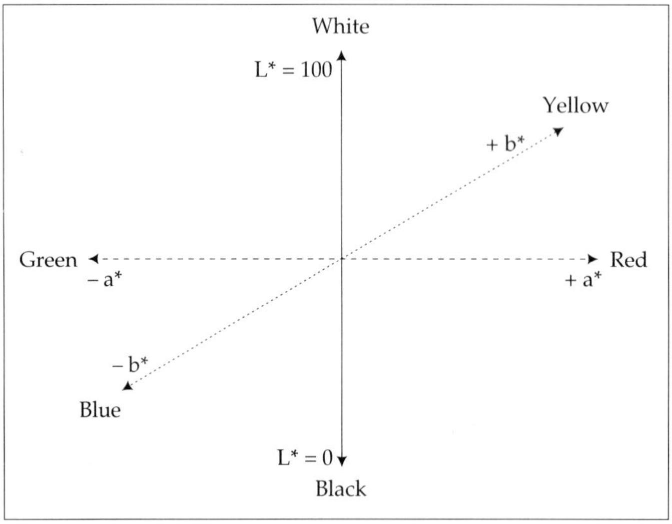

CIELAB colour coordinates (L a* b*) representation

References

1. American Association of Textile Chemist and Colourist [AATCC Technical Manual], (1994), Volume 69, Research Triangle Park, North Carolina.
2. Anon. (1988). *Handbook of Textile Testing,* SP 15 (Part 4): 1988, Bureau of Indian Standards, New Delhi, pp. 99-107, 115-119, 137-146, 157-158, 193-194.
3. *ASTM Standards on Colour and Appearance Measurement.* (1994). Fourth edition, ASTM; 1916 Race, St; Philadelphia, PA, 19103.
4. CIE (1932). *Commission International de l' Eclariage Proceedings*, 1931. Cambridge: Cambridge University Press.
5. CIE No. 15.2 Publication (1986). *Colourimetery,* Second edition, Commission International de l' Eclairage (CIE), Central Bureau of the CIE, Vienna, Austria.
6. Commission International de l' Eclairage (CIE) (1978): " Recommendations on uniform colour spaces, colour Differences equations, Psychometric colour Terms". Supplement No. 2 to CIE Publication No. 15, Colourimetric, Bureau Central de la CIE, Paris.
7. Cummings, W et al. (1956). *J.Soc. Dyers Col*; 72, 373.
8. Gebhard. (1909). *J. Soc. Dyers Col*; 25, 305.
9. Guild, J. (1932). "The colourimetric properties of the spectrum." Philosophical Transactions of the Royal Society of London Series A, Containing Papers of a Mathematical or Physical Character 230, 149-187.
10. Gupta, D. (1999). Fastness properties of natural dyes. *Colourage,* 35–38., 41-46 (August issue).
11. Hunter, R. S. (July 1948). "Photoelectric Colour-Difference Meter."*JOSA*, 38 (7), 661 (Proceedings of the Winter Meeting of the Optical Society of America).

12. Hunter, R.S. (December 1948a). "Accuracy, Precision, and Stability of New Photo-electric Colour-Difference Meter" *JOSA*, 38(12), 1094 (Proceedings of the Thirty-Third Annual Meeting of the Optical Society of America).
13. Saito, M et al. (1988). *Text. Res. J*; 58(8), 450.
14. Smith, T., Guild, J. (1931-32). "The C.I.E. colourimetric standards and their use". *Transactions of the Optical Society*, 33(3), 73-134.
15. Solar Constant of ground level (http://www.new port. Com/Introduction-to-Solar-Radiation/411919/ to 33/ content. aspx)
16. Textile Fading Card example 1 (http://www.gaylord.com/adblock.asp?abid=14017), Gaylord. Retrieved 2010-10-18
17. Textile Fading Card example 2(http://www.preservation equipment.com/store/Products/equiptments-$4-Tools/Instruments-$4-meters/Textile-Fading-Cards), Pel. Retrieved 2010-10-18.
18. Wright, W.D. (1928). "A re-determination of the trichromatic coefficients of the spectral colours". *Transactions of the Optical Society*, 30(4), 141-164.

9 Theory of Dyeing

INTRODUCTION

The dyeing process has been used since ancient times and has undergone numerous changes for its betterment. The hypothesis put forward in 1740-41 indicated that fibre possesses pores, in which the dye particles penetrate at high temperature of the dye bath, which on cooling contract, thus causing the retention of the dye. However, different dyes have different behaviours toward different fibres, which is difficult to explain using the hypothesis. The Bancroft paper on the theory of dyeing states that the dyeing in general is an adsorption process, which results in the formation of definite compounds in special cases (Bancroft, 1914). The fibre tends to adsorb all things in the solution in varying amounts, depending upon the concentration, nature, and temperature of the solution and nature of the fibre. The acid solution is appropriate for dyeing with an acid dye; however, dyeing can be done in a neutral or alkaline solution also. The dye uptake is decreased or increased by the presence of a readily adsorbed anion and cation, respectively. For example, addition of sulphuric acid will cause reduction of acid dye uptake by wool in comparison to the addition of an equivalent amount of hydrochloric acid. More levelled dyeing of the textile is obtained by the addition of sodium sulphate in the dye bath. According to the Witt's theory of dyeing of 1890, the dye dissolves in fabric or mordant, forming a solid solution. He pointed out that the colour of the dyed fabric is that of the dissolved dye and not of the starting solid dye (Bancroft and Ackerman, 1931). For example, the colour obtained by using magenta dye is red and not metallic green.

DYEING PROCESS

The dyeing process is an interaction between the fibre and the dye solution at a particular set of conditions and occurs due to selective dye adsorption by the fibre (Bancroft, 1914a). The low dye adsorption capacity of the fibre from the dyeing bath leads to weak colourfastness properties, indicating the non-suitability of the dye for dyeing the fibre. The dye adsorption on the fibre (wool, silk, or cotton) can be increased by using some chemical substances known as mordants. A mordant fixes the dye on the fibre, which of itself has little or no affinity for the fibre. The mordant can be defined as a chemical compound (substance) that is strongly adsorbed by the fibre (substrate) and, in turn, adsorbs the dye (colour) strongly. The dyeing of mordanted fibre means mainly dyeing of mordant rather than the fibre. Dyeing of mordant without fibre produces a lake. There are two types of mordants, namely, metallic mordants (salts of aluminium, copper, iron, tin, and chromium) and organic mordants (tannins and tannic acid). Metallic mordants can be strongly fixed on the fibre by treatment with silicates, arsenates, phosphates, tannin, or fatty acids. Lime, magnesia, and hydrous oxides of zinc, nickel, and cobalt are helpful in conjunction with other fixing agents.

Knecht, Rawson and Loewenthal in 1910 stated that mordant name is derived from the French word "mordre" (to bite or corrode), because the French dyers believed that the treatment of the fibre with them opened the pores of the fibre, thus making it capable of absorbing the dyes. Later on, it was recognized that the so-called mordants involved into a chemical combination with the dyes by forming insoluble compounds or lakes, and therefore, these chemical compounds (mordants) were considered fixing agents for dyes. Thus, the following is generally believed regarding the mordants:

(i) Mordant forms a definite compound with wool and silk, but not with cotton.

(ii) Mordant forms a definite compound with fixing agent which in turn forms a compound with the dye.

In general, the fixing of acid mordants may be done by metallic mordants, while tartaremetic is the most appropriate fixing agent for tannin. Mordants may be classified as acid or basic in nature. The tannins and oil (fatty acid compounds) are the most important acid mordants used for fixation of basic dyes and metallic oxides. These are also used for fixing basic mordants on the fibre, which later act as mordants in dyeing. These are mainly used in the dyeing of cotton and linen, and for increasing the weight of the silk. The basic mordants consist of the heavy-metal hydrated oxides as their active constituents. The compounds (salts) of aluminum, chromium, copper, iron, and tin are some of the commonly used basic mordants for fixing the acid or phenol-like dyes. The natural dyes are mainly acidic in nature and therefore fixed on the fibre using the basic mordants.

THEORY OF DYEING

The general dyeing theory involves the interaction between the dye solution, substrate (fibre or textile), and the dye auxiliary. Water, the universal solvent, is normally used for preparing the dye solution. However, according to the latest developments, the dyeing mechanism probably depends on the chemical nature of both the fibre and dye. The textile fibres may be covered under two main groups, namely, vegetable fibres and animal fibres. The former includes cotton, linen, flax, hemp, and jute, while the latter involves wool, silk, and leather. The polymer synthetic fibres include polyester, polypropylene, cellulose triacetate (non-ionic), polyamides (nylon, cationic), and polyacrylonitriles (acrylics, anionic). All the textile fibres are polymeric in nature and are made up from very long thread-like molecules. For example, wool and silk consist essentially of polypeptide chain molecules produced by the linear condensation of α-amino acids; cotton and viscose rayon are composed of polycellobiose; and nylon is a polyamide. The average molecular weight of these long-chain molecules is 10,000-20,000, and the molecules are generally arranged along the length of filament in the fibres and may have varying orientation in different fibres. The textile fibres on their X-ray examination produced regular diffraction patterns similar to those produced by the rotation of crystals (Astbury and Wood, 1930; Kratky and Mark, 1937). This indicates that in certain parts of the fibre, the molecules are arranged in orderly and regular manner. These oriented portions have been called "crystallites" or "micelles" and cannot constitute the whole fibre due to a large variation of physical properties of the fibres in comparison to those exhibited by the crystals. It has been found that these micelles are not separate but merge gradually into the amorphous portion of the fibre having the same chemical nature as the crystallites (Hermans, 1938; Kratky, 1938). It is also accepted that one single-chain molecule may pass through and be a part of two or more crystallites. Thus, the fibre comprises closely packed and adjusted regions interspersed with less-compact amorphous regions, and the latter may be accounted as making a system of interlocking pores throughout the mass (Vickerstaff, 1950).

The swelling of fibre takes place in water due to the presence of hydrophilic groupings in the fibre and development of osmotic pressure internally. X-ray examination indicated that swelling occurs in amorphous regions, leading to the opening of pores of convenient size for the passage of dye molecules. The diffusion of the dye molecules then takes place through these pores, followed by the parallel channels in the micellar portions. In the dyeing process of cellulose fibres, the dye molecule may be oriented parallel to micellar and is attached by hydrogen bonding. Because of the weakness of hydrogen bonds, an appropriately long-sized dye molecule is needed for providing a sufficient number of points of attachment so as to attain good fastness properties of the dyed fabric (Tewari, 1994).

The dyeing of fibre appears to be a simple process but is not so easy to explain. The coloured molecules or ions transfer from the dye solution (liquid phase) to the fibre (solid phase). Dyeing is not solely a physical process like the application of paints and pigments; in the dyeing process, some difficulties are faced for obtaining optimum colourfastness, colour yield, level dyeing, colour retention, and other parameters.

The use of auxiliary chemicals (additives) and after-treatments are of great help. The fibre consists of polymeric chains containing different functional groups (OH, COOH, C=O, etc), and the dye is taken up by the fibre as a result of the chemical reaction between them. Numerous dye molecules used for dyeing are water-soluble and form positively and negatively charged ions in solution, just like simple electrolytes. The nature and chemical constituents of the dye molecule are responsible for taking up of the dye by the fibre. The most prominent dye-fibre attachment is through covalent bonds, followed by electrostatic attraction between the oppositely charged dye ions and the fibre ions. The other attractions involved include van der Waals forces, hydrogen bonds, and hydrophobic interactions. Total strength of the molecular interactions is known as the dye affinity toward the substrate. The specific conditions are required for selective attraction of dye by the fibre. The selection of the dyes of appropriate classes along with the dyeing conditions is essential for obtaining the optimum result. Some of the main dye classes include acid, azoic, basic, direct, disperse, metal complex, mordant, reactive, sulphur, and vat dyes. The dyes belonging to different classes can be used to dye a fibre by using varying conditions. For example, wool dyeing can be carried out by using the acid, metal complex, mordant, and reactive dyes.

The chemical reactions involved during the application of the dye to the substrate (fibre) render the dye molecule difficult to be removed from the substrate (fibre), which, in turn, affects the physicochemical equilibrium established during the adsorption of dyes onto the fibres. The effective dyeing process involves the diffusion of the dye in the dye bath and then to the surface of the fibre, followed by the adsorption of the dye at the surface of the fibre and the final diffusion of the dye from the surface of the fibre to its interior. All the dyeing processes in the initial stages involve the transport of large coloured (or potentially coloured) ions or molecules from the dye solution in water (aqueous phase) to the solid fibre phase under suitable conditions. The decrease in the colour of the dye solution and increase of the fibre colour give an indication of dyeing of the substrate. In case of imbibitions, the fibre becomes saturated with coloured liquid due to staining, while there is no change in the concentration of dye in the solution. Dyeing of fibres occurs due to some driving force that is responsible for the transfer of dye from the solution to the fibre, and the solutes behaving in this way are said to be substantive to the fibre used for dyeing. According to colour terms and definitions, substantivity may be defined as "the attraction between a substrate and a dye or other substance under the conditions of test, whereby the latter is selectively extracted from the application medium by the substrate (Anon, 1973). It has been observed that substantivity of different dyes is not the same for a particular fibre, and also, same dye exhibits variations on different substrates. The substantivity of a dye for a substrate (fibre) may be obtained by measuring the degree of exhaustion, as given below (Johnson, 1989).

$$\%\ \text{exhaustion} = \frac{\text{Dye amount at dyeing end}}{\text{Dye amount at dyeing start}} \times 100$$

However, the exhaustion in a dyeing process is dependent on a number of factors, viz., presence and concentration of other substances in the dye bath (including other dyes), the applied dye concentration, the pH, the temperature, the dyeing time, and the material-to-liquor ratio. The problems connected with exhaustion may be reduced to some extent by using substantivity ratio, which is independent of material-to-liquor ratio and dyeing time.

$$\text{Substantivity ratio (SR)} = \frac{[A]\ \text{at equilibrium on fibre}}{[A]\ \text{at equilibrium in solution}}$$

where, [A] = Concentration of dye

Thermodynamic treatments of the dyeing process are almost regularly applied to accurately reversible systems at equilibrium. The studies of dye uptake have been mainly done on the reversible process, viz., on acid dyes on wool and polyamide fibres, basic dyes on acrylic fibres, direct dyes on cotton or viscose rayon fibres, and disperse dyes on polyester or cellulose acetate fibres, but not on the irreversible processes. This type of dyeing trials are normally carried out to equilibrium by keeping all the variables constant, except the concentration of dye in the solution and on the fibre. At equilibrium, the concentration of dye in the fibre is in equilibrium with the concentration of dye in solution. It is also essential to know the contents of initial dye bath and the pure dyes, and carefully washed fibres should be used for dyeing. The dyeing time should be minimum, and this can be achieved by using the material having the largest surface area per unit mass by avoiding closely packed systems or tightly woven material. Dyeing can be carried out by immersing the undyed fibre in a dye bath at the predetermined material-to-liquor ratio for the optimum duration of time. The concentration of dye in the solution and on the dyed substrate should be determined. A number of dyeing processes, namely, azoic coupling of naphthols within the fibre, the formation of covalent bonds between dye and the fibre in reactive dyeing, mordant dyeing, and the oxidation of leuco vat and sulphur dyes, are mainly irreversible ones.

ADSORPTION ISOTHERMS

The data obtained from dyeing trials carried out to equilibrium are normally presented in the form of adsorption isotherms, representing the change in the concentration of dye on the fibre with that in the solution by keeping constant all other variables, including temperature. The adsorption isotherms obtained are close in shape to one of the three typical forms, namely, the partition isotherm, Freundlich isotherm, and Langmuir isotherm (Johnson, 1989; Welham, 2000).

(i) The partition isotherm

This is very simple model. An organic compound or a non-ionic dye dissolves in different organic solvents due to their affinity for them. Similarly, in the dyeing process, a dye can be considered as dissolving in the fibre because of its affinity for the fibre. It is due to this reason that the dyeing process is complying to a solid solution model. A linear relationship is obtained by plotting the concentration of dye in the substrate (fibre) $[D]_{sub}$ (g kg^{-1} or mole kg^{-1}) against the concentration of dye in the dye bath $[D]_{sol}$ (gl^{-1} or mol l^{-1}), which suddenly changes to a solid static state after the period of increase. The advancing, linear part is defined by Equation (1).

$$[D]_{sub} = C[D]_{sol} \tag{1}$$

Where C is a constant and is known as the distribution coefficient or the partition ratio. The flat line in the curve indicates the saturation of the dye in the solution.

(ii) The Freundlich isotherm

Freundlich studied the adsorption by animal charcoal and predicted that the adsorption isotherms could be defined by Equation (2).

$$[D]_{sub} = C[D]^{x}_{sol} \tag{2}$$

This adsorption isotherm is largely observed in practice. Equation (2) can be rewritten in the linear form, as given below.

$$\text{Log } [D]_{sub} = \log C + \times \log[D]_{sol} \tag{3}$$

A linear relationship obtained by plotting $\log[D]_{sub}$ against $\log[D]_{sol}$ indicates the agreement with the Freundlich isotherm.

(iii) The Langmuir isotherm

It was derived by Langmuir to explain the adsorption of gases on metal surfaces. The Langmuir isotherm was found to be suitable to describe dye adsorption by certain fibres by assuming that the dye adsorption takes place on specific fibre sites and no further adsorption takes place after their saturation. It is also assumed that all the sites are identical, with no interaction between them or between the dye molecules occupying the sites. Let us assume that a substrate has $[S]_{sub}$ potential dye sites per kg, and at equilibrium, if a substrate contains $[D]_{sub}$mol kg^{-1} of the dye, with a dye bath containing $[D]_{sol}$ mol l^{-1}, then the rate of desorption is proportional to the dye concentration in the substrate, where C_1 is constant.

$$\frac{-d[D]_{sub}}{dt} = C_1[D]_{sub} \tag{4}$$

During the same time, dye will be entering the substrate at a rate depending on the dye concentration in the solution and number of unoccupied sites in the substrate $([S]_{sub} - [D]_{sub})$. Therefore, the rate of adsorption will be as given in Equation (5), where C_2 is constant.

$$\frac{-d[D]_{sub}}{dt} = C_2[D]_{sol}\ ([S]_{sub} - [D]_{sub}) \tag{5}$$

Thus, at equilibrium,

$$C_1[D]_{sub} = C_2[D]_{sol}([S]_{sub} - [D]_{sub}) \tag{6}$$

On rearrangement, Equation (6) becomes

$$[D]_{sub} = \frac{C[S]_{sub}\,[D]_{sol}}{1 + C[D]_{sol}} \tag{7}$$

where $\qquad C = C_2/C_1$

Equation (7) represents Langmuir isotherm.

Reversion of Equation (7) gives the following.

$$\frac{1}{[D]_{sub}} = \frac{1}{C[S]_{sub}\,[D]_{sol}} + \frac{1}{[S]_{sub}} \tag{8}$$

The saturation value can be obtained using the graphical method.

Equation (6) can be rearranged by dividing it with $[D]_{sol}$ and keeping $C_2/C_1 = C$ as

$$\frac{[D]_{sub}}{[D]_{sol}} = C([S]_{sub} - [D]_{sub}) \tag{9}$$

The equation can be rewritten as (10) if $[D]_{sub}$ is very small in comparison to$[S]_{sub}$

$$\frac{[D]_{sub}}{[D]_{sol}} = C[S]_{sub} \tag{10}$$

Equation (10) becomes a linear relationship, with C and $(S)_{sub}$ remaining constant under a given set of dyeing conditions. In this case, it can be confused with partition isotherm. The Langmuir approach with definable dye sites has been used successfully by many dyers. However, it is to be noted that the dye concentration is normally measured instead of the activity itself. Also, the complete ionization of the dye (100%) is not possible in the solution and the aggregation of the dye molecules may occur in the solution or on the fibre, and the fibre may also undergo chemical change or degrade during the dyeing process.

Dyeing

The mathematical equation used for the calculation of affinity of a dye for a particular substrate (fibre) is applicable to reversible dyeing systems and reveals that they can be classified as non-ionic and ionic dyes systems.

(i) Using Non-ionic dyes

The disperse dyes constitute the major class of non-ionic dyes and may be used for dyeing of hydrophobic fibres made from polyester, secondary cellulose acetate, and cellulose triacetate. Dyeing with disperse dyes can be explained by taking into account their existence in two forms in the dye bath, namely, solid dye and dye in solution. The following equilibrium exists in the presence of fibre:

$$\text{Dye in solid form} \rightleftarrows \text{Dye in solution} \rightleftarrows \text{Dye on the substrate}$$

The overall equilibrium position is ascertained by the solubility and affinity of dye at dyeing temperature. On the basis of experiments, it has been settled that dyeing trials using disperse dyes involve the transfer of dye molecules from a molecular dispersion (i.e., solution) into the substrate.

(ii) Using Ionic dyes

The dyeing of the substrates (fibres) is also carried out using the ionic dyes. A substrate (fibre), having charged groups, on immersion in water gets an electrical potential, which leads to the attraction of the oppositely charged ions and repulsion of the similar charged ions. As a result of this, an electrical double layer is formed, one of which is about single solvent molecule thickness fixed on the surface and the other one (diffuse layer) at some distance into the water solution. Also, the ions possess the same charge as substrate but have the force of attraction for the substrate greater than the electrical forces of repulsion over short but not over longer distances. The distribution of the ions having opposite charge to that of substrate, with the force of attraction greater than the force of repulsion, also occurs. The potential difference between the fixed and free mobile portion is called the zeta potential. The Langmuir approach is applicable when working with acid dyes (negatively charged dye ions) on wool or nylon (with positively charged groups under acidic conditions) or with basic dyes having positive charge on acrylic possessing negatively charged groups on the fibre.

The dyeing behaviour can be readily explained and predicted by taking into consideration the concept of dye sites. There may be variation at times and the results may not agree with the prediction. It is rather difficult to explain the accumulation of anionically charged dyes on cellulosic fibres, which take up a negative charge. The "Donnan Membrane Theory" is helpful in explaining the dye uptake by the cellulosic fibres. Sumner published a number of papers describing the use of Donnan theory to explain the ionization of hydroxyl groups in cellulose and measure for the presence of multivalent anions in the dye bath. The experimental results obtained by Sumner were found to be in agreement with those predicted. Very good results were obtained by applying the approach to the dyeing wool and nylon and accounts for shortcoming of the treatments based on dye sites (Johnson, 1989; Welham, 2000).

The dyeing process is not so simple as it looks and involves complex mechanism. A brief description about the theory of dyeing as described by different researches has been provided above. Two important books entitled *Theory of Colouration of Textiles*, edited by Alan Johnson, and *The Physical Chemistry of Dyeing*, by Thomas Vickerstaff, are very useful to know about each and every detail about the theory of dyeing. Welham has provided a concise description of the theory of dyeing (Welham, 2000). The advancements in the theory of dyeing is taking place continuously.

Part A

COLOUR AND CHEMICAL CONSTITUTION

(i) Material colour

When white light descends on a material, it may be completely reflected or completely absorbed. In case of total reflection and absorbance, the material appears white and black, respectively. The absorbance of certain portion of light and reflection of the remaining portion lead to exhibiting the colour of the reflected light by the material. The single band absorption causes, the material to have the complementary colour (of the absorbed band). In case a material absorbs whole visible light except one band, it will have the colour of the reflected band. Thus, a material appears yellow because of the sole absorbance of the blue portion of the spectrum, or because it absorbs whole visible spectrum, except yellow. Many colourless materials have absorption spectra but the absorption takes place in the infrared or ultraviolet region and not in the visible region of the spectrum. The part of the electromagnetic spectrum visible to humans falls in the wavelength range of 400 nm to 700 nm.

(ii) Correlation between colour and chemical structure

Witt in 1876 pointed out that generally a coloured compound contained a group with multiple bonds. These groups were known as chromophores, e.g., N=N, NO, NO_2, quinonoid structure, and other groups. Witt also noticed that some groups, although not chromophores, intensified the colour when inserted into a coloured molecule. Such groups were known as auxochromes, e.g., Cl, CO_2H, NH_2, and OH. The bathochromic groups in the dyes cause deepening of colour, while hypsochromic groups cause lightening of the colour. The colour of a compound will also be deeper in comparison to other compound if wavelength of the absorption of former is longer than the latter. The Armstrong theory, put forwarded in 1885, pointed out that all colouring matter may possess quinonoid form(o-or p-), or else will be colourless. This is true in case of benzene (colourless) and benzoquinone (coloured) but not for all the colouring compounds.

The Witt and Armstrong theories put forward for explaining the relationship between colour and chemical structure were found to be empirical. The colour concept can be explained by two modern theories, namely the Valence Bond and the Molecular Orbital. The two theories have some common attributes. Light comprises electromagnetic waves, which are made up of a magnetic and an electric component, both oscillating in planes perpendicular to each other and perpendicular to the direction of propagation. The electric component alone is normally involved in interaction with material, and this interaction is suitably explained in terms of light functioning as a train of discrete energy packets called photons or quanta. The molecule on absorption of light undergoes transition from a state of lower energy (E_a) to a state of higher energy (E_b). The change in energy (ΔE) is given by:

$$\Delta E = E_b - E_a$$

The absorption of energy in monoatomic molecules takes place when the photon has the energy equal to ΔE. As the energy of the photon is given by $E = h\nu$,

Therefore,

$$\Delta E = E_b - E_a = h\nu = hc/\lambda$$

h = The Planck's constant and c = Velocity of light.

The total energy state of a polyatomic molecule is the sum of the electronic, vibrational, and rotational components. As ΔE is large for electronic transition relative to vibrational and rotational transitions, therefore the frequency of the photon needed for electronic transitions is high (short wavelength). Thus, the electronic transitions take place in the visible and ultraviolet parts of the spectrum. An electronic level is associated with a large number of vibrational and rotational levels, therefore a transition from one electronic level after absorption of light may be attended by vibrational and rotational transitions. Thus, a range of photons will be absorbed, resulting in a spectrum of closely spaced lines merging into a broad band (Finar, 1973).

(iii) Valence bond application to colour

The colour of a compound can be explained by the Valence Bond (VB) and Molecular Orbital (MO) theories. A prominent difference between the two is that the electrons are treated within pairs in the VB theory, while in MO theory, these can be treated singly. According to the VB theory, a molecule in the ground state possesses electrons in a state of oscillation that increases on absorption of a photon of appropriate energy from a beam of light. The molecule shifts to the excited state and the wavelength of photon absorbed depends on the energy difference between the excited and ground states. The smaller differences lead to longer wavelength. According to VB method, ethylene may be observed as resonance hybrid of (A) and (B).

$$\underset{(A)}{CH_2=CH_2} \longleftrightarrow \underset{(B)}{\overset{+}{C}H_2-\overset{-}{C}H_2}.$$

The ground state may be represented by (A) and excited state by (B). High photon energy is needed to excite ethylene molecule due to large energy difference between two states, implying the very short wavelength. The small molecules normally behave in this manner. The presence of the extended conjugation in a molecule leads to greater contribution of charged structures to the resonance hybrid, thereby requiring longer wavelength of the photon to excite the molecule, e.g., ethylene (λ_{max}= 175 nm), butadience (λ_{max}= 217 nm). Also, the molecules having +I/+R and –I/–R groups at the end of a conjugated system will extend the conjugation as well as increase the contributions of charged structures to resonance hybrid, leading to longer wavelengths, e.g., benzene (λ_{max}= 204 nm), aniline (λ_{max} = 230 nm), and nitrobenzene (λ_{max} = 270 nm). The steric inhibition of resonance causes diminishing of the colour (Finar, 1973).

(iv) Molecular orbital application to colour

According to the MO theory, the transfer of an electron from a bonding to an anti-bonding orbital on absorption of lights leads to the excitation of the atom or molecule. The electronic transitions can happen in various ways. Excitation of a bonding σ-electron to an anti-bonding σ-electron orbital is known as a σ→σ* transition. Similarly, the transition of a bonding II-electron to an anti-bonding Π-orbital is expressed by Π→Π* and an electron of a non-bonding pair of electrons to anti-bonding Π-orbital by n→Π*. Such type of transitions happen with the compounds having double bonds comprising hetero-atoms, e.g., carbonyland others.

These transitions are brought about by the absorption of varying amounts of energy. The allowed transitions in order of energy difference is given below.

$$\sigma\rightarrow\sigma^*>n\rightarrow\sigma^*>\Pi\rightarrow\Pi^*>n\rightarrow II^*$$

Some of the examples of these transitions are given in Table 9.1.

Table 9.1 Transitions in different compounds

Compound	*Transitions*	*Photon energy required for transition*	*Wavelength of absorption $\lambda_{max,}$ nm*
Ethane	σ→σ*	High	135
Ethylene	σ→σ*and II→II*	Lower than ethane	175
Butadiene	II→II*	Lower than ethylene	217
Methanol	σ→σ*, n→σ*	Lower than ethane	150,183
Crotonaldehyde	II→II*, n→II*	Lower than methanol	220, 321
Benzene	II→II*	Lower than ethylene	184,204,254

The energy of a photon required to effect the transition (σ→σ*) is high, thus the absorption takes place at lower wavelength. The presence of unpaired electrons in free radicals changes the electronic energy levels in comparison to the parent molecule. The energy difference between the highest occupied and the lowest unoccupied molecular orbitals is lowered and absorption may fall in the visible region, leading to colouring of free radicals.

The chromophore may defined as a structural feature, the presence of which produces light absorption in the ultraviolet region or colour in the visible region. Some of the examples are keto (>C=O), nitroso(–N=O), nitro (–NO_2), and azo (–N=N–). The compound containing a chromophore is called a chromogen. Any group that brings about a red shift on attachment with a chromophore is called an auxochrome. Hydroxyl (–OH), amino (–NH_2), carboxylic (–COOH), and sulphonic (–SO_3H) acid groups are some of the examples. Thus, a chromophore in combination with an auxochrome behaves as a new chromophore. The shifting of the absorption band to longer and shorter wavelengths is known as bathochromic effect (red shift) and a hypsochromic effect (blue shift), respectively. The increase and decrease in intensity are covered under the hyperchromic effect and hypochromic effect, respectively. The conjugated systems of double bonds give strong bands in the longer wavelength region in comparison to isolated double bonds (Zollinger, 1991). The shifting of absorption bands due to introduction of auxochromes and carboxyl chrompohore is given in Table 9.2.

Table 9.2 Shifting of absorption bands due to introduction of auxochromes and carboxyl chromophore

Compound	*Transition*	λ_{max} *nm*
Alkanes	σ→σ*	140-150
Presence of auxochrome		
RCl	n→σ*	170-175
RBr	n→σ*	200-210
RI	n→σ*	255-260
ROH	n→σ*	180-185
RNH_2	n→σ*	190-200
Presence of carboxyl chromophore		
Aldehydes	Π→Π* n→Π*	180 290-295
Ketones	Π→Π* n→Π*	190 270-280
Saturated mono-carboxylic acid	Π→Π*	200-210 200-205
Esters	n→Π*	205-220
Amides	n→Π*	184,204
Benzene	Π→Π*	254

References

1. Anon. (1973). The Society of dyes and colourists. *J.S.D.C*, 89, 411-422. [Van Beek, H.C.A and Heertjes, P.M. (1963). *J.S.D.C*, 79, 661]
2. Astbury, W.T. and Wood, H.J. (1930). *Nature*, 126, 913.
3. Bancroft, W.D. (1914). The theory of dyeing I. *J. Phys. Chem.*, 18(1), 1-25.
4. Bancroft, W.D. and Ackerman, J.W. (1931). The solid solution theory of dyeing, *Proc. Natl. Acad. Sci*, USA, 17(8), 480-482.
5. Bancroft, W.D. (1914a). The theory of dyeing, III. *J. Phys. Chem*, 18(5), 385-437.
6. Finar, I.L. (1973) (Reprinted 1996). *Organic Chemistry*, Vol. 1, pp. 868-878, Longman Singapore Publishers (Pvt) Ltd, Singapore.
7. Hermans.(1938). *Kolloid Z.*, 83, 71.
8. Huebner. (1934). *J.Soc.Dyers Col.* Jubilee Volume, 1.
9. Johnson Alan. (1989). *The Theory of Colouration of Textiles*, 2nd edition. Society of Dyers and Colourists on behalf of the Dyer's Company Publication Trust.
10. Kratky.(1938). *Kolloid Z.*, 84, 149.
11. Kratky and Mark.(1937). *Z. Phys. Chem.*, 36 B, 129.
12. Knecht, Rawson, Loewenthal. (1910). *A Manual of Dyeing*, 2nd edition pp. 4, 20, 179, 189, 224, 234.
13. Tewari, D.N. (1994). *Tropical Forest Produce.* International Book Distributors, Rajpur Road, Dehra Dun, India, pp. 317-360
14. Vickerstaff Thomas. (1950). *The Physical Chemistry of Dyeing.* Published for Emperical Chemical Industries Limited by Oliver and Boyd, London, Edinburgh.
15. Welham, Arthur.(2000). *The Theory of Dyeing* (and the secret of life). *JSDC*, 116, 140-143.
16. Witt, O.N. (1890). *Chem. Centralbl.* (4)2, II, 10-39.
17. Zollinger, Heinrich. (1991). *Color Chemistry*, Weinheim, New York, Basel, Cambridge.

Index